PREFACE: Prelude to Action

The sun had just appeared over the horizon, as I sat down to my first cuppa java for the day. My eyes caught the calendar on the wall, *"October 23, 1978"; "another Monday."* The brisk, morning breeze was refreshing on my sleep-wrinkled skin and blood-shot eyes. Gradual signs of life were manifesting in the early morning sky; and small motions in the distance heralded the waking of our neighbors and their cats and dogs.

Golden shafts of sunlight sliced their way through gently swaying leaves outside the picture windows overlooking our front gardens, their projections forming animated montages of light on the carpet and the walls. There, on the carpet by the windows, lay our three pets - basking in the warmth of the Sun. Two jet-black miniature dachshunds flanked our elegantly outstretched feline. I remembered how unusually light-colored she appeared for a red Burmese. The Sun heightened the contrasts between their prostrate forms of contentment.

The animated sounds and smells of eggs frying, bread toasting, and coffee brewing filtered into my peaceful reverie. There was a brief break back into reality as my bubbling better-half, re-filled my coffee cup and told me breakfast was due momentarily. For a Monday, it was an incredibly good day, "so good to be alive, free and ..." *Ring-ring, ring-ring.*

She shouted from the kitchen, *"Don't answer that! We're not up yet; and besides, civilized people don't ring this early!"*

Secretly, I had to agree; but I knew that early calls usually meant something important was happening on the east coast somewhere; so I made tracks for the phone in our attached office - just in case I needed to tape-record the details of the conversation *(ring-ring, ring-ring)*. *"I'm going to answer it in the office,"* I yelled as I bounded for the back room.

The multiple "beep-beep-beep-beep" of the STD line blurted into my ear. *"Long distance,"* I thought to myself. I cleared my throat to give my voice that sound of reserved politeness most people expect on a business phone; and then, with a little Irish impishness I answered, *"Good morning! It's Perth - the other side of Earth!* (pause)."

"Is a Mr. Stan Deyo there please?" came the somewhat amused voice on the other end.

"You gottim, chief", I replied.

"I beg your pardon?", he queried again.

"I said, you've got him. It is me you wish to speak to."

"Oh, sorry, uhh, My name is Brian (I couldn't hear the last name as the line faded) *...you don't know me; but a mutual friend of ours in Melbourne suggested I ring you over this morning's UFO incident off King Island near Cape Otway ..."*

As he continued, I thought, *"Why me? I'm missing a perfectly good breakfast over a perfectly typical 'I-saw-a-colored-light-in-the-sky' phone call."*

"Have you heard the details, yet?" he asked.

"No, I haven't read the paper yet; we were just starting breakfast when you rang."

"Oh! Sorry, mate; forgot about the time difference!" he apologized.

"Don't worry about it," I said as I thought how many times I had been roused from a perfectly good sleep by similar long distance calls. *"Can you give me a brief rundown on it?"*

"Sure. The media over here are going wild over a possible UFO-abduction case. It seems that last Saturday night, around 7:10 or so, a young pilot from Melbourne was grabbed - plane and all - right outta the air off the south-eastern coast of Australia. Ground control at King Island had radio contact for about six minutes while the pilot described the damn thing. He reported four very bright landing lights moving at an incredible rate overhead at about 5,000 ft. altitude."

My wife came in with a frown and a 'wind-it-up' gesture. *"This is important!",* I whispered, *"I'll tell you why in a minute; go get the newspaper off the lawn."*

He continued, *"He told ground control it was four 'metallic' lights on a long shape that seemed to be orbiting his plane. It had a green glowing light all over the outside. He said it was coming for him and that his engine was rough idling and coughing. His last words were: 'Unknown aircraft now hovering on top of me'. A loud, metallic noise followed; and then the channel was silent. Whaddaya think? The official position said the pilot was flying upside down and flew into the drink."*

"Wow," I half-whispered, as he waited for my response, *"have they found any oil slicks or wreckage ?"*

"Nothing positive, yet. They say the RAAF spotted an oil slick in the area on Sunday; but that it was too big to have come from a light aircraft. He was flying a Cessna 182, y'know."

"What was the pilot's name?" I asked, almost automatically.

"Something like Valentine or something", his voice trailed off... *"oh, here it is, Frederick Valentich, age twenty; spent three years in the Air Training Corps; then became an instructor pilot for them. Sounds like a level-headed bloke to me."*

"Look, Brian", I interrupted, *"it's too early to make any positive statement on the situation; but I do strongly suspect that this case is going to be a catalyst for renewed public interest in the UFO situation. Tell our mutual friend thanks for me; and I'll keep in touch. Thanks for your trouble. Bye now."*

"No worries, mate", he piped. *"Here's my contact number in Melbourne.'* He rattled off his telephone number and extension as I jotted both onto the phone pad. There was a sharp click followed by two more clicks somewhere along the line as I was hanging up. I mused to myself about the possibility of wire taps. Monday morning's reverie was shattered by that time.

Reading the paper over somewhat crisp eggs and brittle toast accented the Monday syndrome as I explained the situation to my wife - who quickly forgot her pseudo-grizzle. Suddenly, as I was finishing the UFO story in the *West Australian*, a paragraph leapt out of the page at me!

"He (Valentich) *believes in UFOs and he told me (his father) he had seen classified material at the Sale RAAF base which confirmed his beliefs.*"

"*Hey! Listen to this!*" I shouted; but as I read her the paragraph, I could see that she had not yet remembered my own earlier discovery of the existence of the RAAF film library of actual UFO sightings and testing over Australia; so I explained it to her before retiring to the study to do just that - study. Thoughts flashed through my mind with electric speeds: "*Astronauts see UFOs; Iranian encounter hushed up; cattle killed by electrical UFO; Army man takes five-day trip in UFO in only fifteen minutes.*"

My mind raced from one possibility to another. The pieces of an incredible conspiracy were rapidly falling into place. For the first time, I knew why I had lived long enough to see all these events take place. "*Poor Valentich,*" I thought to myself, "*he discovered their little game, too; and look what it cost him.*"

My thoughts drifted back to my own training at the U.S. Air Force Academy sixteen years before. I remembered its rigid discipline and the beautiful, snow-covered landscapes that surrounded those of us who had lived there 7,000 feet up the slopes of the Rockies overlooking the world below. It was there that our fertile, young minds had been impregnated with post-hypnotic suggestions and crammed with subliminal data banks at speeds of up to 200 pages per second. When the 186 of us who had been 'programmed' by mind-control experts were prematurely released - ostensibly for cheating on our final exams - everyone had believed it. In my case, I had been lucky; because, for some reason, one of my instructors had 're-programmed' my mind just before I had left the Academy. I was reasonably certain that he had been a man of good report, a Christian, now, according to 'old friends'. It was partially his interference in my conditioning that had allowed me to eventually remember things that had been locked-away in my subconscious mind - even without knowing the proper access key sequence.

It had been quite spooky at first when I started having those incredible dreams. Three years had passed since I left the Academy; and I had not immediately associated my newly-found 'dream library' of science with the previous subliminal training. 'Visions' of new types of circular-shaped aircraft and spacecraft along with their associated technology had paraded through many of my early-morning 'dream times '.

As I sat and pondered the weight of the years of discovery and understanding that had led me to that moment there in the study, I suddenly felt very tired, very old for my age of thirty-three. A few moments of self-pity passed until I remembered some of the others who had carried the burden of this information. "*Dr. Jessup must have known,*" I thought, "*but they killed him with that pathetic suicide hit. What about Prof. McDonald; he must have known, too. Wasn't it strange about him? a leading atmospheric physicist, champion of the civilian UFO research effort, arch enemy of Dr. Hynek, suddenly, he discovers the CIA involvement in the UFO cover-up and, presto, McDonald suicides under most peculiar circumstances.* "My mind felt like a suitcase-crammed so full it couldn't be shut.

"*Hynek.. yes... what about him?*" I mused, "*..wasn't it odd how he of all people replaced Professor McDonald in the civilian UFO research society?*" I remembered my encounter with Hynek over in Melbourne, "*Let's see, when was it?, '73?, Yup, had to be. I wonder if he still fronts for the CIA boys?*"

A bird flew past the study window and, briefly, I wished I could fly again. Those were the days. My favorite instructor pilot (I could never forget his name) had been Captain Miracle. I had really loved those hours in the sky - soaring and sailing in that tiny silver speck of a T-33 jet. It had seemed like being in another world. Up there, the scene had always been crystalline - especially above cloud level. My gaze fell upon the telephone scratch pad; and my thoughts returned to the conspiracy.

The Cosmic Conspiracy Final Edition 2010 © Stan Deyo 2010

THE
COSMIC
CONSPIRACY
FINAL EDITION 2010

by

Stan Deyo

Publisher

Deyo Enterprises LLC (DEnt)

Pueblo West, Colorado, USA 81007

The Cosmic Conspiracy

FINAL EDITION 2010
2nd printing 2014

Copyright © Stan and Holly Deyo 2010
ISBN: 978-0-9727688-7-0

Previous Printed Versions Copyrighted in:
1978, 1980, 1981, 1982, 1983, 1988, 1992, 1998, 2002
In the following languages: English, French, German and Greek

Deyo Enterprises LLC

P.O. Box 7711

Pueblo West, Colorado, USA 81007

Web Site:

http://standeyo.com/

Email:

standeyo@standeyo.com

ACKNOWLEDGEMENTS

The following persons and organizations have contributed significant amounts of their time and resources to help me research, write and print this book over the last 15 years. So the reader may know that I have not accomplished this work without their help; and so I might publicly thank them for that help, I list their names below with thanks:

ABC Radio	Guy Baskin	Mike Jordan
Advance Press Pty Ltd	Harry Godfrey	Miss D. Marsden
Alan Gibbs	Hewlett Packard Computers	Neil Watson
Alan McGervin	Holly Deyo	Perth Christian Centre
Aviation Studies International	Hugh Kitson	Peter Bruechle
Bill & Shirley Tree	Interavia Magazine	Peter Moses
Bob & Sandy Gregory	Jack & Win Greeve	Peter Nicholls
Bruce & Kathy Webster	John Spall	Rick & Regi
Civil servants & "agents"	John Schuessler	STW Channel 9 Perth
Colin Winter	John Whitfield King	Ted and Shirley Scott
Edward Scott	John X, a "brother" in Perth	Ted Middleton
Frank & Pat Lathlean	Larry Hannigan	The Jayes Family
George Spall	Louise Deyo	Tom Stevens
Greg Hunting	Malcolm & June Dawes	Vic & Jenny Longbon

To the "Groom" - Credit for the work of His "Bride."

FOREWORD

This final edition of the book contains the first 200 pages as printed in 1978 (complete with warts) followed by the 1992 update pages, the 1998 update pages and the 2010 update pages and an expanded word index.

Stan Deyo
10 February 2010

TABLE OF CONTENTS

[Code: **THE.WED** are the 1st and 6th]

I remembered how word had passed down the FBI ranks to me and my ilk who had acted as patriotic informers on various international companies' unconstitutional activities in Middle-East -related affairs. *"There's an intelligence war going on between Hoover and the CIA. You guys had better make tracks and wait till you're called back."* I had then been sent to Australia to 'keep a low profile' until Hoover recalled us. My last instructions had been, *"stay put; but if Hoover is either replaced by the president or dies suddenly, you will be on your own from that time forward until the 'victors' find you."* The rest had become history.

I grabbed a pad of paper and a pen. I began to write. My 'life insurance' began to emerge in the words that flowed onto the pad. The hours passed swiftly as my notes formed into three categories. Firstly, I knew that the real origin of many of the so-called UFOs had to be explained along with a lot of other suppressed, super technology. Secondly, I knew that the suppressive organization, itself, would have to be defined. Thirdly, with all the darkness of the previous two topics a little 'light' would have to be shed. I leaned back in my chair; took a deep breath; said a brief prayer and wrote the three words which would begin a most incredible and, as yet, unfinished adventure for my wife and myself, **"The Cosmic Conspiracy."**

This book which you now hold is the first fruit of that incredible adventure. It is written in three levels of understanding or codification. Kabbalistic gematria and symbolism have been used to veil certain messages from some while enlightening others. I am sure the reasons for this will become most obvious as the world events of the approaching times and seasons unfold. Let us begin with a 'precise mechanism ':

Interested, I looked upon that sculptured place...

Challenged to *find the* one piece *missing*:

Hailed the topmost third - that *capstone* grace.

This I knew: *it was* not *lost.*

Hark: soon returns the *Word*,... and Order.

Until that hour, this is my *chair*:

Service to some *degree* like: *Commander Noah.*
signed: Stan Deyo
on the 28th of the 11th, 1978.

CHAPTER 1: Sightings and Suspicions

It is not this author's intention to waste the reader's time in a deluge of reported UFO sightings and contact cases. Although reference will be made to some five such incidents, it is felt that the majority of this discussion would best be spent probing the real evidence, in an analytical manner. This author has been asked many times, *"Do you believe in flying saucers or UFOs ?"* He has always answered, *"No, I don't* **believe in** *them. I believe they exist, but they aren't my God."*

Yet, some things are certain: Electric "flying saucers" are a real phenomenon and they are intelligently-controlled; some are man-made; some are not; some are modern and some are ancient. Many government agencies have covered-up major findings concerning the UFO situation while many allied agencies have never known anything has actually happened; some multinational corporations have even participated in various stages of research and development of the advanced, technological processes which have ultimately produced electrically-propelled, circular-winged aircraft, submarines, and spacecraft. Contacts with extra-terrestrial beings have been reported by some of the most credible sources on the planet. These things are documented in the following pages.

The Australian Incident

21 October, 1978: The incident occurred on a flight between Melbourne, Australia and a small island to the south of Melbourne. An experienced instructor pilot for the Air Training Corps was alone on a routine, night flight to King Island. His aircraft, a Cessna 182, was flying at an altitude of 4,000 feet. The pilot, Frederick Valentich, was only flying to the island to pick up a load of lobsters for some of the officers of the A.T.C. His flight plan gave his estimated arrival time at King Island as 7:28 p.m.

Valentich had planned to make the round trip that same night and to be back in Melbourne by 10:00 p.m. He never made it to King Island nor, much worse, back to Melbourne.

The official transcript of the last voice contact held between the missing pilot and Melbourne's flight service unit was released twenty-four hours later by the Australian Department of Transport. It is self-explanatory: 7:06 p.m. - The pilot asks Melbourne flight control if there is any known traffic in his area below the 5,000 foot altitude,

Melbourne control responds: *"No known traffic."*
Valentich: *"Seems to be a large aircraft below 5,000 feet."*
Melbourne control: *"What type of aircraft ?"*

Valentich: *"I can't confirm. It has four bright lights. Appear to be landing lights. Aircraft has just passed over me about 1,000 feet above."*

Melbourne control: *"Is large aircraft confirmed?"*

Valentich: *"Affirmative. At the speed it's traveling, are there any RAAF aircraft in the vicinity?"*

Melbourne control: *"Negative."*

7:08 p.m. Valentich: *"Melbourne, it's approaching from east towards me. It seems to be playing some kind of game. Flying at speed I cannot estimate."*

Melbourne control: *"What is your altitude?"*

Valentich: *"4,500 feet."*

Melbourne control: *"Can you confirm you can't identify aircraft ?"*

Valentich: *"Affirmative."*

7:09 p.m. Valentich: *"It's not an aircraft. It's...."* (apparently, the transmission was broken here, but only long enough to miss the magic words).

Melbourne control: *"Can you describe aircraft ?"*

Valentich: *"It is flying past. It is a long shape. Cannot identify more than that. Coming for me right now. It seems to be stationary. I'm orbiting and the thing is orbiting on top of me also. It has a green light and sort of metallic light on the outside... It just vanished."*

Melbourne control: *"Confirm it has vanished."*

Valentich: *"Affirmative. Do you know what sort of aircraft I've got? Is it military ?"*

Melbourne control: *"No. No military traffic in the area."*

7:12 p.m. Valentich: *"Engine is rough-idling and is coughing."*

Melbourne control: *"What are your intentions?"*

Valentich: *"Proceeding to King Island. Unknown aircraft now hovering on top of me."*

Melbourne control acknowledged Valentich's last message before they heard a *long metallic noise on his radio frequency*. Then it was silent. Neither Valentich nor his aircraft has as yet been found. Subsequent interviews with the pilot's father, Guido Valentich, have revealed that his son had told him, *"he had seen classified material at the Sale RAAF base which confirmed his beliefs* (in the existence of UFOs)." The author's comments on the classified RAAF UFO files will be made in the next chapter.

The Time Slip

25 April, 1977: A Chilean Army corporal, Armando Valdes, was leading a routine border patrol northeast of Santiago on the Bolivian frontier when he and his men saw an intensely bright light about a quarter of a mile away. It appeared to move too rapidly for any normal source; so Corporal Valdes told his men to wait for him while he went to investigate the flying light.

A few minutes passed and his men became a bit anxious for his safety; so they searched for him and found he had apparently vanished from the area. Suddenly, the bright light reappeared nearby, and then it, too, vanished. Fifteen minutes passed as the soldiers searched for their missing leader. Then, just as suddenly as the flying light had vanished, Corporal Valdes reappeared in the midst of his men. His haggard and surprised face had suddenly sprouted several days' growth of beard on the chin.

His watch showed the correct time, which was fifteen minutes after the light had left - but it's date was wrong; as it showed five days had passed! He collapsed after recognizing his men with, *"Muchachos!"*

A few minutes later, he remembered nothing of his ordeal. Further comments were not offered by the military authorities at Arica on the Peruvian frontier where the Corporal was later transferred. How can a man live five days in fifteen minutes?

The Iranian Incident

18 September, 1976: A UFO was spotted somewhere over Iran. Two interceptor jets armed with missiles were scrambled to bring it down. Both jets found the target and fired. Neither direct hit had any discernible effect upon the UFO. Brief moments later, an explosion was heard. The details of its cause and its relationship to the deaths of the two pilots and the destruction of their planes are not available to the general public. The official U.S. Air Force policy on the incident was stated by Captain Kenneth A. Minihan of U.S. Air Force Intelligence on 27 May, 1977, *"A copy of an intelligence report concerning the incident was found within this Headquarters; however, the Air Force does not have disclosure authority for this report."* He went on to say that any further dialogue on the incident would have to be held with the **Defence Intelligence Agency**. The main clue here is that the issue comes under the **Defence Department**. Defence from what? It is interesting to note that Colonel Benedict L. Freund - Commander of the U.S. Army Research and Development Group in Europe had previously stated on 18 March, 1975, that the U.S. Air Force had been doing an ambitious research effort into *"anti-gravity propulsion"* under the code name **Project Bluebolt**. He added that he thought the Air Force was no longer conducting such research and that the **Advanced Research Projects Agency** might possibly be supporting such research for the **U.S. D.o.D.**. Colonel Freund felt it even more probable such research was the *"responsibility of either ERDA, NASA, or NSF."*

The Electrified Cattle

8 March, 1975... By this date, following a period of only a few months the states of Minnesota, North Dakota, South Dakota, Nebraska, Kansas, Oklahoma and Texas, along with various regions of Canada reported the deaths of over 200 cattle due to some very bizarre causes. Some of the cattle had been surgically mutilated with such precision that no blood had spilled in the process; others had been killed by having all their blood drained out of them, without a trace of where it went. The areas around these macabre killings were not, however, without some other signs of abnormalities. Tree branches were found sheared-off in some spots while other areas had traces of abnormal radiation. Still other areas sported strange-looking spots that appeared to have been caused by intense heat of some sort. The key clues in these events were the bloodless corpses and the *"strange burn marks on the ground."*

In 1905, the famous Dr. Nikola Tesla (whose work will be discussed in Chapter 4) published a paper on the effects of low-frequency, high-voltage, electrical discharges on human beings. The following is an excerpt from his discourse:

"My arm is now traversed by a powerful electric current, vibrating at about the rate of 1,000,000 times a second... All around me the electro-static force makes itself felt, and the air molecules and particles of dust flying about are acted upon and are hammering violently against my body. So great is this agitation of the particles, that when the lights are turned out you may see streams of feeble light appear on some parts of my body.

"When such a streamer breaks out on any part of the body, it produces a sensation like the pricking of a needle. Were the potentials sufficiently high and the frequency of the vibration rather low, **the skin would probably be ruptured under the tremendous strain, and the blood would rush out with great force in the form of fine spray or jet so thin as to be invisible.**"

Astronauts See UFOs

February, 1976... Astronauts General James McDivitt and Colonel Gordon Cooper released statements to UFO investigator, Lee Speigel, for his record entitled, **UFOs: The Credibility Factor.** McDivitt gave a detailed account of how he had photographed a white, cylindrical UFO with a white pole sticking out of it while he was orbiting Earth in his Gemini IV capsule during June of 1965. He maintained that he had taken photographs of the object with two cockpit cameras. When he had landed, they had checked with North American Air Defence Command to see if any *"bogies"* had been co-tracked with his capsule. None had. McDivitt later reviewed every photograph he had taken during the mission, finding no trace of his UFO. Who doctored the film?

Colonel Cooper's experiences with UFOs occurred some years before he joined NASA. Cooper was stationed with a fighter group in Germany. According to him, they witnessed numerous UFO sightings with incredible performance characteristics over a three-day period. The objects were roundish and very *"metallic-looking."* What is it that metal *"looks like?"* Is it dull? Is it shiny? Does it glimmer in the light? Cooper went on to say that these devices which flew in weird patterns zoomed so high overhead that none of their fighters could get really close to them. They made instant stops and sudden 90-degree turns.

The Recurrent Clues

7 June, 1967: The following is an extract from the late Professor James E. McDonald's statement to the Outer Space Affairs Group of the United Nations concerning the international scientific aspects of the problem of UFOs:

"I also emphasize that there are innumerable facets of the UFO phenomena which I can only describe as utterly bewildering and inexplicable in terms of present-day scientific and technological knowledge. I would also remark that if these objects are not of extraterrestrial origin, then alternative hypotheses that will demand consideration will be even more bizarre, and perhaps of even greater scientific interest to all mankind."

"A wide range of electromagnetic disturbances accompanying close passage or hovering of the UFOs is now on record throughout the world - despite this record not yet being admitted into what one would ordinarily call the 'scientific record'. ***Disturbance of internal-combustion engines coincident with close passage of disc-like or cylindrical unconventional objects*** *is on record in at least several hundred instances... Often the disturbances are accompanied by broad-spectrum electromagnetic noise picked-up on radio devices.*

In many instances compasses, both on ships and in aircraft, have been disturbed. Magnetometers and even watches have been affected. All these reports, far too numerous to cite in detail, point to some kind of electromagnetic noise or electromagnetic side-effects."

Many close contact reports have also reported lingering headaches. Some of them reported humming or buzzing sounds. A few reported seeing the discs rotate counter-clockwise in the northern hemisphere; heat-distortion waves have surrounded other UFOs while still others have left scorch marks behind them. Nearly all have reported extremely bright lights; and some of those appeared as "metallic lights."

For reasons which will become more obvious as this discussion progresses, it is suggested that the evidence given points to a craft with the following characteristics:

1) It is powered by an electric process.
2) The power generation or energy storage process gives-off scorching heat.
3) The electric process involves resonating, magnetic fields.
4) The electric process generates a field that spirals around its vertical axis - giving rotation to the disc as it interacts with the Earth's magnetic fields.
5) The resonating magnetic fields also produce external inductive heating and current flows in nearby objects.
6) The field effect of the electric process does not harm the occupants - who mostly appear humanoid.
7) The electric process is of such high voltages that the air is ionized around the craft giving a bright metallic lustre to it.
8) The electric process can also be used to inject pain, subliminal suggestion, and conscious communication to living creatures.
9) The electric process enables the craft and the crew to execute high-velocity maneuvers without the normal structural fatigue.
10) Because of the strength of the recorded magnetic effects, the audibility of the field resonance, and the ionization of the air only one conclusion can be drawn: the electric process is a high-voltage, high-current, resonating field of low frequency.

A detailed discussion of a device exhibiting most of the preceding characteristics forms Appendix No. 6, which may be considered too technical by some to be incorporated in the main text; however, it and Dr. Nikola Tesla's work hold the key to fully understanding the processes which generate the above factors. This may be the reason that Tesla's papers were confiscated by the U.S. Government at his death in 1943. The only other detailed public source of his work is located behind the Iron Curtain in Belgrad, Yugoslavia; but that is not as limiting a situation as it may sound; for other sources have also continued Tesla's work and have actually published their work on his electric propulsion processes in 'free' countries.

The Cosmic Conspiracy Final Edition 2010 © Stan Deyo 2010

CHAPTER 2: Electro-Gravitic Propulsion

The Talbert Series

Dateline: 20 November, 1955; Mr. Ansel E. Talbert, military and aviation editor for the *New York Herald Tribune* began a series of three articles covering the, then current, world-wide research efforts to conquer the secret of gravity as a means of obtaining energy to propel various descriptions of both aircraft and spacecraft. The following excerpts are from the unabridged original which forms Appendix 1.

"Conquest Of Gravity Aim Of Top Scientists In U.S."

"The initial steps of an almost incredible program to solve the secret of gravity and universal gravitation are being taken today in many of America's top scientific laboratories and research centres."

"...the current efforts to understand gravi0ty and universal gravitation both at the sub-atomic level and at the level of the Universe have the positive backing today of many of America's outstanding physicists.

*"These include **Dr. Edward Teller** of the University of California, who received prime credit for developing the hydrogen bomb; **Dr. J. Robert Oppenheimer**, director of the Institute for Advanced Study at Princeton; **Dr. Freeman J. Dyson**, theoretical physicist at the Institute, and **Dr. John A. Wheeler**, professor of physics at Princeton University, who made important contributions to America's first nuclear fission project."*

Dateline: 21 Nov., 1955; Mr. Talbert's second article gave even more startling commentary. The following are also from the unabridged version in Appendix 1:

"Speeds Of Thousands Of Miles An Hour Without A Jolt Held Likely"

*"Scientists today regard the Earth as a giant magnet. Many in America's aircraft and electronics industries are excited over the possibility of using its magnetic and gravitational fields as a medium of support for **amazing 'flying vehicles' which will not depend on the air for lift.***

*"**Space ships capable of accelerating in a few seconds to speeds many thousands of miles an hour and making sudden changes of course at these speeds without subjecting their passengers to the so-called 'G-forces' caused by gravity's pull also are envisioned.** These concepts are part of a new program to solve the secret of gravity and universal gravitation already in progress in many top scientific laboratories and long-established industrial firms of the nation.*

William P. Lear, *inventor and chairman of the board of* **Lear, Inc.**, *one of the nation's largest electronics firms specializing in aviation, for months has been going over new developments and theories relating to gravity with his chief scientists and engineers.*

...He is convinced that it will be possible to create artificial electrogravitational fields 'whose polarity can be controlled to cancel out gravity.

"...'All the (mass) materials and human beings within these fields will be part of them. They will be adjustable so as to increase or decrease the weight of any object in its surroundings. They won't be affected by the earth's gravity or that of any celestial body.

'This means that if any person was in an anti-gravitational airplane or space ship that carried along its own gravitational field - no matter how fast you accelerated or changed course - your body wouldn't any more feel it than it now feels the speed of the earth'.

"...Eugene M. Gluhareff, president of Gluhareff Helicopter and Airplane Corp. of Manhattan Beach, Calif., has made several theoretical design studies of round or saucer-shaped 'vehicles' for travel into outer space."

It might also be of some interest to the reader that this author witnessed the late William P. Lear making another nationwide statement on a daytime, American television program in about 1969-70. This later statement was made in response to a question from the emcee who wanted to know what Mr. Lear envisioned the next twenty years producing in new technology. Mr. Lear told him that a person would be able to, say, walk into a New York "travel" booth - somewhat similar to a telephone box in shape; deposit his fare, push a button, and walk out the other side of the booth in San Francisco - having been "teleported" across America in seconds! The studio audience automatically laughed at Mr. Lear, much to their uninformed discredit. Mr. Lear just gaped at their performance in utter amazement. How painfully sad and lonely he must have felt at that moment when he realized the great gulf that separated the viewing audience from the realities he had already witnessed in the laboratory... He was a kind and sincere man, and this author, for one, feels a great loss at Mr. Lear's recent death.

22 November, 1955; The following statements are from Ansel Talbert's final article in the series which is fully reproduced (in the public interest) as Appendix-1:

"I firmly believe that before long man will acquire the ability to build an electromagnetic, contra-gravity mechanism that works," stated Mr. Grover Loening - formerly special scientific adviser to the U.S. Air Force on aircraft design and construction.

"George S. Trimble, a brilliant young scientist who is head of the new advanced design division of **Martin Aircraft in Baltimore** *and a member of the sub-committee on high-speed aerodynamics of the National Advisory Committee for Aeronautics, believes that it could be done relatively quickly if sufficient resources and momentum were put behind the program.*

"I think we could do the job in about the time that it actually required to build the first atom bomb if enough trained scientific brain-power simultaneously began thinking about and working towards a solution," he said.

"Dudley Clarke, president of Clarke Electronics Laboratories of Palm Springs, Calif., who years ago worked under Dr. Charles Steinmetz, General Electric Company's electrical and mathematical 'wizard' of the 1930s - is sure that this successful harnessing of gravitation will take place sooner than some of these 'ivy tower' scientists believe.

"Mr. Clarke notes that the force of gravity is powerful enough to generate many thousand times more electricity than now is generated at Niagara Falls and every other water-power centre in the world - if it can be harnessed. **This impending event, he maintains, will make possible the manufacture of anti-gravity 'power packages' which can be bought for a few hundred dollars. These would provide all the heat and power needed by one family for an indefinite period.**"

The Interavia Leak

Can anyone seriously believe that all these men were deluded dreamers with no concrete facts to build upon at that time? One science fiction story which pre-dated the above articles put another viewpoint forward. Written by Raymond F. Jones for **Astounding Science Fiction** magazine, the story, entitled, "*Noise Level*," described how American scientific and military leaders gathered the nation's best brains together, demanding that they develop 'anti-gravity' similar to that (supposedly) previously developed by an unknown inventor who had recently destroyed both his invention and himself in the process of demonstrating his process to the authorities.

The whole story about the unknown inventor had been a ruse to remove the "it-can't-be-done" syndrome from the minds of the assembled scientists. By convincing them someone else had already done it, they succeeded in getting the gathering to develop a crude, but functioning, anti-gravity disc. Were the Tribune articles such a ploy? If they were, the ruse continued into 1956 when **INTERAVIA** magazine published the article, "*Towards Flight Without Stress or Strain... or Weight*" in Volume XI, No. 5, p. 373 and 374 by the American author, **Intel**, in Washington, D.C. The article (see Appendix 2) - datelined, "*Washington, D.C. - March 23, 1956*" began by stating:

"*Electrogravitics research, seeking the source of gravity and its control, has reached a stage where* **profound implications for the entire human race begin to emerge...**

...And towards the long-term progress of mankind and man's civilization, a whole new concept of electro-physics is being levered-out into the light of human knowledge."

"*There are gravity research projects in every major country of the world. A few are over 30 years old. Most are much newer. Some are purely theoretical. Some projects are mostly empirical, studying gravitic isotopes, electrical phenomena and the statistics of mass.*" (see Appendix 3 entitled, **The Gravitics Situation** - published by Gravity Rand in London in 1956 - for more detail on the empirical approaches.)

Some of the companies involved in this phase include **Lear Inc., Gluhareff Helicopter and Airplane Corp., The Glenn L. Martin Co., Sperry-Rand Corp., Bell Aircraft, Clarke Electronics Laboratories, The U.S. General Electric Company.**

The article went on to say that various empirical tests on metals like steel, barium, aluminium, etc. had so far proven that metals could be given reduced weight at the same mass.

In fact, it appeared that the "energizing process" could produce negative weight-to-mass ratios! The steady-state, weight reductions had already reduced some materials to 70 per cent of their original weight.

The work of Dr. Townsend T. Brown on his electrogravitic discs was (at that date) the result of thirty years research according to the author. He had succeeded in building electrically-propelled disc airfoils which he had patented in the U.S. the following extract from this Interavia article illustrates the amount of success that the governments involved were willing to discuss, openly:

"A localized gravitic field used as a ponderamotive force has been created in the laboratory. Disc airfoils two feet in diameter and incorporating a variation of the simple two-plate electrical condenser charged with fifty kilovolts and a total continuous energy input of fifty watts have achieved a speed of seventeen feet per second (a little over eleven miles per hour) in a circular air course twenty feet in diameter. More lately these discs have been increased in diameter to three feet and run in a fifty-foot diameter air course under a charge of a hundred and fifty kilovolts with results so impressive as to be highly classified. Variations of this work done under a vacuum have produced much greater efficiencies that can only be described as startling. Work is now under way developing a flame-jet generator to supply power up to fifteen million volts."

Now then, it is necessary to flash-back to April of 1955 when the magazine, **Scientific American**, printed an article dealing with both the history of and current developments within the field of electrostatic. On page 110 of that discussion a most enlightening dissertation on the effects of vacuums and high voltages on mechanical forces within atoms was presented. It discussed the research of **Professor John G. Trump** (of M.I.T.'s electrical engineering department) on electro-static power generation techniques:

"Professor Trump illustrates how the power-generating capacity of electro-static machines may be stepped up by asking you to consider two metallic plates, 100 square inches in area, facing each other and separated by an insulator. If a voltage amounting to an electric field of 300 volts per centimeter is applied between them, the plates will be attracted to each other with a force of one 2000th of a pound. Increase the field to 30,000 volts per centimeter and the attraction becomes half a pound." (this meant that 1000 times the force was generated with only 100 times the voltage!) *"Now immerse the plates in a high vacuum - a good insulator, though one difficult to maintain, and increase the field to three million volts per centimeter. The force of attraction jumps to 5,700 pounds!"* (this meant that 11,400 times the previous force was generated with, again, only 100 times the voltage!!!) Returning, now, to the Interavia article:

"Such a force raised exponentially to levels capable of pushing man-carrying vehicles through the air, or outer space (a vacuum of very good insulation properties) at ultra-high speeds is now the object of concerted effort in several countries. Once achieved it will eliminate most of the structural difficulties now encountered in the construction of high-speed aircraft. Importantly, the gravitic field that provides the basic propulsive force simultaneously reacts on all matter within that field's influence.

The force is not a physical one acting initially at a specific point in the vehicle that needs then to be translated to all other parts. It is an electrogravitic field acting on all parts simultaneously.'

As much as the idea of simultaneity appeals, it is more correctly stated that sudden changes in the field's direction rapidly transmit the new inertial moment to all parts of the craft so that the elastic rebound of its atomic lattice is not distressed to the point of structural fatigue.

Note: Apologies to the reader who has found some of the preceding too complex. There was no other obvious way to document the facts to the necessary degree for convincing the "die-hard" sceptic. If this chapter becomes too involved for the reader, do not hesitate to skip it, as its main purpose is to document the conclusion that "electric flying saucers" are at least being manufactured on Earth - if nowhere else.

According to the **Interavia** article, using the electric field system to propel a saucer-shaped craft would enable the craft to accelerate to speeds of thousands of miles per hour; stop suddenly; or change direction almost instantly. Such manoeuvres could be achieved by altering the intensity, polarity and direction of the field's charge. Further detailed were some of the other implications of man's conquest of gravity:

"In road cars, trains, and boats the headaches of transmission of power from the engine to wheels or propellers would simply cease to exist. Construction of bridges and big buildings would be greatly simplified by temporary induced weightlessness, etc., Other facets of work now under way indicate the possibility of close controls over the growth of plant life; new therapeutic techniques; permanent fuel-less heating units for homes and industrial establishments; new sources of industrial power; new manufacturing techniques; a whole new field of chemistry" and *"communications possibilities..."* which *"confound the imagination. There are apparently in the ether an entirely new unsuspected family of electrical waves similar to electromagnetic radio waves in basic concept. Electrogravitic waves have been created and transmitted through concentric layers of the most efficient kinds of electromagnetic and electro-static shielding without apparent loss of power in any way."*

The Gravitics Situation

In December of 1956, Gravity Rand Ltd. - located in London, published a discussion entitled, **The Gravitics Situation**. The first page read, *"Theme of the science* (of gravitics) *for 1956-1970: Serendipity.'* It also quoted Professor Einstein's view on gravitics:

*"It may not be an unattainable hope that some day a clearer knowledge of the processes of gravitation may be reached; and the extreme generality and detachment of the relativity theory may be **illuminated** by the particular study of a precise mechanism.'*

The Gravitics Situation is the one piece of documentation that the reader will not be able to acquire without great endurance or incredible luck. It puts the finger on so may nerve centres of the body of those scientists who sired "anti-gravity" that there can be no doubt to the enlightened reader: **Mankind has developed "anti-gravity."** As a result, this document has been totally reproduced in the public interest as Appendix 3.

To fully appreciate the implications of the document, one should read the entire thing; however, certain key phrases from its forty-three pages of technical discussion have been selected to give a brief picture of its purpose. They follow below:

page 3: *"This point has been appreciated in the United States and a program in hand may now ensure that **development of large sized disks will be continued. This is backed by the U.S. Government**, but it is something that will be pursued on a small scale. This acceptance follows Brown's original suggestion embodied in Project Winterhaven. Winterhaven recommended that a major effort be concentrated on electrogravitics: based on the principle of his disks."*

page 4: *"..aims were re-written around a new report which is apparently based on newer thoughts (than Winterhaven) and with some later patents not yet published - which form the basis of current U.S. policy. It is a matter of some controversy whether this research could be accelerated by more money but the impression in Gravity Rand is that the base of industry is perhaps more than adequately wide. Already companies are specializing in evolution of particular components of an electrogravitics disk. This implies that the science is in the same state as the ICBM - namely that no new breakthroughs are needed, only intensive development engineering.'*

page 4 (con't): *"...The power of the device to undermine the electro-static forces holding the atom together is a destructive by-product of military significance. In unpublished work Gravity Rand has indicated the possible effect of such a device for demolition."*

page 7: *"If a real spin or rotation is applied to a planar geoid the gravitational equipotentials can be made less convex, plane or concave. These have the effect of adjusting the intensity of the gravitational field at will: which is a requirement for the gravity absorber."*

page 13: *"Again the principle itself will function equally in a vacuum - Townsend Brown's saucers could move in a vacuum readily enough - but the supporting parts must also work in a vacuum. In practice they tend to give trouble, just as gas turbine bits and pieces start giving trouble in proportion to the altitude gained in flight."*

The document also has a glossary of new terms on pages 17,18, and 19 and a 'gold mine' of additional reference papers listed on pages 20 and 21 which are followed by a summary of Dr. Townsend T. Brown's original patent application for his 'electric flying saucer' beginning on page 22.

Professor F. Mozer contributed a brilliant discussion on the existence of negative mass particles and their utilization in the construction of neutral-gravity bodies. It begins on page 30 of the document (Appendix 3).

Also of special interest to a few will be the brief discussion by Dr. Deser and Dr. Arnowitt which begins on page 39 of the same document. It is entitled, *"A link Between Gravitation and Nuclear Energy."* It uses Einstein's usual structures found in General Relativity as a building stone for a so-called "creation tensor" to convert gravitational energy to nuclear energy. Their field equations are difficult and are not in a readily workable state. Still, they do represent an interesting aspect for some.

This incredible document was compiled in 1956! Can there be any doubt that such information has been superseded by even more incredible developments in the last twenty-two years?... No!

The Cosmic Conspiracy Final Edition 2010 © Stan Deyo 2010

Aeronautics And Electrogravitics

The date was 30 December, 1957. The article was published in Product Engineering, Volume 28, No. 26 on page 12. It was entitled, *"Electrogravitics: Science or Daydream?"* The article gave even more clues to the then-current state of the art of electrogravitics:

A few weeks from now, at a special session of the **Institute of the Aeronautical Sciences** *(New York City, Jan. 27-31), a group of dedicated men will discuss what some people label pure science-fiction, but others believe is an attainable goal. The subject: electrogravitics - the science of controlling gravity.'*

*...**David B. Witty** noted in his award-winning essay for the Gravity Research Foundation,* **gravitational screening is crucial in all theories of gravitation.'**

*...**E.M. Gluhareff, Pres. of Gluhareff Helicopters,** suggests much progress might come if gravity were considered as 'push' rather than 'pull' - with all matter being pushed toward the centre of the earth by a sort of 'electronic rain' from outer space.'*

Such recently discovered atomic fragments as hyperons and K-particles (Dec. 2, p. 16) appear to interact with nuclear matter in ways not explained by present theories. Scientists are now suggesting that these interactions may explain, or be explained by gravity. It is even possible that gravitational energy may prove to be transformable into particles of this type."

Perhaps British aeronautical engineer A.V. Cleaver (see "Electro-Gravitics ": What it is - Or might be" by A.V. Cleaver, F. R. Ae. s., Fellow B.I.S., published in the British Interplanetary Society Journal for April-June of 1957 in Vol. 16, No.2, pages 84-94) is right in insisting that if any anti-gravity is to be developed the first thing needed is a new principle in fundamental physics, not just a new invention or application of known principles... Nevertheless, the **Air Force is encouraging research in electrogravitics, and many companies and individuals are working on the problem.** *It could be that one of them will confound the experts."*

Comments By Keyhoe

In his book of 1957, entitled, **The Flying Saucer Conspiracy**, Donald Keyhoe also mentioned some rather indicting news on pages 200-201:

On 2nd February, while visiting Bogota, Columbia, William P. Lear manufacturer of aircraft and electronic equipment, told a news conference that the flying saucers are real. When Lear's story was flashed to the United States by the AP, it was a hard blow for the UFO censors. But this was only the beginning. Within twenty-four hours Lear amplified his first statement: "I feel the flying saucers are real ', he said, 'because of four points'.

1st, *he said, there have been numerous manifestations over long periods of time.*

2nd, *many observations have been made simultaneously by reliable observers.*

3rd, *there are great possibilities linked with the theory of gravitational fields.*

4th, there are now serious efforts in progress to prove the existence of anti-gravitational forces and to convert atomic energy directly to electricity.'

"This new AP story **dismayed the Pentagon, for it could easily disclose our top-secret research to duplicate the UFOs propulsion.** *There had already been*

one hint despite Pentagon precautions. *During a meeting of aviation leaders in New York, on the 25th of January, G.S. Trimble, vice-president of advanced design for the Glenn L. Martin Aircraft Company, had made an amazing disclosure...* '**Unlimited power, freedom from gravitational attraction, and infinitely short travel time are now becoming feasible'**, *he told the press. Then he added that eventually all commercial air transportation would be in vehicles operating on these fantastic principles.'*

Can one really believe that such keen interest and such wide-spread research over twenty-five years ago was all for nothing? Can one really believe that, even if the concepts of "electrogravitics" were proven false, no papers have been printed which discussed the errors? No...

The situation is clearly this: There are two sources of "UFOs" or "Flying Saucers." One is man-made from the mid-fifties; and the other has been with mankind since the ancient days of the Old Testament and the Epic of Gilgamesh. One wonders why neither source has identified itself to mankind. In addition, one wonders if the "elder source" did not infiltrate and take control of mankind's fledgling flying saucer research and development programs of the last two decades... Yes, one wonders.

Other Gravitic Research

The Russians Have Never Been One To Be Left Out. They Have resorted to many types of subterfuge to obtain the secrets of anti-gravity propulsion systems. To this end, they enticed one of the west's most brilliant physicists, a former "member of the fathers of the hydrogen-bomb group," to defect to Russia in 1950. As it turns out, he was a KGB employee. His name: Dr. Bruno Pontecorvo.

Since at least 1961, it has been known by the CIA that Pontecorvo had successfully demonstrated a gravitic aircraft (which had no 'engine') for the Russians. It has also been passed to this author by former intelligence personnel that the chief of all Russian antigravity research is Dr. Andrei Sakharov. His American counterpart has been identified to this author by American intelligence agents as Dr. Edward Teller who has worked in close association with more than fifty U.S. anti-gravity research programs since the early 1950s...(more on him later).

In the October 1961 issue of *Practical Mechanics* , an article was published by I. A. Van As, entitled, *"Anti-Gravity The Science of Electro-Gravitics"* which observed: *"An anti-gravity machine is not impossible and many countries including Russia are at present investigating this new approach to aviation. Canada has its 'Project Magnet' which is the production of an anti-gravity machine using the electrogravitic principle. Many American aircraft manufacturers are spending millions of dollars on the use of gravity as applied to their industry. A number of universities are also going into the problem, which, incidentally, is not a new one. An actual flying model using this principle was made in England before the war."*

Three years passed quickly, and yet another startling announcement was made by a Major Alexander P. de Seversky in the October 1964 issue of the *Popular Mechanics* magazine. Pages 34-37 and 121-122 disclosed details in (including photos and illustrations) a news item entitled, *"Major de Seversky's Ion-Propelled Aircraft."* Major de Seversky was a former WWII consultant to the U.S.. Chiefs of Staff in the formulation of basic, U.S. air -strategy concepts.

The article stated, *"He also contributed to the designs of the p-35 and p-43 which led to the development of the p-47 Thunderbolt, one of America's most effective wartime fighter planes.'*

He had developed a working, light-weight model aircraft that "flew with electrons" for his employer: Electron-Atom, Inc., of Long Island City in New York. His aircraft which was a two-ounce, saucer-shaped model required 90 watts of power (30,000 volts at 3 milliamps) to fly. The power-to-weight ratio was .96hp per pound as compared to the .065hp per pound of the Piper Cub aircraft. However, just raising the voltage level to 3,000,000 volts would have produced a power-to-weight ratio of better than .0000192hp per pound if the previously mentioned Dr. Trump's figures were even 50 per cent achieved!

Even assuming that the fuel-consumption **rate** would increase by 100 times, the craft would travel between 50 and 100 times further than the Piper Cub... But, one must remember, *"UFOs or flying saucers, as the masses call them, simply do not exist; and, furthermore, no government of the world could be responsible for such devices"...* Oh really?

Pine Gap, Australia

Nestled in a shallow little valley at the southern foot of the Macdonnell Ranges about twelve miles by air from the dead centre of Australia is one of the modern wonders of the world. The **apparent, surface entrances** to this super-technological retreat are located in the vicinity of 23 degrees 48 minutes south by 133 degrees 43 minutes east. It is one of the top three of several very-secret, U.S. Government - financed 'bases' in Australia.

Note: It is not this author's intention to start another hue and cry for the removal of these secret facilities; it is simply to illustrate how advanced modern technology may have become. Once a person understands how very advanced these secret discoveries may have become, he is forced to realize how futile it would be to try to argue with those who possess such knowledge, such power, without at least an equivalent support power... (such is available, but this will be discussed in Section III).

The Pine Gap facility as it is informally known, is officially called the **Joint Defence Space Research Facility**; however, Australians refer to it as *"Pine Gap."* The Pine Gap facility has had several functions. Its original function was to execute research and development of space defence technology. The primary responsibility for the facility has been controlled by the U.S. Defence Advanced Research Projects Agency (DARPA) located in the U.S. at 1400 Wilson Blvd., Arlington, Virginia 22209, in the Architect Building.

During the early stages of the establishing of Pine Gap, the Director of DARPA was Stephen J. Lukasik. Key members of his staff were a) Kent Kresa - special assistant for Undersea Warfare Technology, and director of tactical technology, b) Lawrence G. Roberts - director of information processing techniques, c) Eric H. Willis - director of nuclear monitoring research, d) David E. Mann - director of strategic technology, and e) Lt. Col. Austin W. Kibler - director of the human resources research division.

Pine Gap has what is believed to be the deepest and straightest 'water bore' in Australia drilled beneath it. The bore is at least 28,000 feet deep (5.3 miles).

This bore could also be used as an underground antenna for very low frequency electricity broadcasts. As Pine Gap is supposed to be involved in both upper atmospheric and sub-surface research, it is quite feasible, if for no other reason, that the 'bore antenna' could be used to tune a gigantic 'standing wave' field around the entire planet!

Such a system might easily be tuned to frequencies from, say, 9,000 cycles per second to 14,336 cycles per second. Enough to set up a resonating, electric field to an altitude of 250 miles above the earth! However, this possibility will be explored in a bit more detail in Chapter 4 of this section.

It is also rumored that Pine Gap has a very large nuclear facility used to power its enormous 'transceiver.' Other rumored projects include high-voltage, high-energy plasma accelerators, possibly for use in new methods of power generation, a 'death ray' or plasma cannon, and even specialized power broadcasts to fuel 'electric submarines' as far away as the Indian Ocean. It is almost certain that the earlier version of Pine Gap's very low frequency transceiver which is located at the North West Cape near Exmouth Bay in Australia was and still is used to transmit very powerful undersea electric currents to U.S. submarines which trail long antennae behind them. It is also known that electricity transmitted in this way can be 'strong' enough to recharge on-board 'high-voltage batteries' known as 'plasma-dynamic storage cells.'

The Nation Review - a national Australian newspaper, had some interesting things to say about Pine Gap in their May 17-23 issue of 1974:

"The Pine Gap research facility near Alice Springs has managed to keep secret, until now, one of the most unbelievable research projects in the world."

"The United States has been carrying out continuous research into electromagnetic propulsion (EMP for short) at Pine Gap since it was established in 1966.... Nixon (former U.S. President) *last year announced 1975 as a target date for the completion of the project. At that time it was to relieve the petrol crisis.'*

"I understand that last minute flaws in the design and operation of the EMP vehicles have probably put the completion date back by four years" (author's note: that means a 1978-1979 public release date..... *Security aspects of the EMP project have included hypnotic and post-hypnotic keys implanted in personnel prior to their acceptance into the project. It is likely, however, that this technique has been replaced now that it is known that a side effect of LSD and other hallucinogenics is to remove partial hypnosis effects.'*

This author wrote to <u>The Nation Review</u> seeking additional information from the author of the previous article, William H. Martin. A reply was sent by his secretary which sought more information that it gave. Apparently, 'William H. Martin' is a nom de plume. No further dialogue was attempted.

According to several eye-witnesses, white disks about 30 feet in diameter with 'U.S. Air Force' markings have been ferried into Australia inside large, military air transports, **which have landed at one or the other of the two airports servicing Pine Gap**. Other eye-witness reports have seen these same air transports unloading incredible amounts of modern furniture, food, and other provisions which one would ordinarily expect to see in a very plush hotel. Could it be that nearby to the obvious Pine Gap facility in an underground, man-made city of multiple levels is... the real 'Pine Gap facility?'

Could it be that Pine Gap is also a so-called 'bolt-hole' to be used to shelter key U.S. personnel in the event of a natural weather catastrophe or a full-scale nuclear attack code-named **Noah's Ark**?... One does wonder.

Should anyone doubt the degree of orbital surveillance that is possible today; a 1973 press release about one of Pine Gap's other functions should convince him. The release said that Pine Gap and its sister station in Guam supported photographic satellites as part of what is known as *Operation Big Bird*. Each 'bird' weighed eleven tons; was fifty-feet long by ten-feet wide; could scan the entire surface of the earth every 24 hours; and could vary its altitude from 100 to 200 miles for 'close-look' photographs of 'interesting areas.' Whenever ground control instructed the 'bird' to take a 'close look' or 'high-resolution' photograph of some strategic area, a huge Perkin-Elmer camera would be used by the 'bird' to take a low-altitude photograph of the objective.

The resulting pictures would be so sharp that objects of only twelve inches across were identifiable. The 'birds' as well as the early-warning satellites of *Programme 647* use infrared sensors and films. It is almost certain that the Russians have equivalent systems in operation... (is 'big brother' at home tonight?').

A similar station to Pine Gap is located in the Transvaal, South Africa; but it is difficult to obtain much information about it. Most of the employees there are disguised as U.S. consular employees (I think, however, that twelve hundred consular staff is a bit excessive). My sources reported the station was located near a place called 'Krugersdorp' or 'Koedespoort ?'. It is, apparently, also linked to another VLF station at the South Pole, possibly 'Operation Deep Freeze.' Is it not a strange coincidence that the two 'grids' mentioned by Capt. Bruce Cathie have 'poles' located at the South Pole, too?

Some of the major contractors and suppliers for the Pine Gap facility have been Collins Radio, Ling-Tempco-Vought (L.T.V.) - both of Texas, McMahon Construction and I.B.M. It is also rumored that there are 'super IBM computer systems' on a floating platform - 'down the well' underneath the facility.

A VLF power transceiver, and 'electric flying machines' are not so hard to believe. Remember, L.T.V. is an aerospace company formed as a conglomerate of electronics and aircraft manufacturing subsidiaries. IBM has, also, long-ago developed mammoth computers with super-cooled, crystalline, main-memory units. These computers can recognize both voice and visual patterns. Their main-memory sizes are said to be in excess of 2,000,000,000 bytes (characters)! Is it not possible that there have been some secrets which the 'invisible government' of planet earth has kept from its subjects? Remember, there were over 100,000 people working on the *Manhattan Project* to produce the first A-bomb. Did that secret escape in time to help the Japanese?

If Pine Gap is an electric power broadcasting facility, then the disappearance of Valentich may be related. Suppose that Captain Bruce Cathie's hypothesis of 'power grids' circumnavigating the earth is correct; but that the locations which he picked from his 'grid poles' were only part of several other grids, themselves. If Pine Gap were then chosen as a 'grid pole,' one would be able to make a very interesting observation. A compass centred on Pine Gap and extended to Perth will, when scribed about the Pine Gap centre, form a circle around Australia which intersects the following areas: Perth, the VLF transceiver at Exmouth, Brisbane, Canberra, Sydney, Melbourne (along the Mornington Peninsula and **an area just off Cape Otway between the Cape and King Island!**

Was the reason Valentich said the UFO was playing a game and making passes because he had inadvertently been caught in the craft's field like an iron filing is attracted by a magnet swung past? Such an event would have given Valentich the impression the UFO was moving - when in reality Valentich's aircraft itself had been sucked into the field so rapidly (through an undetectable uniform acceleration) that his plane had passed under the UFO and had oscillated from side to side in smaller and smaller arcs until it finally stuck onto the underside of the UFO. The final metallic noise could have been either the actual contact of the two craft before Valentich dropped the mike or the radio squeal produced when the high-density, electromagnetic field of the UFO inductively burned-out his radio which was transmitting at the time.

One must also consider the possibility that Valentich had seen too much of the top-secret RAAF -UFO records at Sale AFB; and had become a potential security problem to 'those in the know.' He could have been grabbed on purpose. The suggestion is rough to say so soon after he has disappeared; but it is a distinct possibility.

Incidentally, Darwin and Adelaide fall on a circle around Pine Gap with a radius of some 800-odd miles while the previously mentioned areas fell on a circle of some 1200-odd miles. The Department of Transport (formerly called the Department of Civil Aviation) lists Pine Gap as 'R233' on the navigation maps for the area. 'R233' is listed as a restricted air space reserved for space research (i.e. do not fly over the area as shown on the map). The 'R233' space is a circle around Pine Gap with a radius of five nautical miles.

Who Was Mendelov?

The Mendelov Conspiracy by Martin Caidin was published in 1972. Martin Caidin has six other titles on the shelves which are selling quite well. The only one of his seven books which has either been withdrawn from sale or made extremely difficult to obtain is *The Mendelov Conspiracy*. The reason would be most obvious to anyone who has had the good fortune to read the book. It is thinly disguised fact in the form of science fiction about a plot to take over the world. The conspiracy is headed by a Dr. Vadim Mendelov (a physicist whose biographical sketch closely fits either Dr. Edward Teller or the Russian Dr. Andrei Sakharov ...'Sakharov' .....'Mendelov '..hm - m - m ...). The conspirators are discovered by a U.S. newsman who writes for a big paper... (remember Ansel E. Talbert ?). The newsman - named Brady - writes a series of articles on 'anti-gravity,' UFOs, and 'electrogravitics.' He is fired by the newspaper after the third article... (Talbert, too, wrote only three articles - even though his first articles were originally listed as 'the first and second in a series!'). The incredible fact is that 'UFOs' of the book were built on a principle that is an existing technology - right now. The names of the key characters and the large aircraft corporations in the book were changed ever so cleverly to conceal the real names of the conspirators - whom Caidin, himself, had obviously already identified. Why did Martin Caidin write the book? Why has it been made so scarce when the demand for it is so intense?

The points of this chapter are beacons illuminating the path for those who pursue the truth and, hence, wisdom. They will lead to an understanding of the second greatest mystery in human affairs. Pursue their lead, and the related details entwined in the remaining fabric of this book will not only leap from the pages with incredible agility, but will also lead 'he - who - seeks' to an understanding of that which is surely the greatest mystery in all human endeavor.

Sleep tight tonight your Air Force is awake!

CHAPTER 3: CONSTRUCTS OF REALITY

The Melbourne Meeting

The letters from Dr. James R. Maxfield arrived in Melbourne on the fifteenth of May 1972. They had been dictated six days earlier in Dr. Maxfield's radiation research clinic in Dallas, Texas. One of the letters instructed the recipient (this author, Stan Deyo) to have a visit with Sir John Williams. Apparently, Dr. Maxfield had previously contacted Sir John about my coming to Australia... (see copies of this letter and the other as Appendix - 4). The same letter also stated that the Chief Superintendent of the Aeronautical Research Lab at Melbourne would be contacted by both Sir John and Dr. Maxfield on my behalf (so that I might seek employment there to continue my research into 'anti-gravity'). Dr. Maxfield's letter went on to say that he and Dr. Edward Teller were planning to come to Australia in October and hoped they might 'get together' with me at that time.

The second letter was a copy of the one which Dr. Maxfield had sent to the A.R.L. Superintendent. It had told the superintendent that, although I was then working as a computer systems analyst for a well-known tractor firm in Melbourne that he (Maxfield) hoped the A.R.L. could find a position for me - **as I had been working in a field that he (Maxfield) and Dr. Teller had been *'interested in.'*** I knew what that 'field' was... 'antigravity.'

In my last meeting with Dr. Maxfield in America in 1971, I was told about various other research projects in America which had been or were under the watchful eye of Dr. Edward Teller. It was an incredible moment! It was like meeting the real Santa Claus... finding out that the mysterious 'they' really did have names and faces, and super technology. Dr. Maxfield told me how he and *'Ed Teller'* had *'sponsored other young minds'* (like mine) in the pursuits of the secrets of gravitational energy. It was mind-boggling. He went on to say that there had been over fifty 'anti-gravity' research projects in the U.S. since 1948! (Some of the results of these projects and their accompanying contract numbers from the issuing authorities will be discussed in the next chapter).

I then prepared two preliminary papers on electrogravitic propulsion for the A.R.L. as per Dr. Maxfield's instructions. After a few weeks had passed, Dr. Tom Keeble - the director of the mechanical engineering division of the A.R.L. - called me into the facility for a critique of my preliminary papers. Dr. Keeble with two of his research staff attended the closed-door meeting. A short time was spent discussing the papers and some of the somewhat embarrassing mistakes I had made when preparing the material. After this, Dr. Keeble asked why I had not stayed in America to finish my research.

I then related the long story of my FBI involvement, my training at the U.S. Air Force Academy, and my subsequent and somewhat peripheral involvement in the U.S. 'anti-gravity' research program. Dr. Keeble looked as though he wanted to say something which he felt he could not because of certain 'restrictions.' His furrowed brow framed his bushy eyebrows as he finally said, *"Look, we know your theory works; your design is not the best for a fully-operational model; but it will work. What has puzzled us the most is how you found out about it... about the project..."*

He went on to say, something like, *"We knew that the Yanks, no offence meant, and the Canadians did some mind control or tuning experiments in the early sixties; but we thought they had abandoned it because so many of the test blokes had gone mental or suicided under the subliminal effects of the conditioning. Yet, here you are as living proof that they did succeed."*

I interjected saying, *"Yes, that could be quite true; however, if it is, then you have apparently made the assumption that I am not one of those who cracked-up under the strain, haven't you?*

"Yes," he said as he smiled. *"We have made that assumption here. Furthermore, I, personally, feel that your mind is one of those that they tuned to tap into - now, don't laugh - other people's sub-conscious minds."*

"You're joking!" I interjected. *"Look, don't patronize me; if you do think I'm nuts, just say so, and let's be done with it."*

This must have convinced Dr. Keeble and his staff to chance trying to tell me some things which might ordinarily have been too risky for them to have said at that moment. The room was apparently bugged by ASIO (Australian Security and Intelligence Organization) - or so they were trying to tell me with their silent gestures toward the book shelf behind them followed with questions as to whether or not ASIO had, as yet, contacted me. In fact, one of the chaps in the meeting suggested that, because of the very quiet manner I had used to enter Australia, ASIO might not have known what information I had stored sway in my sub-conscious data bank.

Dr. Keeble leaned forward and said, ***"There are extensive motion-picture libraries of these flying saucers taken right here in Australia. The RAAF have control of these libraries.*** *I've seen them. Good stuff you Yanks have put together on that project!"* As he finished speaking, I asked him if I could also see these filmed records.

Dr. Keeble and his colleagues all said, *"Oh no, you couldn't possibly... they require a clearance that you as a Yank cannot obtain."* Yet, as they were all saying these words, with their hands, they were frantically pointing to the book shelf and in some cases miming the word, *'maybe'* with their lip movements. Their message came through..., *Whoever was bugging their room was not of the same philosophy as those people present; however, one or more of those present would make a later attempt to show me those film records.*

Dr. Keeble tried to say something more, *"You see, there's something like a group of us who... ah..."* I interjected, *"by group, do you mean club or a formal organization?"*

"No," he said, *"..scientists, engineers around the world... we feel... well...I"* and his voice trailed off leaving the sentence unfinished as he allowed one of his staff to

add further comments on the earlier subject of 'mind training' before he, himself, joined into the same discussion. "

...So you see," Dr. Keeble resumed, *"your mind can theoretically eavesdrop on the collective knowledge of all those people in the world who study or practice any of the subjects your mind might ever wish to, consciously, address as a means to solving any problem requiring conceptual knowledge which you haven't previously gained by any other scholastic or practical means..."*

The very thought made my mind race with the possibilities - the very probabilities, that what I had just heard was true!... *"But then,"* I thought, *"who would believe me, if I told them I have already experienced this school-of-the-mind effect?... Who indeed?"* One wonders what Valentich really got to see on the RAAF base at Sale; was it the actual UFO films ?....

He Went 'Thattaway'

Months passed after the Melbourne meeting and I was not taken into the employ of the A.R.L. In fact, both papers previously submitted to the A.R.L. had even been 'classified' and removed from the receiving authority at the A.R.L. Strange events began to manifest. Someone broke into my home in broad daylight; and touched nothing. When the local police were called in and shown the forced entry, their answer was, *"Sorry, mate, this is a political situation and we can't do anything about it..."*

Weeks passed and Dr. Alan Hynek came to town. I was summoned to see him by an intermediary. The meeting took place in an attorney's penthouse in Melbourne. Dr. Hynek questioned me on my knowledge of the UFO situation. Copies of the previously-mentioned A.R.L. documents were given to Dr. Hynek. After a lengthy discussion Dr. Hynek smiled and reached in his pocket producing a hidden pocket tape recorder, which he then turned off. He asked if I had ever discusses the UFO-coverup situation with the American comedian, Dick Gregory. The answer was *"No, but why do you ask?"* Hynek then said that Dick Gregory had made the same claims and had named the same responsible parties to Hynek, himself, just before Hynek had left the U.S. for Melbourne. Hynek advised me to 'keep in touch' at a particular address and telephone number in the U.S.

It was not until months later when a phone call from a 'reliable source' in Auckland, New Zealand informed the author that both Hynek and an attorney in Melbourne were CIA operatives that the situation became a bit clearer. Apparently, I had caused quite a stir in certain local and international intelligence organizations by releasing basic theories about suppressed 'flying saucer' technology to the civilian populace of Melbourne and Auckland. It became more and more obvious with subsequent 'unofficial' visits by A.S.I.O. operatives and employees from the Australian Defence Standards Lab., that I was under the 'eyepiece.'

As the pursuit of those who wanted me either dead or very tightly muzzled warmed-up, I managed to 'disappear' into the Australian scene. Although the elusive methods which I used were not unique, it may be that I will need to use the same techniques again in the near future. As a result, it must suffice to say that I 'went bush' for around a year's time, finally surfacing in the most remote city on Earth: Perth - finding it the most delightful hiding place I could have chosen, had there been a choice.

The low-profile was, however, soon discovered by the normal processes of meeting new people. It became all too obvious that the pseudonym I had adopted was no longer a 'cover' of any magnitude. It was because of this realization, that I finally agreed to make a statewide radio broadcast on the national Australian network, the ABC.

I felt that one of two things would result from doing the show: 1) I would either have, subsequent, a very short duration in life-span, or 2) The information I could suddenly release all at once might just make me too noticed, thereby insuring my extended longevity. There was nothing to lose; so when Neil Watson of the ABC asked me to do the show - live, I surfaced with a 'bang.' The switchboard jammed for three hours over the broadcast which only lasted an hour and a half. Many people made cassette recordings of the program from their radios. Hundreds of copies seemed to appear overnight in the state. The interest was so high, that the ABC authorized two more shows with only a few 'restrictions' as to what subjects could be discussed on the air. The gamble had paid off.

A year passed with the public interest climbing steadily because of the 'underground' tapes which seemed to traverse not only Australia but also many other nations. Then, Channel 9 - the local favorite in documentary television, asked me to do a thirty minute spot as a 'surprise trailer' to their forthcoming, special UFO production by director, Guy Baskin. The production was entitled, **UFOs Are Here**. I was quite willing to do the show, realizing the more coverage I could get, the longer I could expect to live on Earth...

Again, the public response was overwhelming. As a result, two more Channel 9 television specials were produced in which I was allowed to say almost everything I desired (although some items had to be omitted with my permission for the legal protection of Channel 9). The resulting ratings of the shows were a sign that the 'plot heard around the world' might not be as impossible an objective as previously thought.

In the months that have followed the Channel 9 UFO documentaries, I have also been further assisted in my efforts to 'spread the word' by Brisbane's Radio Station 4IP. Subjects such as Biblical prophecy, suppressed 'flying saucer' technology, the conspiracy to overthrow the established governments of the world, and a host of other subjects which are to be found in this book were first broadcast there at 4IP by Alan McGervin and Greg Hunting in an effort to educate their listeners not only to the many problems facing today's society, but also to the solution which I have accepted for those same problems.

Conservative estimates now indicate that 8,000 original cassette recordings of a variety of my radio broadcasts, private lectures, and the recent series of public lectures given in Brisbane are currently circulating in Australia. The second and third generation copies of these recordings raise the figure to an estimated 20,000 tapes. Assuming that each tape has been heard by at least ten persons and the average family size is four people, it is quite reasonable to say that one in every fifty families in Australia has already heard about at least a few of the topics I have discussed. Although the public interest has been gratifying, it has also become so overwhelming, that I have had to write this book to avoid needless repetition of the same facts and deductions.

Although there are yet many details of my personal adventures remaining to be told, they must now wait for some other time, and some other space; for much more important observations now need to be related to the reader...

Time And Space

One of the hottest debates in physics today is over the true nature of space: Is it a 'luminiferous ether' (see Appendix 5) or is it some abstract, ten-dimensional, Reimannian construction like Dr. Albert Einstein proposed in his *Theory of General Relativity*? If it does, indeed, require a super-dimensioned construction to explain the physical laws of the universe, one must wonder why this construction could not be replaced by one using real and observable dimensions like width, length, and depth.

If, on the other hand, space is a 'luminiferous ether' or some tenuous 'fluid,' then one wonders why the functions of the physical laws of the universe cannot be observed; and, hence, translated into a mathematical construct of reality, of three real dimensions with time expressed as a ratio of relative distances and vectors.

Strange as it may seem, space has already been properly described right here on Earth as far back as 1954! Space is a 'fine structure '.. a 'tenuous medium, fluid or field.' All gravitational, electromagnetic, and electrostatic phenomena occur as results of various interactions of energy - 'waves' in this 'fluid space.' In pages 172-174, 176, 178 and 180 of Scientific American in 1954, a brilliant discussion giving three-dimensional explanations of many nuclear phenomena (based largely on previous discussions written by Douglas Crockwell) was conducted by Albert G. Ingalls. Crockwell's explanations offer the only real solution to the apparent paradox which certain nuclear events present to the researcher: A particle sometimes behaves like a wave. The discussion stated:

*"It seems reasonable, as a first thought to accept each particle-field relationship as an inseparable something, which is perceived sometimes in one fashion and sometimes in another. We might also think of the **particle portion of the effect as that which is experienced radial to the course or potential course**. We know that some relationship of this sort exists, whether or not it is exactly as stated. Variation of one effect is accompanied by a reciprocal variation in the other effect. In other words, the more the particle-field manifests itself as a particle, the less it manifests itself as a field, and vice versa."*

*"..We also know that charged particles in motion exhibit a 'sense' or quality of right- or left-handedness which characterizes their charges. From this we can infer a kind of **tangential motion in space around the course of a particle - a motion which differs between particles of unlike charge**.'*

The discussion went on to say, *"...It is important to remember that the field does not rotate as a unit. The areas of the field vary only in the diameter and the phase of translation. As the field is explored from the centre outward, the phase of rotation lags progressively. **Hence, its structure can be considered as a series of concentric phase shells**, each 360 degrees out of step with adjoining neighbors."*

*"**The field and particle are one, and at all points the action is similar. The diameter of translation is greatest when the particle is at relative rest. An increase of particle-field velocity is accompanies by an increased rate of rotation but a smaller radius of rotation."***

The summary statements of the discussion brought out a very interesting point about James Clerk Maxwell - the Einstein of the nineteenth century:

"I submit a line from the great James Clerk Maxwell's preface to his theory of electromagnetic radiation: 'In several parts of this treatise, an attempt has been made to explain electromagnetic phenomena by means of a mechanical action..' "

A very common phenomenon illustrates Crockwell's model 'particle-field' concept. If one blows a weak 'smoke ring,' it moves slowly away in a rapidly-widening ring. If, however, one blows a strong 'smoke ring,' it moves away rapidly - maintaining a very small diameter. If a person nearby were struck by the latter smoke ring, the impulse or particle effect would be more obvious than the tangential expansion pressure on the ring.

On the other hand, if that same person were struck by the first smoke ring of less translational energy, it would not be felt as a direct impact so much as an expanding crawl over the individual's person. Although it would prove a bit difficult in practice, one could, theoretically, shoot two smoke rings at each other so that their encounter would produce either mutual annihilation or mutual enhancement dependent upon the rotation vector applied to each ring as it left the issuing orifice. The annihilation would produce a visual effect like a 'barred-spiral galaxy' while the enhancement would produce a visual effect like the 'Sombrero galaxy.' The reader who is keenly interested in the mechanics of gravity and electromagnetism must pursue the preceding lines of thought to properly understand the 'missing link' which unites the physical laws of the microcosm with those of the macrocosm.

If one views 'space' as an infinite existence - a continuum - comprised of endless levels of sub-nuclear particle-fields forming atoms, forming planetary systems which form galaxies that, in turn, form galactic cells and, etc. ad infinitum, then one can easily visualize that the 'ether' of man's particular level of existence is a 'fluid' comprised of ultra-small 'particle-fields' which, in turn, are made-up of relatively equally small 'particle-fields.' 'Time,' as such, in a continuum of such magnitude is equally relative.

'Time' is not an absolute dimension in reality. The only absolute is energy. The distribution of energy within the various levels of the hierarchy of existence creates the phenomenon called, 'time.' As the distribution of energy is not uniform, 'time' itself, is not uniform in the universe. When a person says it took him five seconds to walk across a room, he is really saying a clock pendulum moved or changed its energy-distribution level five times as compared to his own, single change of energy-distribution made by his walk across the room. Time is a ratio of changes in energy-density. 'Time' on an atom passes much faster than 'time' at the Earth level does. If a person's body were to be 'pumped' with resonant energy, it would make him age several days in only a few relative minutes to someone watching him. If, however, the person were to be 'drained' with resonant energy, it would lower his energy-density causing him to age only a few minutes in several relative days of the observer's time. How incredible it would be.

Suppose a group of scientists had to solve a very time-dependent problem fast. If they were to take their pencils and paper with themselves into a 'field' which harmonically 'pumped' their energy-densities to a higher level, 'time' would extend for them. They would have several relative days to solve their problem while only a few relative minutes of time had passed to the world outside their 'field '... fascinating! If the American base at Pine Gap could be used to 'pump' resonant, low-frequency energy into certain circles of the country, the effects could be mind-boggling. Why, in just a few days of time relative to the rest of the world, certain parts of Australia could pass several years of time relative to its occupants.

Has the reader ever had those days that seem to 'fly by?' On the other hand, if the same facility could be used to 'drain' energy from those same circles of influence, the days would seem to 'drag by' to those so influenced. If an electric air - or space-craft based on the same principle of resonance were to be suddenly accelerated into a new vector at speeds which would normally break its molecular lattice apart, a 'relative' or 'apparent' 35g acceleration could be easily amortized over a relative 'time-dilation' of 1:35 inside the field of the craft, giving the craft and its crew the relative acceleration of only 1g...! If the reader has been able to grasp the preceding dissertation on time and space, he now knows why 'UFOs' have such incredible performance characteristics. They are only relatively incredible.

The 'Gravitational' Effect

The reader is asked to examine figures (3-a), (3-b), and (3-c). All three figures represent the same 'system' in different energy stages. The system consists of a rectangular fish-tank, sealed on all sides. The pump (c) pumps water into the tank through hole (a) and extracts the water from hole (b). A small trolley car (d) with one, upright end resides on the track (f). Assuming that no air bubbles appear in the circulation patterns, when the pump is activated, the trolley car will move away from the water inlet (a) toward the end (e) with no visible means of acceleration to an observer outside the tank. The trolley car will come to rest flat against the end as shown in figure (3-b).

Now, let the same process be repeated with several holes having been drilled through the trolley car's upright as shown in figure (3-c). The trolley car will not come to rest flat against the end (e). Instead, it will stand back from the end a certain distance. To the observer, the trolley car would have oscillated toward the end (e) very briefly before coming to rest in the position shown. The reason is plain: The water from the inlet struck the upright driving it toward (e); however, some of the water passed through the holes in the upright striking (e) and rebounding back into the upright as it approached (e) creating an 'energy-cushion' (h) between the end and the upright. So, as long as the observer could not see or feel the energy 'waves' moving the trolley car he could only deduce some 'force' was at work which had either 'pushed' or 'pulled' the trolley car to the end.

If the observer could then somehow place his hand into the tank while the system was operating, without destroying the pressure seal, he could take the upright in his fingers and move it toward the end (e). Upon releasing the trolley car, he would see it move away from the end (e) returning to its former position - however, this time the observer would have 'felt' the force that tugged at the trolley car. Yet, that 'force' was actually the resultant of one force acting upon its own partial reflection from the end (e).

If the observer could then move the trolley car toward the end of the water inlet (a), he would find that the 'tugging' force (g) had apparently reversed direction as it would then be pulling *toward* the end (e). He would then realize that releasing the trolley car would allow it to '*gravitate*' to the point along the track (f) where the force of the incoming water was being balanced by the force of the reflected water.

In reality, all mass on the surface or outer shell of the Earth is matter whose nuclear density (like the holes in the upright) has caused it to 'gravitate' to that distance or radius which is the resultant of an energy-input-wave passing through its nuclear holes, meeting itself in the centre of Earth, and reflecting back toward itself.

The Cosmic Conspiracy Final Edition 2010 © Stan Deyo 2010

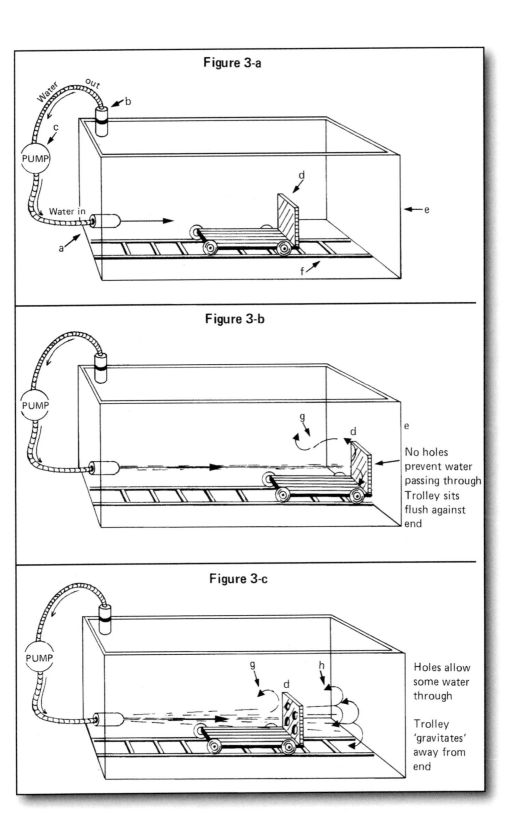

Figure 3-a

Water out

b

Water

c

PUMP

Water in

a

d

e

f

Figure 3-b

PUMP

g

d

e

No holes
prevent water
passing through
Trolley sits
flush against
end

Figure 3-c

PUMP

g

d

h

Holes allow
some water
through

Trolley
'gravitates'
away from
end

The centre of the Earth is like the end (e) while that spot where the trolley car 'gravitates to' is like the surface of Earth.

Certain conclusions follow from these observations. Firstly, 'gravity' is the resultant of a bidirectional pair of forces. Secondly, 'gravity' has higher oscillation frequencies the closer the observer approaches to the centre of the input-energy wave reflection point. Thirdly, a mass whose nuclear 'hole-spacing' gives it a greater 'particle-field' density than another with more 'holes' per unit area will not reside at the same radius from the 'centre' as the other will. It will, in fact, reside closer to the centre. This effect can be detected on the Earth's surface, by a study of the so-called 'specific gravitational nature' of the elements - which will show that 'gravity' and 'specific gravity' are the same phenomenon.

It is quite simple. Even though acceleration due to gravity is basically a constant on Earth, when a mass takes on a 'gravitational' radius from the centre of the Earth it is related to the ratio of its atomic 'particle-field' areas divided into the atomic spacing areas (or 'holes'). Therefore, those masses of the higher ratios of 'particle-field' densities to atomic 'holes' reside closer to the centre; and are referred to as more dense objects. That mass whose ratio approaches infinity converts to pure, radiant energy from the centre of the system.

The 'gravitational' effect is, in reality, the resultant of more than just two opposing forces. As any object must have spin around its own axis to have the effect, it follows that the 'gravitational' effect is the resultant of the spin:anti-spin moments and the convergent:divergent moments... This is the reason that 'gravitational' fields vary by the fourth power of their angular velocity in free space.

Such a theory of 'gravity' also predicts that 'gravitational' shells exist in any 'gravitational' field. These shells would give the illusion that 'gravity' reversed its direction as one approached the centre of the generating field. In this theory, if one were to drill a hole to the centre of the Earth, and were to drop a stone down that hole, one could expect the stone to fall only a portion of the distance to the centre before reversing direction and 'falling' back up the hole until it found its shell of balanced forces. Recent tests have even visually shown the Sun has at least one such inner shell which can be 'seen' through the sunspots. This inner shell spins so fast, that for the first time it appears that the hyperbolic equation which Laplace once formulated for orbital radius versus relative velocity is totally acceptable!

The 'Magnetic' Effect

"Magnetism" is the same type of phenomenon as 'gravity.' All things in normal states can be 'magnetized' to some degree. 'Gravity,' also, normally effects all things in its field. 'Gravity' has frequency and so does 'magnetism.' 'Gravity' has 'shells' of balanced forces and so does 'magnetism.' Where is the difference? It is quite simple. If a mass (like the Earth) spins around its own axis, the resulting reductions of pressure in the 'fluid of space' (see Appendix 5) produce 'gravity.' As the sub-particles of this mass (like electrons) orbit a chain of atoms in spiraling courses, thus reducing the pressure at right angles to the particle path in the 'fluid of space,' an interlocking series of vectors give the illusion of 'magnetism.' The motion of Earth around the Sun could be called a 'magnetic' moment while its motion around its own axis is a 'gravitic' moment.

The frequency of a nuclear magnetic field is very high when compared to the frequency of the gravitic field that contains the Earth. 'Gravity' is a source of power and so is 'magnetism.' Both are results of balanced, but dynamic forces.

By introducing shielding or a method of diverting the energy density of one 'shell' to a lower state 'shell' one can draw energy from both fields. There is more - much more to say, but it would prove to be beyond the scope of this book.

In a forthcoming publication, the author will thoroughly explain 'gravity' and 'space.' The mystery of the so-called 'North' and 'South' poles of magnets will be visually explained as will be the 'right-hand' twist of magnetic fields. A method of utilizing 'gravity' as a means of transport will also be detailed by a discussion of the use of resonating, low-frequency, high-voltage electric fields. But, these things must wait for the present moment.

Some Secret Sciences

In 1971 Dr. Oleg Jefimenko, a scientist at West Virginia University announced the world's first practical motor to be powered by the Earth's electric field. He claimed his motor could be used to avert the power shortage in the Western world. He was correct; however, the details of his process were quickly silenced. It is known that his process used an electrostatic motor to generate motion and, hence, power. His first generator apparently used a balloon-lofted wire to 'short-out' the differential voltages between earth's surface and 1500 ft above the surface.

His unit produced about one-tenth of a horsepower of continuous energy with only the earth's electric field as a power source. The only solid leaks this author could find which illustrated Dr. Jefimenko's technique were printed in the Scientific American magazine. Unfortunately, the magazine would not give this author permission to reprint the articles or the drawings of Dr. Jefimenko's generator; so the reader must follow-up these leads himself: Scientific American, October 1974, p. 126-131; and March 1972, p. 57...(Do it).

The reader should also secure a copy of the following document from either the Wright Air Development Centre at Wright-Patterson Air Force Base in Ohio or the University of North Carolina Physics Department at Chapel Hill, North Carolina: **Conference on the Role of Gravitation in Physics**, WADC TECHNICAL REPORT 57-216 (March 1957) or ASTIA document No. AD118180. This report gives 167 pages of summary on the proceedings of a conference attended by the world's 44 leading physicists who came from London, Copenhagen, France, Turkey, Stockholm, Japan, Switzerland, and America.

In 1972, a Yugoslav-born physicist, Mr. Bogdan Maglich, announced a new power source for the home. Mr. Maglich made his announcement while on leave from Rutgers University where he was a professor. His new power source was detailed in a paper to the American Physical Society. The power source was about the size of a domestic air-conditioning unit. It used colliding beams of deuterons (the nuclei of heavy hydrogen) enclosed in a magnetic field as the power source. It could 'breed' its own fuel while it generated electricity, directly, with no steam turbines or any other sort of the usual intermediary devices. What ever became of Mr. Maglich's process?

A wealth of information on classified research projects is available to the reader for a fee of around $100 from the Smithsonian Science Information Exchange located in Washington D.C. (20036) at Room 300 of 1730 M Street, N.W. All the reader need do, is join the organization (for a very small fee) and then ask for a "Notice of Research Report" on whatever subject he wishes. This author asked for, *Electro-Gravitic Propulsion Systems For Aircraft and Spacecraft* since 1975. A book of thirty project summaries was sent in response. Some of the highlights follow:

a) Pulsed Electromagnetic Gas Acceleration - supported by an unspecified unit within N.A.S.A. at Washington D.C.; annual funds in excess of $160,000; research being performed by the School of Engineering and Applied Sciences at Princeton University, Box 430, Princeton, New Jersey 08540. *The research is studying plasma propulsion devices for space flight as well as for plasma dynamic lasers.*

b) Properties and Behavior of Explosion Fields and Plasma Flows - supported by the U.S. Dept. of Defence for the Air Force under contract number DF024930 and F44620-75-C-0021; funding of $30,730 for fiscal year 1977; development by Air Vehicle Corporation of San Diego, Calif.; *the research is developing an advanced plasmadynamic electric propulsion engine with very high rate energy conversion devices. These techniques are to be used for satellite surveillance and close air support.*

c) Magnetic Field Annihilation of Impulsive Plasma Current Sheets - supported by the U.S. Dept of Defence for the Air Force under contract number DF029200 and F44620-71-C-0031; development by Thomson Ramo Woolridge Inc. of 1 Space Park at Redondo Beach, Calif. 90278; funding of $60,508 for fiscal year 1976; *the project is pursuing the conversion of magnetic energy to plasma kinetic energy through the annihilation of anti-parallel magnetic fields. Furthermore, it is pursuing the development of large, total-impulse, acceleration devices using pulsed plasma thrusters for orbit acquisition and stationkeeping.*

d) Mechanisms of Magnetoplasmadynamic Arc Jet Acceleration Processes - supported by U.S. Dept. of Defence for the Air Force under contract number DF037860 and F44620-74-C-0017; annual funds of $15,475 in fiscal year 1977; development by Techion Inc. of Newport Beach, Calif.; *with the main purpose of using the atmospheric environment as an arc jet propellant source for a magneto-plasma-dynamic, arc-operated space thruster, which would be code-named SERJ (for space electric ramjet).*

e) Experimental Research on Fluid Dynamic Energy Conversion and Transfer Processes - supported by the U.S. Dept. of Defence for the Air Force under contract number DF136500 and F33615-73-C-4053; with fiscal year funds of $313,098 in 1977; development by Universal Energy Systems, Inc. of Medway, Ohio, 45341 under the eye of the supporting agency's Flight Dynamics Laboratory at Wright-Patterson AFB also in Ohio; *purpose: to design new aerospace propulsion systems and electromagnetic weapons.*

f) New Horizons in Propulsion - supported by N.A.S.A.'s Office of Space Science Jet Propulsion Lab. at the California Institute of Technology in Pasadena, Calif. 91109; funds undisclosed; *this project was designed to explore several new areas of propulsion:*

1) the production and storage of anti-matter along with the use of matter/anti-matter mutual annihilations as a propulsive source.
2) the study of new methods of obtaining energy from the interaction between an electrically-conducting fluid in turbulent motion aboard a spacecraft and the fluctuating magnetic field of a nearby planet.
3) the utilization of various planetary atmospheres as propulsion systems,
4) the design of a practical process for converting high energy-density sources into thrust, and
5) the use of lasers as propulsion systems.

There is more.... much more. Why is it that the U.S. Government stated that they had no 'death-ray' during the 1977 controversy? One must wonder since it is common knowledge in Australia that as early as 1973 the U.S. had destroyed an Australian drone aircraft over five miles away with a ship-board plasma cannon ("Sky-Cannon ") which could only be photographed with special cameras operating at over fifty thousand frames a second!

Consider the statement of Dr. Stanton T. Friedman - a well-known nuclear physicist, who said that he had personally assisted on various developmental stages of ion-thruster engines for "small companies" like Westinghouse, General Electric, and General Motors. In fact, he stated that Westinghouse had already test-fired an 1100 megawatt ion-thrust engine sometime before 1970. The proof is everywhere... everywhere... but, why hasn't the reader been told? This question must be answered.

CHAPTER 4: RADIANT GENIUS: TESLA

On the stroke of midnight between the 9th and 10th of July in the year 1856 in the small Yugoslavian town of Smiljan, one of the most distinguished inventors of modern times was born. His name: Dr. Nikola Tesla.

Dr. Tesla was unquestionably a prolific genius in several aspects. By 1884 he had acquired a detailed knowledge of some twelve languages while attending various European universities. His mental ability was also quite exceptional in another way because his mind could 'conjure-up' visual images to represent words spoken to him. His photographic memory coupled with his mental 'animation facility' gave him incredible advantages in problem-solving.

Between 1882 and 1888, he patented many devices which employed the use of rotating magnetic fields and which produced a method of transmitting electricity with alternating currents. To those who do not realize what this means, all the world's modern electrical generation and transmission equipment is designed around the original ideas and patents of this man, Dr. Nikola Tesla - who once loved the American ideal so much that he became a naturalized U.S. citizen in 1889. If the reader is like most people, his first question would be, "Who?.." It is a pity, but his name and the credit which belong to him have both been suppressed in the "interests of national security.'

Dr Tesla invented a unique process for generating extremely high voltages between 1889 and 1892. The type of device which he designed to generate these voltages still bears his name: The Tesla Coil. During these same years, he also patented several types of isochronous oscillators. Then, between 1891 and 1893, he patented the Tesla wireless (radio telegraph) system and developed "cool process" electron tubes. Between 1896 and 1898, he published a still valid theory of radioactivity and radiated energy. During those same years, he developed high-potential vacuum tubes (the forerunner of all the fluorescent lights the world uses today). In 1899 he presented a radio-controlled, electrically-powered submarine to the U.S. Navy. They failed to see any practical use for it and dismissed the idea!

However, one of his most incredible achievements was also accomplished in 1899; and it was not rejected by the Morgans, Rockefellers, and Westinghouses of the world. In 1899, he published photographs and papers (to a limited readership) detailing his **discovery of terrestrial resonance and the law of propagation of conduction currents through the globe. His paper also gave details of his high-potential, wireless electrical broadcasting station which broadcast usable power - not radio communications - by transmitting energy in the form of 'standing waves' or stationary waves' in the Earth's crust and upper ionosphere!**

Tesla Electrifies Earth

During the summer of 1899, Dr. Tesla managed to construct and subsequently test one of the most incredible devices mankind has yet to see. Dr. Tesla had convinced Colonel John Jacob Astor (of Waldorf Astoria fame) to gamble $30,000 on this 'summer test.' The dry goods company named Simpson and Crawford contributed another $10,000 to the project while Mr. Leonard E. Curtis who owned the Colorado Springs Electric Company contributed both land and the use of one of his Colorado Springs power-generation plants.

Tesla had chosen the Rocky Mountains for their frequent supply of highly-charged thunderstorms. It was Tesla's idea to tap into these charged clouds from the ground using a tuned circuit. He had already determined that the lightning flashes from these storms seemed to discharge along a specific 'stationary wave' pattern that (he had theorized) encircled the entire globe of Earth. He built a gigantic Tesla coil which used the Earth's surface as one 'plate' of a spherical capacitor to the former. He linked them with a dielectric medium "the atmosphere. To overcome the problem of distance between the Earth and the ionosphere (which ranges from 25 miles up to 250 miles up). Tesla had aimed one end of the secondary coil of his gigantic Tesla coil at the ionosphere - the idea being that the charge-crowding effect at the ball-capped tip of the 200 foot pole above the secondary would raise the effective voltage between the earth and the ionosphere. This would have had the same effect as bringing them closer.

His device consisted of three main points: 1) a secondary of seventy-five turns having one end grounded into the Earth with a large copper plate and the other end attached to a 200 foot tall shaft topped with a three-foot diameter ball; 2) a heavy gauge primary coil of seventy-five feet in diameter which surrounded the secondary and was inductively linked by an air gap; and 3) an electromechanical circuit to provide the high frequency (circa 150 kc) power source to drive the primary.

The interested reader is encouraged to obtain a copy of Tesla's patents numbered: **a) No. 649,621 of 15 May, 1900** from the original specifications filed 2 September, 1897 under serial No. 650,343; **b) No. 685,953 of 5 November, 1901** from the original application of 24 June 1899 under serial No. 62,315; c) **No. 685,954 of 5 November, 1901** from the original specifications filed 1 August, 1899 under serial No. 62,316; **d) No. 685,956 of 5 November 1901** from the original specifications filed 1 August 1899 under serial No. 725,749; **e) No. 685,012 of 22 October, 1901** from the original specifications filed 21 March 1900; **f) No. 568,178 of 22 September, 1896** from the original specifications filed 20 June 1896 under serial No. 596,262; **g) No. 577,670 of February, 1897** from the original specifications filed 3 September, 1896 under serial No. 604,723; **h) No. 583,953 of 8 June 1897** from the original application filed 19 October, 1896 under serial No. 609,292; and **i) No. 593,138 of 2 November, 1897 from the original application filed 20 March, 1897** under serial No. 628,453.

(Note: the reader may be able to obtain all the preceding extracts in addition to thousands of other pages and photos of Tesla's works from the TESLA BOOK COMPANY, P.O. Box 1685, Ventura, California 93002).

Tesla's Colorado experiment was done at a spot very close to where this author's alma mater, the U.S. Air Force Academy, now stands. This author has spent many fascinated hours perched there some 7,000 feet up the slope of 'El Diablo Mountain' (the devil mountain) watching those magnificent thunderstorms of the summer season in the 'Rockies.'

It is, perhaps, for this reason that Dr. Tesla's words of 1904 concerning his highly successful Colorado test of 3 July, 1899 struck such concord in this author's own thoughts... (from the <u>Electrical World and Engineer</u> of 5 May, 1904), *"...I first obtained the first decisive experimental evidence of a truth for the overwhelming advancement of humanity. A dense mass of strongly charged clouds had gathered in the West* (over Pike's Peak) *and toward evening a violent storm broke loose which, after spending its fury in the mountains, was driven away with great velocity over the plains. Heavy and long persistent arcs* (lightning flashes) *formed almost in regular time intervals.... no doubt whatever, we were observing stationary waves. Subsequently similar observations were also made by my assistant, Mr. Fritz Lowenstein* (who later saw) *the true nature of the wonderful phenomenon...The tremendous significance of this fact in the transmission of energy by my system had already become quite clear to me..."*

*"**As the source of the disturbances** (which was the electrical thunderstorm) **moved away discharges came successively on their nodes and loops. Impossible as it seemed, this planet, despite its vast extent, behaved like a conductor of limited dimensions...**"*

Tesla had discovered a process which could transmit almost unlimited amounts of electrical energy to any place on Earth with negligible losses. He had managed to estimate the resonant frequency of the Earth -to-ionospheric cavity at 150 kc. It was a good guess; but one which has later proved not as efficient as possible because the cavity resonance changes from moment to moment - depending on solar wind densities and sunspot activity.

The United States has a very low-frequency transmitter located at the South Pole. It is similar to Tesla's process, except that it can be tuned to a frequency in the range of 7 Hz. Even-numbered super harmonics of this frequency such as 14,336 Hz (two to the eleventh power times 7 Hz) have been used by both the U.S. Navy's VLF transmitter at Australia's Northwest Cape and the American Defence Advanced Research Projects Agency's VLF facility at Pine Gap in the dead centre of Australia.

It may be significant that the $10,000,000,000 North American Air Defence Command (NORAD) is buried one mile underneath Cheyenne Mountain (also in the Rocky Mountains of Colorado). If one views the planet as a round ball of "electrical fluid" as did Tesla, one can see how a resonant circuit using the earth as a conductor would generate very high voltages at the antipodes of such a transmitter. if NORAD has a VLF-Resonance transmitter, then the antipode of its broadcast would be near a very small island group in the Southern portion of the Indian Ocean. Could it be that the still highly-classified, electrically-propelled U.S. submarines 'recharge' their dynamic-plasma batteries there? Furthermore, one wonders even more about the antipodes of the Pine Gap and North-west Cape transmitters. The Pine Gap transmitter sits very near the Tropic of Capricorn and its antipode is very near the Tropic of Cancer on the centre of the Atlantic Ocean's great dividing ridge. The North-west Cape transmitter just slightly north of the Tropic of Capricorn is, however, of extreme interest because its antipode is directly in the middle of the Bermuda Triangle where many strange electrical phenomena have been observed in recent years.. Isn't this significant?...

One must also wonder what purpose the gigantic Russian VLF transceiver at Riga on the coast of the Baltic Sea serves. For over three years now, ham radio operators all over the world have been plagued by a somewhat elusive ten to fifteen cycle "buzz" that suddenly appears across a bandwidth of over 30,000 frequencies. The signal seems to last anywhere from ten seconds to, sometimes, three minutes. Efforts to triangulate its source have always pointed to Riga.

The signal is a Russian product. It seems to do nothing but jam random frequencies at random time; but, is it simply a random jamming device? It is known that the signal travels along the same 'cavity' waveguide that Dr. Tesla's system used; the signal also behaves somewhat like a 'beat' signal between two very low frequency signals; yet, the most amazing 'fingerprint' of this elusive signal is its excessive bandwidths.

The signal must be a type of VLF, high-voltage broadcast. The 'beat' characteristic is formed as it interacts with one or more existing 'resonant-broadcast fields' around the globe. The wide-band static is caused by spurious electromagnetic radiation from 'cross-field' shorts and from leakage to Earth in the main signal. At least one other source of the 'other resonant fields' can be attributed to the United States, whose transmitters adorn the Earth in at least four places... as previously mentioned. It is quite feasible the two 'super-powers' are having a broadcast power 'war.' This may seem a bit far-fetched; but when he is apprised of the other possible applications of such a powerful, resonant, electrodynamic field, he will perhaps join the author in his suspicions.

When a living organism is subjected to very dense and/or very rapidly changing electromagnetic fields, various physiological responses can result - dependent upon a variety of factors. In some cases, genetic malformations may occur; while in others, immediate tissue damage may result in the form of 'burns;' and yet in still others, the effects may be purely neurological causing prolonged psychotic symptoms like paranoia, depression, and hallucination. Perhaps the most 'fascinating' possibility is the use of such fields to 'model' human behavior within the confines of the field. There would, of course, be distance limitations; but it is theoretically possible to induce subliminal suggestion, sleeplessness, and lethargy through very low frequency fields of some 10 to 15 Hz. (which lie in the range of the major resonant frequency of the earth -to-ionospheric-cavity at 14.3 Hz).

As Dr. Tesla once observed... *the Earth is like a gigantic, spinning ball of the electric fluid* (or ether) *of space...* His observation would imply that an electromagnetic 'thump' on one part of that "ball of electric fluid" would send 'splash rings' spiraling around the planet only to meet at the antipode of the initial 'thump.' There, quite contrary to one's first thought, these 'splash rings' would not simply reflect from meeting themselves at the antipodes simply to eventually return to the source point; not so, not so, for they would indeed meet at the antipodes; but the vector sums of their meetings would not be 'head-on' - as the density of the transmitting medium is not compressible. Instead of colliding, the wave fronts would actually produce a return wave front which would then spiral back to the source like a spinning 'smoke ring' (or toroid) of energy. This re-directed wave front would only partially with interact with the next oncoming wave from the source. This would be due to the fact that both wave fronts would be separated by an angle somewhere between 90 and 180 degrees - depending upon the relative motion of the planet's conductive surface in contrast to the planet's spinning magnetic field.

That angle between the send and return waves could also be controlled at the source by enhancing or retarding the wave pulses' spin moment. If one were to then observe the path of the energy pulses, one would perceive *an immense grid system which would be broadcasting power around the entire world.* The grid illusion would be caused by the energy-density spirals from the initial wave being overlapped by the energy-density spirals of the return wave. Two such transceivers could explain the twin 'grid-networks' which Capt. Bruce Cathie discussed in his book, *Harmonic 695.* If such is the case, however, Capt. Cathie's grid-cell sizes appear to be in error - but only by a few miles; because, it the broadcast frequency were 14,327 Hz, then the cell size would be a diamond-shaped 13 miles by 13 miles at an altitude of 158 miles and 12.5 miles by 12.5 miles at sea level.

To tap the energy of such a broadcast requires the construction off a tunable Tesla coil which either operates at 14.3 KHz or some even upper harmonic of it. Of course, the higher the harmonic the coil is tuned to, the lower will be the power received in an inverse proportion to the increased frequency. The Tesla coil's 'primary' would be the high-voltage, inner coil, in this instance, as the voltage will need to be stepped-down. The total length of the wire used in the primary should equal either the full or one-quarter wave-length of the resonant harmonic chosen. The secondary should have the equivalent in ampere-turns. Placing a resistive load (like a filament light globe) across the secondary (which should be vertically movable to allow tuning of the inductive coupling of the coils) should indicate to the observer when the system is receiving by its glow. The main problem will be guessing the incoming voltage level. Depending on the field density, it could be as high as 500 Megavolts... which could produce problems.

However, if the harmonic chosen is high enough, the effective voltage will drop considerably. For now, this phase of the reader's education must finish until a later publication.

For now, the author and many others salute that rare genius, Dr. Tesla, the man who:
- won the Nobel Prize in 1912
- patented over 900 new processes in the field of energy conversion
- received fourteen doctorates from universities all over the world
- resonated the entire Earth in 1899 with over 100,000,000 volts
- broadcast electricity over twenty-five miles (without wires) to light 10,000 watts of filament globes
- devised the system of alternating current power generation and transmission which lights the world today
- designed a 'force field' to shield America from air attack in WWII
- suggested a process for a charged-particle 'death-ray' (which is now a reality)
- discussed electrical weather control in practical terms in 1905
- and devised a procedure which would turn the upper ionosphere into a single fluorescent light source to permanently light the Earth...

Dr. Nikola Tesla, who had crammed several lifetimes of research into one, was to finally expire - a lonely and <u>apparently</u> forgotten figure - on the 7th of January, 1943. Only time will reveal whether he was really 'forgotten' or whether his work was so advanced that it has required two generations of secrecy to 'safely' administer its findings...

CHAPTER 5: EFFECTING WEATHER WARFARE

Many of Dr. Tesla's incredible contributions to the world of physics yet remain to be revealed to the public - for 'security reasons.' If he had only known what horrible misuse would be applied to some of his most promising ideas, one wonders if he would have even voiced them. It is certain that some of his ideas are being used by Russia and the United States alike in producing controlled weather phenomena as instruments of war.

In 1976 during the month of October, a new kind of radio/radar jamming signal manifested. The signal was triangulated to Riga; and, properly, blamed on the Russians - who apologized profusely for a few low-frequency tests. Those 'low-frequencies' were in the range of 143Hz! Sound familiar? About this same time, according to a report filed in June of 1978 by Edward Campbell of the <u>London Evening News</u>, *"the Russians had a nation-wide hunt going on for anybody who had ever known or met Tesla."*

According to the same source, *"Early in 1977 meteorologists reported an extraordinary 'blocking effect' extending down the west coast of America and a similar 'iron curtain' on the east coast, and along the Russo-Polish border up to Finland. These 'blocks' stopped the normal circulation of the weather."*

"Scientific brows wrinkled in alarm when it was discovered that each of these doorstops on world weather seemed to be associated with very large 'standing waves' of electromagnetic energy.' Meanwhile, according to the article, world weather went *"haywire."* *"Snow fell in Miami. Floods swept Europe; and the American nuclear-submarine monitoring system went on the blink after two of the American observation satellites were destroyed by 'electron beam technology.'"*

In December of 1974, Howard Benedict of the Associated Press in Washington filed an earlier report on the subject of weather warfare entitled: <u>*The Weather - As a Secret Weapon?*</u>

Benedict's article said that although both the U.S. and Russian official spokesmen had denied such weather weapons existed at that moment, research was *"under way."* What else could they have said? *"We've got super weather weapons now?"* One could imagine the result in the masses. Even the smallest weather anomaly would have subsequently raised cries of, *"War! It's War!"*

The official position is one of 'preemptive weapons production' over the Russians' probable development of the same weapons. One wonders what sort of weapons they were discussing? Benedict's article states:

".... current Pentagon research - at a cost of $1.52 million annually is limited to rain making, rain suppression, and hail and fog dispersal..."

"...Last October, the Soviet Union introduced a resolution in the United Nations asking the Geneva disarmament conference to outlaw weather research for military purposes."

"The Soviet ambassador, Mr. Jacob Malik, said scientists had concluded that a future weather weapons arsenal might even include the ability to create 'acoustic fields on the sea and ocean surface to combat individual ships or whole flotillas.' "

His article went on to say that by cloud-seeding areas a thousand miles up-wind from a particular country, one could cause severe droughts for years afterward to the target country. How incredible! One can imagine the comments after such a disaster, *"Gee, the weather's sure been bad lately; what a pity, tut, tut."*

His report also mentioned the possible alteration of the ozone layer [found at about 25 to 30 miles altitude] by chemical and physical means to allow excessive ultraviolet radiation to bombard certain areas of the planet. Perhaps one of the most interesting possibilities that his article mentioned was suggested by Dr. Gordon McDonald of Dartmouth College - an internationally known geophysicist and astrophysicist: **"...the release of thermal energy in the Arctic or Antarctic, perhaps through nuclear explosions along the base of an ice sheet, could initiate outward sliding of the ice sheet... 'the immediate effect, would be to create massive tsunamis (tidal waves) that could completely wreck coastal regions."**

Furthermore, according to the same source, **"He** [Dr. McDonald] **outlined a concept in which enhanced electrical oscillations in the earth's atmosphere might be used to impair human brains. He said research indicates that weak oscillating electrical fields can influence the brain causing small but measurable reduction in a person's performance. Lightning research has shown that it might be possible to control lightning to create such low frequency oscillations in the ionosphere!"**

Several points must surely leap out of the page to the 'initiate' reader: 1) the use of VLF, high-voltage fields in the ionosphere, 2) the heating of a polar cap, and 3) selective lightning production. The first point needs no additional explanation; as it has been previously discussed in Dr. Tesla's chapter. The second point may not have been quite as obvious. It bears expansion.

When a VLF field like Tesla's is tuned to a frequency that creates a complete cycle at the antipode to the sending tower, the thermodynamics of the energy-exchange are uniform at both poles; however, when a frequency is used which completes a full cycle at the sending pole, it causes excessive cooling at the sending tower, **and excessive heating at the antipodes!** Could this be used to melt a polar ice cap, thus flooding either Russia or the Canadian coastlines? Is this why both America and Russia have established VLF stations at the South Pole? One must wonder.

The third point is linked to the first point; however, a by-product of the production of those VLF, high-voltage, resonating fields is that one can **direct lightning strikes to specific targets on the other side of the Earth!** [The renascent reader is reminded of the prophecy in Revelation 13:13 which speaks of a future super-power which will have the technology to *"bring down fire from heaven in the sight of men."*]

The Power Struggle

If such technology is as advanced as all the evidence indicates, then the United States and Russia, would be in an intense power struggle for control of those adjacently or diametrically situated areas of the planet which could be used as VLF-broadcasting tower locations to effect weather warfare. Look at the following situations:

1) America has known VLF stations located at the South Pole near McMurdo Sound, South Africa in Transvaal, and in central and north-western Australia. Three of them are nuclear-powered. The antipodes of them are, respectively, in the regions of the central Barents Sea above the Swedish-Finnish border [and Riga], on the Tropic of Cancer midway between Hawaii and Cabo San Lucas, in the middle of the Bermuda Triangle, and on the Tropic of Cancer over the Mid-Atlantic Ridge. If America has other VLF weapons strategically deployed against Russia, they would be in areas like the South Sandwich Islands, Cape Horn, the South-East Tasmanian Cape, the South Tasman Ridge, and the southern portion of New Zealand's South Island.

2) Russia has a known VLF station at Riga and a suspected one at the South Pole location of Vostok. The antipodes of these are, respectively, in the South Pacific Sea near the ice packs at the southern tip of the Albatross Cordillera [58.S by 157.W] and the Baffin Bay Basin ice packs. Either location could produce coastal flooding or tidal wave phenomena under certain conditions. Edgar Cayce once predicted that the Great Lakes would become a part of the Atlantic Sea and the Hudson Bay at some future date. Was he correct?

If the U.S. wants to protect itself from a direct VLF attack over America proper, it must surely be defending the sub-oceanic locality around the mid-Indian Ridge bounded by 23.S to 55.S by 55.E to 120.E in the Southern Indian Ocean. Conversely, the Russians' recent occupation of the Bellingshausen Sea region at the South Pole between the Weddell Sea and tip of Byrd Land must indicate a like-wise defensive posture for potential VLF warfare.

It cannot fail to impress the reader that the draft presented to the August, 1975, Geneva Conference on international disarmament by delegates from both the U.S.A. and the U.S.S.R. contained the ban of nineteen man-made weather catastrophes. **Specifically, they included: the triggering of avalanches and landslides; awakening volcanoes; causing earthquakes and tidal waves; harnessing lightning bolts; guiding hurricanes or cyclones to strategic targets** [Australians must wonder whether or not the unusual behavior of Cyclone Tracy at Darwin was somehow linked with a battle of the Titans]; **melting ice caps to flood nearby coastal states; changing directions of rivers; generating fog, hail and rain; and deliberately destroying portions of the Earth's ozone shield**. What was the real purpose of President Carter's, *"Operation Noah's Ark ?"* The next chapter may be able to assist the reader in forming a fresh opinion.

Note: These words were first written in 1978; and now in 2009 they are a patent reality. Will anyone listen, now? Will anyone do something, now? There is much worse to come.

CHAPTER 6: The Weather Factor

Great glowing galaxies of light hang as if motionless in a timeless continuum. In one of those galactic, spiral vortices of light, a small and almost insignificant dwarf star of spectral class Gl spins quite rapidly in its seemingly slower orbit around the centre of what is known as the Milky Way Galaxy - comprised of some 100,000,000,000 other orbital star-systems of varying sizes and colors. That small star is, itself, orbited by nine small planetary bodies which are constantly bathed in a storm of 'solar particles' emitted by that star, referred to by the inhabitants of its third planet as, "**The Sun**.' Almost 4,700,000 tons of matter are radiated by the Sun each second in the form of charged particles and electromagnetic radiation.

The inhabitants of the third satellite call their abode, "**Earth**.' At the present time, their scientists are in a great state of confusion over the strange and, obviously, unpredicted behavior of the Sun. Their 'long-accepted' mathematical models of the Sun had forecast a relatively long life for it before it was to enter the 'Red Giant' stage, supposedly, some 1,000,000,000 earth-years hence. Recent abnormalities in the Sun's spin-rate, surface-oscillation rates, and sunspot activities have worried them a great deal. Perhaps they will understand in time, perhaps...

The Weather Forecasts

According to an article written by Gary Hughes (in February of 1978) which was reprinted in the Australian Sunday Times newspaper, the world is facing a miniature ice age. His article entitled, *"Mini Ice Age on the Way"* stated: *"The freezing weather which brought south-west England to a crunching halt this week and buried the countryside under mountainous snow drifts is here to stay. For experts say it is part of the same weather pattern that paralysed Scotland and New York in January and caused freak heat waves in Australia and other southern hemisphere countries.'*

Elsewhere in his article, Hughes quoted the well-known astrophysicist, Dr. John Gribbin, from the science policy research unit at Sussex University. According to Dr. Gribbin the cause of the cooling in the northern hemisphere is the Sun, itself. The increased number of sunspots (magnetic holes in the Sun's burning chromosphere) has apparently brought about the beginning of a new ice age on Earth, according to Dr. Gribbin. It was also Dr. Gribbin who pointed out that individual sunspot occurrences are not as vital in the weather changes as are the cumulative effects of a constant increase in sunspot activity. In a public release issue in January of the same year, an American geologist, Madeleine Briskin, expressed the same gloomy prediction: *"An ice age is coming!"*

On the 26th of July in 1977, The West Australian daily newspaper printed: *"Climate in Danger - Experts."* It began: *"Washington, Mon: A panel of American scientists and engineers has warned that continued use of oil, gas, and coal will cause a carbon dioxide build-up that could change the climate and perhaps cause oceans to flood coastal cities."*

Roger Revelle of the University of California at San Diego, told a news briefing that the potential climatic changes might mean increased snows in polar regions and destruction of the west Antarctic ice pack [recall last chapter's discussion?] *resulting in a rise in the sea level of five metres in 300 years."*

*"...if present trends **continue,** global temperatures **will** probably increase 6° C by the 22nd century. Such an increase would far exceed the temperatures of the past several thousand years."*

On the 10th of May in 1976, The West Australian also printed: *"Big Climate Changes Pose Threat."*: *Washington, Sun: Major world climate changes are under way which will cause economic and political upheavals 'almost beyond comprehension,' according to an internal report of the Central Intelligence Agency (the C.I.A.)."*

"'The new climatic era brings a threat of famine and starvation to many areas of the world,' *the (C.I.A.) report says. 'The change of climate is cooling some significant agricultural areas and causing drought in others. If, for example, there is a northern hemisphere drop of one degree Celsius it would mean that India will have a major drought every four years and can only support three-quarters of its present population.* **The world reserve would have to supply 30 to 50 million tonnes of grain each year to prevent the death of 150,000,000 Indians,'"** the report said. [The reader is asked to remember the figure of 150,000,000 Indian deaths by famine in the next section of this book which discusses another side of this situation].

The news article continued, *"The report, which was concerned with possible political and economic threats that the United States could expect from such drastic events, said that starvation and famine would lead to social unrest and global migrations of populations.'*

On the 21st of January, 1977, the magazine, Pacific Computer Weekly published an interesting article: *"A unique and complex modeling project which was used to analyze the effects on Earth's atmosphere after **the Sun was removed**, has created high interest world-wide. The project was undertaken by Barrie Hunt, principal research scientist with the Australian Numerical and Meteorology Research Centre.'*

His study which utilized the *"C.S.I.R.O.'s Cyber 76 in Canberra and the department of Meteorology's dual 360/65s in Melbourne,"* obtained the *"surprising conclusion that activity on the Earth would remain for longer than the text-book conclusions of about 10 days."* if the Sun were to suddenly 'go-out.' His model showed the Earth in some life-like state even after 50 days of total darkness. His next project was to be one to analyze the *"effects of speeding and slowing the rotation of the Earth by a factor of five."* One must ask the obvious, *"**Why has there been such high-level interest in his project - unless there was already some doubt in somebody's mind as to the duration of the Sun's steady-state energy levels?"***

The Daily News, a West Australian newspaper in Perth, Australia, on September 14, 1976 published an article entitled, *"What On Earth Is Happening?"* ... a report written by Angus McPherson in London. It started:

"Planet Earth, it seems, is on the rampage. The earthquake barrage in the Far East probably has done more damage and killed more people in China than a nuclear attack. On the other side of the Pacific with one major volcano erupting and another ready to go, it seems as 'if the whole Caribbean is about to blow-up'.."

"Day after remorseless day of sunshine has burnt England as brown as the Arizona desert - and this is only a part of a world-wide weather aberration that has brought drought, shriveled crops and hungry livestock to Western Europe, the American Mid-West, India, and Australia.'

" This year of drought and disaster, 1976, has come when the spots on the Sun are at a minimum. 'I for one, find it very hard to see that as a coincidence,' says Dr. John Gribbin, one of Britain's most avant-garde astrophysicists.'

" 'Records I have studied do seem to show that bursts of earthquake activity on Earth come when the sunspots - which fluctuate roughly every 11 years, are either at their maximum or their minimum.' "

"In just the past two years, after an exhaustive study, scientists at America's leading atmospheric research centre at Boulder, Colorado, confirmed that they (sunspots) did alter the weather.'

"But what astronomer Carl Sagan has called the 'Cosmic Connection' is well-established, and the effects of the Sun's next rash of spots, due in about 1980, will be watched with fervor."

Earthquakes Have Increased

In 1975, Hawaii suffered two earthquakes and a new eruption of the volcano, Kilauea. One of the quakes which was 7.2 on the Richter scale was the biggest earthquake to hit Hawaii in over 75 years.

In January of 1976, Iceland was struck by a quake measuring 6.5 on the Richter scale. Also in January, more than 20 earthquakes were reported in the Pacific region near the Russian Kamchatka Peninsula. The event was the worst in 25 years.

In October of 1976 earthquakes and tremors hit Scotland and Guatemala. The death toll in Guatemala was over 22,000 while the injured numbered over 74,000. It left 1,300,000 people without homes.

In May, the west coast of New Zealand's South Island suffered a quake at 7 on the scale; while one of only 6.5 clobbered North-eastern Italy. Uzbek in the USSR suffered one at 7.2; and the Chinese-Burma border suffered one at 6.9. In June, Papua New Guinea received one at 7.2; while Irian Jaya was hit at 7.1 on the scale. Bali received one at 7.

In July, China received hers, which was, as previously mentioned, an absolute horror at 8.2 on the Richter scale; and by December of 1976, California residents were preparing for another big earthquake to hit them in April of 1977. What they finally received was a 40,000 square kilometer "bulge."

It is almost certain that California will have the worst earthquake in its history before the turn of the century. It is just as amazing to find that Californians are still living right on top of the fault line like there was no tomorrow. They even have a land development scheme nicknamed, *"The Faultline Estates!"*

America has officially begun to prepare itself for earthquake disasters in the nine most probable regions. Of the nine regions, California is the only one which has spent any large amounts preparing the buildings and other edifices for the eventuality of a horrendous earthquake. It is a matter of public record that Carter's Office of Science and Technology has drawn-up plans for the widespread reinforcement of American skyscrapers in the most prone areas. Incredible effort has been thrown into early earthquake detection and prevention studies.

Between the years 1897 and 1946, the average number of observed earthquakes over Richter 6 was 3 per decade. Between 1946 and 1956, the average jumped to 7. In the following decade, it jumped to 17 earthquakes over Richter 6. Then, in 1967, the yearly earthquake figure for Richter 6 or better was 17! In 1968, it was 19; in 1969, it was 21; in 1970, it was 24; and in 1971 it was 34.

However, during the decade of 1967 to 1976 there were 180 earthquakes over Richter 7 on the scale! Note that in the recorded history of man, an estimated 74,000,000 people have been killed either by earthquakes or indirectly by their attendant fires, floods, landslides, and disease.

Planets And Sunspots

Sunspots are actually holes ripped into the surface of the Sun by magnetic storms deep inside the Sun's multi-shelled core. Various factors can influence the propagation of these 'mysterious' blemishes. When a star is young, its outer shell of hydrogen is usually quite thick. As a result, it very seldom has such deep vortices of magnetic turbulence that visible black spots are made to appear. There are, however, times even in the young and stable state that a star's surface may be so disturbed as to show visible signs of magnetic turbulence.

If a large enough external body interacts or collides with the outer planetary or "gravitic" shells of a young star, it can create relatively brief, but visible, "sunspots.' As a star grows older, it burns up its hydrogen fuel layer in the process of nuclear fusion. This means the hydrogen layer grows thinner and thinner as the star ages. As a result it takes less and less in external forces to create visible signs of the magnetic turbulence.

In the case of those stars which have planets occupying two or more of their 'gravitic' shells or "discrete orbits," the periodic motion of the planets themselves can effect 'sunspot' (or 'starspot') activity after about half of the original hydrogen shell has been consumed. The Sun is one of those stars whose hydrogen supply is over half of its original amount and whose orbital shells contain planets. As the nine charted planets of the Sun all orbit the Sun at different speeds, it is only infrequently that all nine of the planets are somewhere on the same side of the Sun at the same time. It is even more infrequent that all nine form a straight line from the centre of the Sun to the outermost of the nine planets. It is, in fact hundreds of thousands of years between such events. Yet, as partial alignments do occur, certain imbalances in the spin of the entire solar system manifest.

Consider a man standing on a small platform which, in turn, rests on the floor, mounted in such a way that it spins around quite readily when someone spins it. Let the man be given two equally heavy balls with rigid cables attached, so that the man can swing the balls out at his side. Let the man be spun quite rapidly. The balls will 'orbit' away from the man's spinning body. His body will be erect and stable, as the balls will both be traveling at the same speed and at the same distance from his body. If the man then pulls-in one of the balls so that it is closer to his body than the other, the resulting imbalance of forces will cause the man's erect posture to away to the side of the ball with the longest tether; and, subsequently, to vibrate in an awkward fashion until he stops spinning.

If one were to replace the man and the platform with a pole that was anchored to the floor, one could create a very good analogy to the processes which cause planets to periodically 'shake-up' their star (or the Sun, in this case). One could affix nine rigid bars each of a different length to the upright pole so that they could be easily spun around the pole by the observer. To the ends of these nine bars, nine balls of varying weights could be attached. If the observer were to, then, spin all nine balls around the pole by thrusting each in turn with his hand, he would see a peculiar sight. As the balls orbited the pole it would gyrate violently for a time; then it would 'settle down' and sway briefly in an arc; then it would 'straighten-up' and momentarily appear as erect and as stable as the man did when his balls were 'balanced;' and then it would gyrate madly, again.

Such is the case with the Sun. The nine planets are connected by invisible "rods" of gravity to both centre and the surface of the Sun. As the planets line-up in varying numbers, attitudes, and times, the Sun is shaken to various degrees, the more so, the older it becomes.

In March, 1951 issue of RCA Review, John H. Nelson commented on planetary orbital-patterns and sunspots and the correlation of both to the heavy radio frequency storms that even now plague Earth communications systems with a growing regularity:

"It can be readily seen from these graphs that disturbed conditions [excessive radio interference] *show good correlation with planetary configuration. It is definitely shown that each of the six planets studies if effective in some configurations."*

Nelson went on to say that short-wave frequencies are disturbed when Jupiter, Saturn and Mars line-up in either a straight line or at right angles to each other. He also emphasized that the phenomenon, *"is not due to gravitational effect or tidal pulls between planets and the Sun.'* His last statement is, technically, correct; however, "gravitational effect" does not mean the same as 'gravitic tensors.' Remember, 'gravity' is that illusion produced by the interaction of two forces hitting each other and forming a zone or shell of equilibrium. The actual connectors of either force to its respective source are 'tensors' like muscles between parts of the body. As it is the balance between these 'connectors' which generates the illusion of gravity, great and powerful changes in the magnitude of both 'tensors' can result in only small changes to the apparent 'gravitational effect.'

Dr. Immanuel Velikovsky felt that these 'tensors' could be better explained as functions of electric charges; thus giving the solar system the appearance of a balanced 'atom' on a large scale. His words are slightly different - but they imply the same as this author's. For those who were unaware, Dr. Velikovsky was a regular correspondent with and visitor to the late Dr. Albert Einstein, who eventually died with one of Dr. Velikovsky's early manuscripts open on his desk. It was also Dr. Velikovsky who quoted a most enlightening statement from an article in the April 15, 1951 edition of the New York Times in his own book, *Earth In Upheaval*. The article reported that:

"...evidence of a strange and unexplained correlation between the positions of Jupiter, Saturn and Mars, in their orbits around the Sun and the presence of violent electrical disturbances in the Earth's upper atmosphere, seems to indicate [that] the planets and the Sun share in a cosmic electrical-balance mechanism that extends a billion miles from the centre of our solar system. Such an electrical balance is not accounted for in current astrophysical theories."

In the same book, (p.259) Dr. Velikovsky went on to say, *"By 1953 the strange fact was established that the solar tides in the Earth's upper atmosphere are sixteen times more powerful than the lunar tides in the atmosphere, a fact in complete conflict with the tidal theory, according to which, the action of the Moon on oceanic tides is several times more powerful then that of the Sun. the fifty fold discrepancy is still without an acceptable explanation."*

An undeniable set of circumstances now presents itself to the people of Earth: The Sun is entering into a transitional stage, which is allowing the increasingly eccentric alignment of its nine planets to effect abnormal magnetic and electromagnetic 'storms' on its surface. These 'storms' - evidenced by the extremely high number of sunspots, are sending increasingly more violent 'sprays' of charged 'particles' and short wavelength radiations outward to collide with all the planets in a form called the 'solar wind' by the N.A.S.A. technicians. This 'solar wind' of great energy density can increase or decrease the spin rate of the Earth in its orbit; hence, it can change the length of an Earth day.' This 'solar wind' has already been responsible for creating ionospheric 'tornadoes' which caused the Skylab space station to prematurely drop from its correct orbit to one which eventually resulted in its crashing down to Earth. The increased solar wind was blamed on *"abnormally high sunspot activity accompanying the 1980 arrival of the peak of the current, 11-year sunspot cycle."*

The biggest solar flare since 1968 occurred in May of 1978. Its size was over fifty times the surface area of the Earth! The resulting radio frequency and high-speed particle bombardments of Earth caused widespread short-wave, radio communications disruptions. The National Oceanic and Atmospheric Administration in Boulder, Colorado reported that the U.S. Coastguard had lost all radio communication with its ships in the Atlantic while the initial flare effects continued.

Only two and a half months later, a huge sunspot formed on the Sun. The previous flare had receded leaving a 'thin' area on the Sun's surface. Then, the relatively high 'planetary alignment effect' in the July-Sept. period of 1978 triggered such an intense magnetic storm on the Sun that a sunspot of over 3,900,000,000 square kilometers formed, becoming the biggest since 1947, by five times the size of the one in 1947!

This sunspot steadily grew in size; and almost doubled its initial size in only a few short months. It temporarily reduced the visible light and heat effects on the Sun by two percent on the visible disk! Over 20,000,000 highly-charged particles were bombarding Earth each second as a result of this huge sunspot. It was surely the major factor in causing the 1978 Indian floods and the absolutely disastrous summers of the late 1970's in Europe, Australia and many other places.

THE HEWLETT-PACKARD SYSTEM 9845S COMPUTER

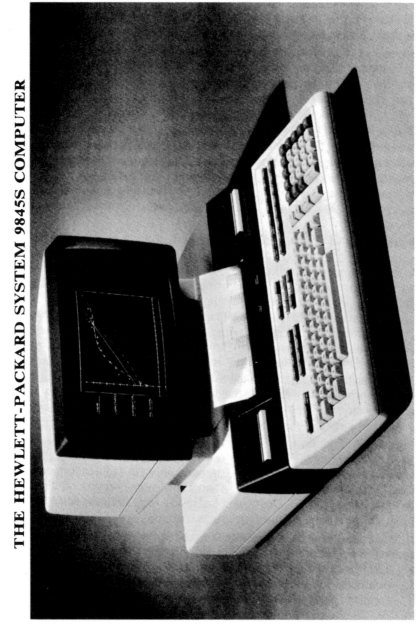

The Hewlett-Packard Series 9800 System 45, features the most powerful central processor and the largest mass memory ever offered in a desktop computer, a 12-inch CRT display, enhanced BASIC language, an optional graphics package with high-speed hard-copy printing and applications software.

A news report by Bruce Sandham in the Brisbane Sunday Mail Color of 5 November, 1978 had these things to say about the sunspot:" These enormous emissions from the Sun can have a profound effect on the way humans react,' says Dr. Michel Gauquelin, Director of cosmic research at France's Strasbury University.

This sunspot activity is causing massive ionization which can have dramatic influence on our behavior,' he explains. 'Some areas of the human body contain a positive charge of electricity and others a negative one. So the arrival of electromagnetic waves from the Sun can seriously upset the delicate balance between the two."

Dr. Gauquelin continued on the subject of possible sunspot effects on world diseases, *"The invasion of a human cell by a virus is virtually an electrical reaction. Normally the virus and the living cell carry a charge of negative ions, thus the virus is repelled and can't enter the cell. When something happens to the body and the cell's charge becomes positive, then the virus is free to enter. This 'something' could be the disturbance caused by solar activity.'*

The article also noted the high correlation of suicides, industrial mishaps, divorce, and insanity during heavy sunspot peaks like this one. Dr. Gauquelin also told Mr. Sandham (the journalist) that these effects could last for up to four more years; however, this was a gross understatement, as the reader will soon discover.

DISASTER IN 1982?

The reader is invited to study the computer generated graph in Chart No. 1. The graph is a correlation of relative-planetary-positions against time. If all nine planets were to be aligned in a straight line from the Sun outward, the graph would show a dot on the line labeled '9.' If the planets were to be unaligned as much as possible, the graph would show a dot on the value '4.55.' The dotted reference lines help the reader to quickly assess what months and years have had and will have very high relative-planetary-position quotients. If (please note the conditional tense) this method of approximating the timing of sunspot activity is valid, then the next seven years may hold eighteen periods of solar activity that will be far more disastrous for mankind than any of those in the past 3,563 years!

This graph is only one of some six hundred and thirty produced by the author on a Hewlett-Packard computer. The computer, and 'HP System 45' 'crunched' through 2,000,000,000 calculations and drew these graphs in a little under two and a half days. The calculations were simply those of running a mathematical model of the solar system in reverse motion for six thousand years, and then in forward motion for one thousand years. These figures are only monthly position checks; so slight variations in some of the figures will occur on a daily position check.

The reader will note that the highest peak shown on the graph occurs in February, 1982. **IF** Dr Gribbin's hypothesis is correct, then the cumulative effects of sunspot activity triggered by certain planetary positions may not return to an acceptable level until after 1986...

The 'dropped lines' were generated by the computer whenever the relative-planetary-position quotient exceeded a statistically critical value of '6.'

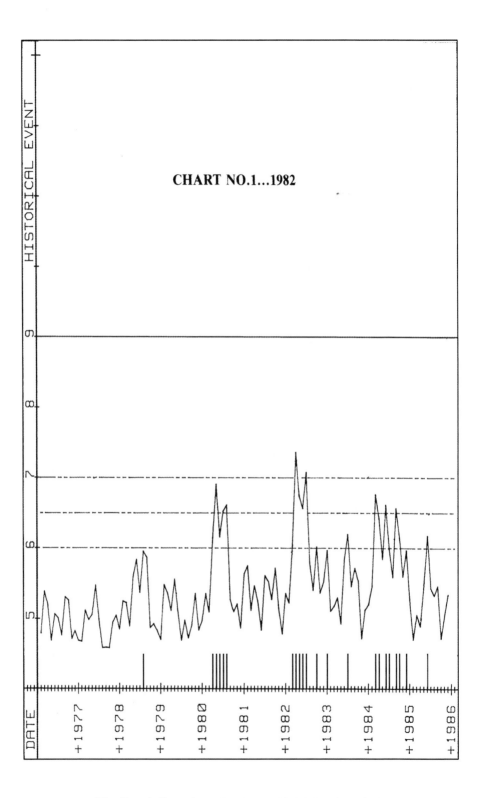

CHART NO.1...1982

The reader will note that both the solar flare of May and the giant sunspot of July-Sept. occurred or were initiated on the only two near-critical peaks in 1978. The author has also made a preliminary study of ancient history in correlation to the last 6,000 years of planetary motions as shown by the other six hundred charts. In forty-nine out of the fifty major revolutions and wars of the last 3,500 years, the graphs' high-peaks matched the event to the year. The graphs' peaks also agreed to the year with the corrected Biblical dates for the Flood of Noah and the Exodus of Israel from Egypt amidst great geological distress and plagues of vermin. It is this author's intention to publish a sequel to this book which will contain all the charts mentioned and his correlations at a later date; so that the reader may analyze the results for himself. *[Note: those charts are now included in the 1992 additions attached to this book]*

For the computer-users in the readership a copy of the author's program listings written in the System 9845's basic language will be in that book; so that cross-checks and 'fine-tuning' may be effected by those who wish to pursue the subject... if there is any time left.

Project 'Noah's Ark'

Early in 1977, the President of the United States, Mr. Carter, announced a few details of a top-secret Pentagon project code -named: ***"Noah's Ark***.' It is, supposedly, a system of some 96 'bunkers' and 'bolt-holes' which have been established at various places on or near the Earth to house approximately 6,500 key officials in case of a nuclear war.

Many of these 'bolt-holes' are underground cities complete with streets, sidewalks, lakes, small electric cars, apartments and office buildings. One such 'city' is carved out of a mountain near Washington. It is called Mount Weather. One other such 'city' is most probably located at each of the super VLF broadcast stations around the planet.

One wonders why the project was code-named "Noah's Ark.' Any Biblical student knows that "Noah's Ark" has to do with a weather calamity - not a nuclear war. Carter would have been better advised calling it something like "Project Gomorrah," "Project Brimstone," or their ilk; yet he chose a weather calamity. **Could it be that he, like many other senior statesmen and key scientists - knows the Earth is heading for a severe weather catastrophes**? If so, one can understand why the masses have not been informed, can't one? It appears that a new solution is needed, one which requires a Copernican revolution in human philosophy. Pray that it comes soon.

SUGGESTED READING LIST FOR SECTION 1

1) Caidin, Martin; *The Mendelov Conspiracy*:* 1974 by Granada Publishing Ltd., Mayflower Books Ltd.

2) Cathie, B.L., *Harmonic 695*; 1977 by A.H. & A.W. Reed Ltd., 65-67 Taranaki St. Wellington.

3) Gribbin, John and Plagemann, Stephen; *The Jupiter Effect*; 1977 by Fontana/ Collins.

4) Hoffman, Banesh; *Einstein*; 1975 by Granada Publishing Ltd., Frogmore, St. Albans, Herts AL22NF.

5) Kraspedon, Dino; *My Contact with UFOs*; 1977 by Sphere Books Ltd.

6) Lucas, George; *Star Wars*; 1977 by Sphere Books Ltd.

7) Velikovsky, Immanuel; *Ages in Chaos*; 1973 by Abacus/Sphere Books Ltd.

8) Velikovsky, Immanuel; *Earth in Upheaval*; 1974 by Abacus/Sphere Books Ltd.

9) Velikovsky, Immanuel; *Worlds in Collision*; 1977 by Abacus/Sphere Books Ltd.,

10) Watkins, Leslie; *Alternative Three*; 1978 by Sphere Books Ltd.

*   Now reprinted under new title, *Encounter Three*, 1979 by Martin Caidin, Pinnacle Books Inc., Book available by mail from: Pinnacle book Mailing Service, P.O. Box 1050, Rockville Centre, N.Y. 11571.

SECTION II: Mysticism And Numerology

Mystery Babylon

CHAPTER 1: Centuries Of Darkness

To properly ponder those mysterious, unseen forces which direct modern social dynamics, one must look to the past, to the ancient past, to that obscure time when the gods and god-like supermen of classical legend were some form of reality upon Earth. It is in the history of those ancient days that the key to understanding lies; for it is from the ancient secret teachings of so-called 'enlightened or illuminated' masters of wisdom that the modern mystics draw their blueprint for world peace in the form of a dictatorship.

Thoth Hermes Trismegistus

Most of modern mysticism is based upon the teachings of a legendary character called, 'THOTH' by the Egyptians, 'HERMES' by the Greeks, and 'THOTH HERMES TRISMEGISTUS' by the Illuminati (or enlightened ones). The latter title means, 'the messenger of the gods who has been thrice illustrious master.' Legend states that Hermes was the source of all basic knowledge man possesses in physics, geometry, medicine, astronomy (including ancient 'astrology'), and many other sciences and arts. He is reputed to have contributed between 24,000 and 36,000 scrolls of information to the ancients. He is even credited with the origin of Egyptian hieroglyphics. Whether or not he personally wrote those scrolls is another issue, however. It is quite possible that he made the information known to man from a great library, not of Earth origin, as his full title suggests.

The Greeks not only credited Hermes with their knowledge of commerce, invention, and the esoteric teachings; but also, with their somewhat less desirable traits of cunning and theft. Obviously, they credited most of their ancient sciences to Hermes, because they are known today as the 'hermetic' sciences.

The Illuminati not only credit Hermes with the preceding; but also with being the original authority on the 'Word' which is not 'lost.' To the Illuminati the 'lost Word' was once God in a human form. They represent this 'human god' with the word of four letters 'I.N.R.I' (not really meaning Christ) and the symbol of a red rose inside a cubic stone. Supposedly, using the three virtues of Faith, Hope, and Charity, the Illuminati search for this 'lost Word' until that day when the 'lost Word' will be found and, as they claim, **will again become flesh and blood.** His function **will be,** supposedly, to unite all human philosophies and religious disciplines into one, very-worldly order. However, it appears to this author that the Illuminati are somewhat misinformed on this matter; since the Bible plainly states that the real 'Word' was never lost, has already become flesh and blood once before, and is due to return to Earth soon as God only - not some hybrid, human god as they suggest.

Pathways Of Mystery

Analysis of ancient history is difficult, at best; however, trying to analyze the spread of the secretive Hermetic teachings, which by definition are not for public knowledge, can prove to be an almost impossible task. If it were not for a few somewhat opportune 'leaks' in history, this author could not have deduced even the following pathways...

As previously stated, Hermes first appeared in the ancient writings of the Egyptians as their god, Thoth. It would appear that Thoth or Hermes was not an earthman. He would most probably have appeared to the Egyptians between 2442BC and 200BC because he would have manifested after the great flood of Noah and before the Old Kingdom of Egypt.

The Hermetic teachings next surfaced in the first Babylonian Empire founded by Hammurabi the Great around 1760BC. Hammurabi's empire was even named as a function of the Hermetic school of thought. Babylon, the ancient capital of Babylonia, was called 'Bab-ilu' in the ancient tongue. 'Bab-ilu' translates to 'gate of the god.' As ancient Babylonia covered a portion of what is thought to be the gateway to the hidden Garden of Eden, it might well be that 'gate of the god' was originally speaking of that place where Lucifer (Being of Light) first appeared to early man. If so, then 'Bab-ilu' might be more appropriately written as, 'Bab-ilu Thoth' or 'Bab-ilu Lucifer,' but more on this will come in Section III.

Hermetic teachings flourished in the ancient kingdoms, and young minds were constantly being initiated into various phases or schools of those mysteries which were handed-down from that ancient superhuman, Hermes. Biblical history reveals that one of those young initiates was an adopted member of the Royal household of one of the Egyptian Pharaohs, possibly Seqenenre II or Ramses II. It was that young initiate who was later to be a great leader of a new nation of his own people; his name? Moses, the great, Hebrew lawgiver.

Moses was unique in many aspects, yet one aspect stands out more than the others when viewed in the light of mysticism. Moses was an initiate in both the Hermetic mysteries and the mysteries of the 'One-God' of Israel (or 'Kabbala'). This explains to a certain degree how the royal Egyptian magicians were able to duplicate some of the plagues that Moses had produced under the instructions of the 'One-God of Israel.' It is significant, however, that the Kabbalistic teachings of the 'One-God of Israel' allowed Moses to not only defeat the Hermetic scholars but also to utterly destroy order in Egypt.

A definite divergence of purpose manifested between the two schools of mystery or 'illumination.' Those teachings which Moses passed-on to Aaron and Joshua were called the 'oral Kabbala' or the unadulterated instructions of the 'One-God of Israel.' They were not the teachings of Thoth, Hermes, or Lucifer; whose teachings were not of the 'true light' and prospered mostly in Egypt, Babylonia and Assyria.

Centuries passed as the two diametrically opposed mystery schools continued their development. Then, in Israel, during the reign of King Solomon a rather bizarre incident occurred which ultimately disrupted the Kabbalistic mystery school which Moses had initiated. The incident centred around the construction of the Great Temple of Solomon at Jerusalem.

The Widow's Son

Since the time of Moses, no one man has ever known all three of the great secrets of the mystery school of Moses. In King Solomon's time the three secrets (sometimes called the three 'Words') were known only by King Solomon, King Hiram of Tyre (ancient Phoenicia), and Hiram Abiff also of Tyre (sometimes called the widow's son by modern Freemasons). Each of these three men was the master of his own craft in the Kabbalistic mystery school, and as such was titled a 'Master Mason.' Each man knew only his secret or 'Word' and not the others'. The three began the construction of Solomon's Great Temple with the help of their brother 'Freemasons'. It was the craft of Hiram Abiff, the artisan of Tyre which was to decorate the Great Temple. In addition to his work being ornamental it was also to serve as both a secret instructional aid to subsequent initiates and as part of the functional process whereby God would communicate more power and wisdom to Israel. However, when Hiram Abiff was not quite finished with his part of the Temple, a group of three of his brother initiates of lower rank tried to force him to yield his secret 'Word.' In the process, Hiram Abiff, the widow's son from Tyre, was brutally murdered by the three. It was a black day for more reasons than one might immediately see. The three were found and executed in a manner similar to that by which they had murdered Hiram Abiff. The Temple, however, could not be finished because Abiff was the only one who knew the details of the 'Word' which was lost at his untimely death.

Three serious shocks had killed many of the initiates' motivation. Firstly, one of the only three Master Masons alive had been murdered by 'brothers' in the Temple; secondly, his secret 'Word' had apparently been lost for all time; and lastly, the Temple of Solomon was not able to be completed to the degree that only Abiff's secret would have produced.

Before his death, however, Hiram Abiff was able to produce a complete set of scrolls which detailed the construction of Solomon's Great Temple. For five centuries after Abiff's death these 'Temple Scrolls' lay in a secret place. They were found again lost as the Temple was destroyed in 70AD. In the last four years these scrolls have been re-opened and translated by the very elite of modern mystics - the Illuminati - who now know the location of the two hollow pillars of Solomon's Temple, known as 'BOAZ' and 'JACHIN.'

Darkness Sneaked In

In the wake of this great tragedy, the Kabbalistic Brotherhood was infiltrated by the Hermetic mystery school initiates from Egypt, Assyria, and Babylon. It was a slow, but deliberate, process which allowed the subtle injection of deceptive Babylonian teachings into the ancient Mosaic order of Israel. The most obvious blasphemy was the introduction of a new and secret name for God. This 'new' name was a composite made from each of three cultures' names for their god - whether he be pagan or not. The name was 'Jah-Bul-On.' In the Hebrew alone, this name is misleading, for it means: 'The Lord of withering or failing strength.' 'Jah' represents the Hebrew name of God, **Yahweh;** 'Bul' represents the Assyrian name for their God, 'Baal;' and 'On' represents the Egyptian god, 'Osiris.'

Ask any European Freemason who will secretly discuss this matter. This name is sacred within almost all the European Masonic orders affiliated with the 'Grand Orient Lodge.' (Not all the Masonic order - just, some). Even the original name of Solomon was changed. Originally his name was pronounced: 'Shel-ou-mouh' which meant 'his peace.' However, the infiltrators changed his name to a combination which represents the 'eternal light' in the language of three pagan cultures: 'Sol,' in Latin, 'Om,' in Hindi dialect, 'On,' Hamitic or Egyptian.

Time passed; and Israel was captured by King Nebuchadnezzar of Babylonia, which isn't really surprising, since the Kabbalistic society of Israel had been infiltrated over several hundred years since Solomon's time. During the 'Babylonian Captivity' a young Hebrew names Daniel was taken to Babylon to be instructed in Nebuchadnezzar's court. While Daniel was under the instruction of Babylon's 'Hermetic initiates,' the Lord God of Israel gave Daniel the direct instruction or 'enlightenment' which would restore the true Kabbala to the minds of the chosen, wise men.

These revelations, then, explain why much of modern Freemasonry is based upon ancient pagan ritual and philosophy. The true teachings which were given to Moses are not lost; they are preserved in the 'Oral Kabbala.' In the teachings of the Oral Kabbala are secrets which will soon inspire the largest number of true Christian conversions of all time during the coming period to be known as 'The Great Tribulation' or 'The time of Jacob's Trouble.' Two 'ancient witnesses' will personally supervise 144,000 new 'initiates' who will use the ancient teachings of the God of Israel to evangelize the troubled world of the 'Great Tribulation.'

Now, begins the end of a long descent down steps in darkness for those Hermetic initiates: the Jacobins and the Illuminati. Let the mind of wisdom comprehend this.

CHAPTER 2: Orders of 'Illuminism'

Words Of Caution

It is a common human failing to blame every member of an ethnic, social, or religious group for the unpleasant actions of the more active or vocal group members - who, quite frequently, do not represent the majority of the group. So, it is emphatically stressed that no allusion has been made or will be made in this book which in any way suggests that the ills of the modern world are due solely to a Jewish, Masonic, Jesuit, capitalist, communist, or fascist conspiracy. Some, but not all, members of these groups, however, are members of various interlocked conspiracies to form a new world order. Do not form an automatic dislike or suspicion for anyone who happens to be a member of any of the above groups, especially the Jewish and Masonic groups, which seem to be most popular targets of late. Let each individual person's actions defend or indict himself. Please remember this.

The Unseen Hand

The preceding chapter outlined the early history of the 'unseen hand' of those ancient mystery schools which so mercilessly dictated human affairs. This chapter will concern itself with the re-emergent, modern versions of those same ancient Hermetic and pseudo-Kabbalistic orders.

Benjamin Disraeli (1801-1884) former Prime Minister of England once observed, *"So, you see,... that the world is governed by very different personages to what is imagined by those who are not themselves behind the scenes."*

A few years later, Dr. Nicholas Murray Butler (1862-1947) was nominated to the presidency of Columbia University by J.P. Morgan and Co. of Wall Street. Dr. Butler who was also an 'insider' or a 'behind-the-scenes man' observed, *"The world is divided into three kinds of people - a very small group that **makes things happen**, a somewhat larger group that **watches things happen**, and a great multitude that **never knows what has happened**."*

Deism Sired Illuminism

In 1624, Lord Herbert of Charbury published "De Veritate" which gave his definition of what he called 'natural religion.' Lord Herbert stated that certain beliefs were so obvious that all men of reason must accept them Among those beliefs was the belief in the existence of the one God who not only dispenses rewards and punishments; but also puts upon man an obligation to worship God in repentant piety and virtue. Such was the foundation of early deism. Yet, even before Lord Herbert's formalization of early deism, other men had begun to form similar concepts.

In Spain, Ignatius Loyola, founder of the Society of Jesus (the Jesuits), had helped to form a secret society from former Jesuits and Franciscans calling it the 'Alumbrados,' which meant 'the enlightened' or 'the Illuminati.' By 1654, France had received her 'society of the enlightened' in the form of the 'Guerinet.' Their main draw cards to the bored young noblemen of the time were 'visionaries' and 'ghost-seers.'

From the great confusion of so-called 'enlightened' societies finally emerged two distinct camps, the pure mystics and the mystics of reason. The former were primarily on the Jesuit side of Illuminism. The latter formed the basis for the German school which now thrives in most of the developed nations. The pure mystics entertained many rituals of unknown origin and true function. They simply retained them in fear of reprisal by their version of God. The German school known collectively as the Bavarian Illuminati would accept no rite, ceremony, or belief that was not, as they said, 'reasonable.' The Bavarian Illuminati felt that Christianity was based on a book of 'groundless historical confusion.' Furthermore, they held the entire Old Testament to be a 'repository of crimes and obscenities' in which God was supposed to have connived with the 'so-called chosen people.' They felt, also, that Biblical prophecy was an offence to the reasonable mind, as they could not envision a God who would have to 'stoop to revelations in written form.' In essence, both groups were deistic, they simply had different limits to what 'was reasonable.' These limits, however, were far enough apart to later precipitate a feud between the two in Bavaria and, eventually, even in France.

The Bavarian Illuminati

In 1731 Benjamin Franklin joined Freemasonry; and it was only a short time before he attained the rank of Grand Master of the Pennsylvania Masonic Lodges. It soon became obvious from the statements of his 'American Philosophical Society' that Franklin was a member of the 'rationalist' or 'realist' school of enlightenment. Other Freemasons in the colonial states of America shared Franklin's opinions which paralleled the European 'Enlightenment' movements.

Among Franklin's Masonic brothers who sympathized with the 'illuminated' form of the emerging continental Masonic reforms were Thomas Jefferson, John Adams, and the Marquis de Lafayette. The four of them formed the nucleus of the American branch of the French and Bavarian enlightenment schools. Years passed; and as the American Revolution came to a finish, a brilliant young student in Bavaria was formulating his own revolution 'for a different reason.'

Adam Weishaupt was born in 1748. At the age of twenty-two, he was elected to the chair of Canon Law at the University of Ingolstadt in Bavaria. Traditionally, this chair had been held by the Jesuits since 1750. However, because Weishaupt, whom the Jesuits had trained, took the chair without joining their Order, they began to pressure him. On the first of May, 1776, just seven years after winning the chair, Dr. Weishaupt, in self-defence, formed his own secret society of illuminists. He named the society 'The Order of Perfectibilists;' but this name soon proved an unwise choice for obvious reasons, so he changed its name to 'The Order of the Illuminati.' The 'Bavarian Illuminati' had been born in the middle of a raging storm of continental enlightenment movements.

Two years passed; and by 1778 Weishaupt had managed to infiltrate the Masonic Lodges as a fully-initiated Master Mason. He then began a deliberate process to selectively pick-off the cream of the Masonic Lodges of Europe in the typically Teutonic fashion that marked his manner.

In the Bavarian mystery schools of Freemasonry, a legend had evolved which led many of the Bavarian Masons to believe that undisclosed 'superiors' were watching the progress of the Lodges all over the world. Furthermore, they believed that these 'superiors' would periodically manifest themselves to such junior Freemasons as had merited their favor, with the specific purpose of advancing that junior in the upper ranks of the mystery schools of Masonry. It was this superstition that allowed Weishaupt to play the part of a secret 'superior.' By inventing his own series of ranks, secret signs, and secret words, he was able to approach the influential nobility of European Freemasons and convince them that he was inducting them into a new level of Masonry. In no time at all, he found his plan successful beyond his own expectations, which had been as high as his own ego was tall. Baron Von Knigge, Judge Zwackh, the Duke of Orleans (eventually the Grand Master of the Grand Orient Masonic Lodges of France), and over 600 other men of influential positions had eventually joined Weishaupt's Bavarian Illuminati by 1783 forming six major 'circles' of illumination.

It soon transpired that the Jesuits as well as the remaining Masons realized what a force Weishaupt had constructed. The battle between these established orders and Weishaupt's 'new order' started in Bavaria, spreading to France where, eventually, it is rumored to have precipitated the French Revolution of 1837... seven years after Weishaupt's death.

However, let us continue with Weishaupt's moves in the early 1780's. He listed six main points in his manifesto which should have left no uncertainty in anyone's mind as to what he was planning: violent world-wide revolutions in the coming generations which were to yield, a new world order, or a benevolent' dictatorship. His six main points were the **abolition of**: *1. Ordered or nationalistic governments in the form of monarchies, 2. private property, 3. inheritance rights, 4. patriotism to national causes, 5. social order in families, sexual laws, and moral codes, 6. all religious disciplines based on faith in God as opposed to faith in nature, man, and reason...* [deism].

Weishaupt wanted a deistic republic of global dimensions. To those who have read the book entitled 'Protocols of the Learned Elders of Zion' it must be obvious, since the book discussed not only Weishaupt's six points of subversive revolution (and 18 more), that neither Jewish Elders nor Masons were entirely responsible for writing it. Neither group would have been stupid enough to so obviously indict themselves in such a document, secret or otherwise. The protocols are real; they do exist; and they have been exercised with alarming precision by some group for more than 100 years. They were truly written by the Illuminati, that same Illuminati whose Hermetic code insists on secrecy, and a 'low profile.' The Jews and Masons have been made the scapegoat for something they have not done, even though some of both groups have at times aided the cause by their own ignorance.

Degrees Of Illuminism

Although Weishaupt established thirteen (13) degrees in his Bavarian Illuminati, the upper nine (9) degrees were the ones from which his key persons were chosen.

Freemasons will recognize some of the following degrees which Weishaupt devised:

Nursery degrees:		1) Preparation
		2) Novice
		3) Minerval
		4) Illuminatus Minor
Masonry degrees:	SYMBOLIC::	5) Apprentice
		6) Fellow-Craft
		7) Master Mason
	SCOTCH::	8) Illuminatus Major or Scotch Novice
		9) Illuminatus Dirigens, or Scotch Knight
Mysteries degrees:	LESSER::	10) Epopt or Priest
		11) Prince or Regent
	GREATER::	12) Magus or Philosopher
		13) Rex, Homme Roi, or Areopagite

The Bavarian Illuminati used secret names for both individuals within the Order and for the locations of their Lodges. Weishaupt adopted the code name, 'Spartacus,' after the slave who led the great slave uprising during the reign of Pompey the Great; Baron Von Knigge was code-named, 'Philo;' Judge Zwackh, 'Cato' Professor Westenreider, 'Pythagoras' etc. Each initiate had to take a classical name for his code name. Their cities and countries took names of ancient cities and regions. Munich became 'Athens' and Bavaria became "Grecia.' In time, the Illuminati were forced by an untimely discovery of their plans by the Bavarian authorities to move to France, England, and America.

After the French Revolution, Weishaupt's Illuminati went underground. Many sources even say the Illuminati movement died out after the Revolution. However, modern 'observers' must surely concede that the most daring and diabolical social experiments in the history of man are presently illustrating the basic tenets of Weishaupt's school in both Russia and China. The Russian 'Society of the Green Glove' and the Chinese 'Society of the Green Dragon' are not dead, nor are the Illuminati who created them.

Rituals And Punishments

The Illuminati at the higher degrees seemed to progressively excel in pagan, orgiastic ritual. The use of red ribbons, cloaks, and girdles accompanied black candles, the ankh, a rose on a cross, secret words and signs, and a human skeleton. In one initiation, the candidate's pudenda were tied up with a string as the ceremony started. His naked body was then painted with crosses using human blood for the paint. In a later part of the ceremony a 'gigantic, semi-transparent' form rose up out of a burning pyre to demand an oath from the candidate. As the ceremony drew toward its end, the candidate was threatened with swift and sure death if he violated his oath. He was then presented with a vessel of human blood with which he had to bathe himself. Following that, he had to drink one glassful of the blood. At the completion, his pudenda were untied and he was placed in a normal bath. Such rituals, however, seemed more popular in the French branch of the Bavarian Illuminati.

Of course, should anyone doubt that modern secret societies could be so strict or macabre in their demands, he need only hear the Masonic Oath to which the modern candidates must swear. In many Lodges, should a Mason at any time prove faithless to the Order by a variety of actions, including breaking secrecy, he is liable to be ritually killed by having his throat cut, his head cut off, or by being severed in two and having his bowels burned to ashes.

Lesser penalties can consist of having his tongue torn out, his left breast laid open, or his right hand cut-off and slung over his left shoulder to wither and die. Modern Masons, however, maintain that these rituals are only symbolic and that a lesser punishment of 'branding the violator as void of all moral worth' is now in common use. This is most certainly the case in most orders of Masonry; however, some incidents in the recent past make one wonder. A case in point was the Masonic, ritual-murder of the American Freemason, William Morgan, in 1826 for breaking his oath by writing the book, "Freemasonry Exposed and Explained." His death triggered mass resignations from Masonry all over America. The reader is urged to read, *"Darkness Visible - A Revelation and Interpretation of Freemasonry"* by Walton Hannah. Although pressures brought to bear upon Hannah make copies hard to get, it is essential reading if one is to fully understand the implications of the following chapter which concerns the secrets of the 'Great Seal of the United States of America.'

Modern Masonic Movements

Reminders of ancient pagan practices and cultures still permeate the North American and European Masonic movements - as some of the following names show:

1) Ancient Egyptian Arabic Order Nobles of the Mystic Shrine: 30,000 32nd degree Masons.
2) Ancient Egyptian Order of Sciots: 3,000 Blue Lodge Masons.
3) Daughters of the Nile, Supreme Temple: 76,000 women.
4) Imperial Council of the Ancient Arabic Order of the Nobles of the Mystic Shrine for North America:100,000,000 members.
5) Red Cross of Constantine - United Imperial Council: 5,850 members.
6) Supreme Council, Mystic Order, Veiled Prophets of Enchanted Realm : 100,000 Blue Lodge Masons.
7) Supreme Council, Ancient Accepted Scottish Rite of Freemasonry: 511,369 members.
8) Supreme Council 33rd Degree, Ancient and Accepted Scottish Rite of Freemasonry.
9) Supreme Shrine of the Order of the White Shrine of Jerusalem: 180,303 Master Masons

Remember: not all Masons, nor all Jews support the Illuminati. Do not judge those who do so... The proper application of FAITH, HOPE, and CHARITY will overcome all darkness.

CHAPTER 3: Seal of Deception

With the exception of the year in which Jesus Christ rose from the dead, the year 1776A.D. must be the most significant date in recorded history. In that mystical year of 1776, one major revolution officially began in America while a philosophical one began in Bavaria. The latter was, however, to eventually affect the outcome of not only the American Revolution, but also the French Revolution, the Bolshevik Revolution, the Chinese Cultural Reformations and countless other bloody, social experiments of the last 200 years.

The Principled Architect

Although Thomas Jefferson disagreed with Weishaupt on several of his protocols of the 'new world order,' he did support Weishaupt's idea of a republic, of abolishing inheritance laws to prevent feudalism, and of basing the new government on strong reason as opposed to pure faith alone. Thomas Jefferson was the principle architect of the Declaration of Independence for the United States of America. Jefferson, Benjamin Franklin, and John Adams worked diligently together on the initial drafts of the Declaration; but it soon became obvious to both Franklin and Adams that Jefferson not only had definite ideas for the Declaration's format; but that he also thought himself much more qualified (which he, not doubt, was) to present the proclamation than they, themselves. In the end, they allowed their 'enlightened' Masonic brother to finish the document by himself. So, on the 4th of July in 1776, the United States of America declared its independence, just two months after Dr. Adam Weishaupt officially began the Bavarian Order of the Illuminati.

During that same year, 'someone' authorized a small group of 'illuminated' Masons to design a seal for the new republic of thirteen (13) colonial states. Six years passed before the design of the seal was completed and unofficially adopted by the predominantly Masonic Congress of the United States of America on the 10th of June, 1782. The actual Seal was to be called, 'The Great Seal of the United States of America.' It was cast in two metallic plaques. Both were delivered to Thomas Jefferson on the 17th of June, 1982 while he stood in the dim candlelight of his Monticello drawing room in Virginia. They were presented to him by a hooded figure who entered quietly from the garden carrying the plaques in a red velvet bag.

Sometime, during that same year, world Freemasonry held a congressional meeting of their own in Europe. It was to be remembered as the 'Congress at Wilhelmsbad.' At this meeting, Continental Freemasonry was merged with Weishaupt's Illuminati. As a result, many English Lodges broke relations with the 'illuminated' Continental Lodges which were collectively called the 'Grand Orient Lodge.' By 1784 Weishaupt found himself under fire from many Master Masons, even his cohort, Baron Von Knigge.

Weishaupt was fortunate, however, that Thomas Jefferson had just been appointed the American Ambassador to France in 1784. Jefferson quickly sized-up the European mood which was quite complex, considering that four major philosophies were competing for dominance in Europe. Seeing that Weishaupt's brand of illuminism stood the best chance of uniting the major elements of Masonry, Jacobinism, and the Jesuit teachings, he quickly sprang to Weishaupt's defence throughout the major strongholds of the 'Enlightenment' in Europe.

Five years passed; and Jefferson returned home to America to become the Secretary of State. Shortly after his return, the Constitutional Congress officially adopted two major constructions on the 15th of September, 1789: The U.S. Constitution and the 'Great Seal of the United States of America.'

The Obvious Seal

The Great Seal of the United States of America is not what it appears to be. It appears to be a seal designed to represent the birth of a new social order from the thirteen colonial states. It is supposed to represent a nation of Christians. It is supposed to represent freedom, strength, and peace. On the surface, it may appear to represent these things, for sure; but it holds many more meanings which are not only secret, but also, sinister and anti-Christian. To prove this last statement, the symbolism of the 'obvious' seal should be explained.

The front side of the Seal (see figure 1) has two Latin phrases around its periphery: *"Annuit Coeptis,"* and on the lower scroll, *"Novus Ordo Seclorum.'* Historically, these phrases have been translated, respectively, as: *"HE (GOD) has favored our undertakings"* and *"A new order of the ages.'* The eye in the triangle is usually viewed as the all-seeing eye of God's wisdom; and the light which streams out of the triangle, supposedly, represents the light of God's wisdom. The date on the lower row of blocks on the pyramid represents 1776 - the date of American Independence, among other things.

The back side of the Seal (see figure 2) has only one Latin phrase: *"E Pluribus Unum"* or *"Out of many, one."* The eagle has been the symbol for divinity in many ancient religious philosophies. It was apparently used here to represent God, strength, and freedom. The thirteen arrows represent strength to defend while the olive branch represents peace. The stars above the eagle's head are arranged in a Mogen David or 'shield of David.' The reason for this has not been publicly explained, not has the reason for a lot of other anomalous symbolism in this Seal.

Seal Of Babylon

When the Great Seal of the United States was designed, it was designed by the Illuminati. This is certain. Whether Thomas Jefferson really knew the secret of this Seal is not certain, even though circumstantial evidence does implicate him as being a close friend of the one man in the world who could have designed such a mystical seal at that time: Dr. Adam Weishaupt.

The front of the Seal is a 'reasonable' place to start this critique. What does a pyramid have to do with the United States of America? Why does the pyramid have no capstone? Why is the 'eye of God's wisdom shown inside a triangle? Why are there 13 steps in the pyramid?...

FIGURE 1

3 points of the triangle

1 great light

1 all-seeing eye

13 letters

13 steps

2 forked tongues

MDCCLXXVI

9 roman numerals

ANNUIT COEPTIS

NOVUS ORDO SECLORUM

THE GREAT SEAL

a)   **MDCCLXXVI** = **1776** = **1110** *(Babylonian)* + **666** *(Arabic)*
b)   **\$1.00** *(Babylonian Sexigesimal money system)* = **\$.60** *(Arabic centesimal money system)*
c)   *Hence:* **\$1110** *(Babylonian)* × **\$.60** *(Arabic)* = **\$666** *(Arabic)*
d)   **1776** *reduced by numerological rules* = **1 + 7 + 7 + 6** *which reduces to* **3**
e)   **MDC.CLX.XVI** = **1600 + 160 + 16**
f)   **1600** *reduced as above* = **1 + 6 + 0 + 0 = 7**
g)   **160** *reduced as above* = **1 + 6 + 0 = 7**
h)   **16** *reduced as above* = **1 + 6 = 7**
i)   **7 + 7 + 7 = 21** *which finally reduces to* **3**
j)   **1776 ÷ 16 = 111**

FIGURE 2

'One out of Many'

The pyramid is the Great Pyramid at Gizeh. The top is missing to show that the capstone or the 'lost word' is missing. This pyramid was not built solely to bury a pharaoh. It was also used to house the great secrets handed down from the ancient times of Ammon, Horus, Ra, Osiris, and Thoth. These secrets form the basis for all the modern Hermetic mystery schools, all of which are pagan societies. They are not Christian. They may say they believe in 'God;' but their Lord is not Christ.

The *'eye of God's wisdom'* is a Masonic symbol. The eye represents *'The Great Architect of the Universe.'* *The Grand Geometrician of the Universe,'* *'God,'* or *'Yod.'* It is enclosed by a triangle which is a mystery symbol for wisdom, strength, and beauty, a trinity. This triangle, however, is upside down if it is supposed to represent the infinite looking down; because the point should be down. This means that the eye is not of God but of a man, of a finite order. The thirteen steps represent something rebelliously evil; because, thirteen is the value that both Biblical and Kabbalistic gematria assigns to represent Satan, sin, murderer, serpent, dragon, Belial, tempter, or rebellion.

"Annuit Coeptis... Novus Ordo Seclorum" has also been translated by some mystics as, *"Announcing the arrival of a new secret order of the age,"* and *"The Established Order of the ages looks favorably upon our endeavors."* The term *"Great Seal,"* itself, is indicative of another major clue to the nature of this seal. The *"Great Seal"* is *"Mahamudra"* in Sanskrit. It stands for, *"the union of all apparent duality and the supreme joy that comes with the realization of mystical union."* To solve the next part of the Seal, mysticism must be employed in the merging of the *"apparent duality of the Seal"...* for it is two pieces.

However, before employing the 'union of the two,' the symbolism of the back side of the Seal should be explained. The eagle is a symbol of ancient, pagan forms of deity, which will be most obvious when the 'union' is effected a bit farther into this text. The golden eagle was the symbol of Rome. In fact, the Jews once tore a golden eagle from the wall of their own, Holy Temple after Rome had mounted it there. The nine tail feathers of the eagle represent the nine beings of the innermost circle of enlightenment in the 'Great White Brotherhood' - or the 'Illuminati.' 'Nine' also represents 'finality or the end' in Biblical gematria, and "the ninth hour of the day Christ died.' The eagle's head looks to the right symbolizing: *'it looks favorably upon the 'right wing'* or *'the established order.'* In contrast, the WWII Nazi bird looks to the 'left wing' as a symbol of revolution or rebellion. The so-called communist doctrines of the world also fall into the 'left wing' with the fascist orders; because both are totalitarian or one-way pyramidal structures instead of 'limited government.'

Whose established order does the eagle looks to? Is it that worldly order spoken of by the ancient Biblical prophets? Again, using gematria, the twelve horizontal lines on the eagle's shield represent 'perfect government.' The six vertical bars of three dark lines each interspersed with seven white spaces on the lower part of the shield yields the following under gematrical analysis: *"Sinful man* (6), *of the dark trinity* (3 dark lines), *in opposition to the Seal of God* or *the seven churches of Christ* (7 white spaces)."* The Latin phrase *"E Pluribus Unum"* means, *"Giving order to chaos by uniting the many into one."* The arrows in the right claw are symbols of war and military might. The olive branch is a symbol of peace and the fruits of peace. The stars above the eagle's head are entirely mystical in meaning. The thirteen small stars are five-pointed. They are more correctly known as 'the Seals of Solomon.

Many people have been taught incorrectly that Solomon's Seal was the interlaced twin equilateral triangle. It is not. Solomon's five-pointed seal is used by initiate witches and warlocks as a means of protection against 'evil spirits.' They stand inside the five-point star which, itself, is inside of a circle, sometimes called 'circle of Satan.' The thirteen five-pointed stars are formed into the traditional 'Magen David' or 'Shield of David.' These interlaced triangles represent the 'union of the two realities.' That is the union of God with man, of the infinite with the finite, or of the spiritual with the physical.

Having thus laid the foundation, it is time to erect the structure. If one makes transparencies of the front and back of the Great Seal it facilitates the breaking of the hidden code. By overlaying the front piece onto the back piece so that the triangle around the eye completely covers ten of the stars in the Mogen David, one can see three remaining Seals of Solomon - each on the exterior of the covering triangle (see figure 3). These three bright stars can represent the three 'emblematical lights' of Freemasons: The Bible, the Square, and the Compass, which can also be referred to as the 'jewels of the lodge.' They can also represent the Satanic trinity. there are six separate positions in which these three 'lights' can be seen outside the triangle, but inside the 'circle of illumination.' In each position are messages to the mystics. The eagle's wing-tips point to some; the arcs of the circles intersect others. The twenty-eight dark lines which surround the Mogen David cluster are reference lines which cross with other extended reference lines from the shafts of the arrows to yield further coded messages both in Latin and symbolism. Unfortunately, some of these messages were still being translated as this book went to press; so the reader will have to do some of the work on his own. One may find another series of coded messages by arranging the pyramid so that one of its base angles fits into the triangular tip of the shield (figure 4). It is significant that so many of the angles and dimensions on one of the plaques so closely match those on the other plaque. Another most unique message is formed by arranging the Seal as in figure 5. One is reminded of another symbolic message in the 13th Chapter of the book of Revelation. The 12th verse mentions a second 'beast' (the new Rome?) exercising the authority of the former 'beast' (ancient Rome?) while in the sight or presence of the 'revived,' first 'beast.' The eye of the 'new world order' does appear to be in the eyesight of the old roman eagle. When the blazing eye of the triangle is viewed in this particular way, it forms the well-known symbol for the ancient, pagan god of the Egyptians: Horus - the winged eye of eternal light. The eagle's head and the triangulated eye oversee the capstone of the pyramid denoting superior strength and wisdom. Many Freemasons and Rosicrucians will instantly recognize this combination, for it adorns their lodges and many official documents. The tip of the shield rests upon the 6th step of the pyramid. The capstone is the rectangular block formed by the top of the shield with the 12 horizontal lines of 'perfect government.' Masons will understand this rectangular block represents the **double cubic stone** found at the sight of the second temple; however, be forewarned, this symbolic message is not from the true Light. It reads: *"The missing capstone and the lost word are both supplied when the perfect government or the cubic stone of the great winged light of the sky takes his rightful place."* Furthermore, as the bird can also represent the Egyptian Phoenix or bird of eternal life, this view could represent the Phoenix rising up from the ashes of a previous chaos or desolation (see Genesis 1:2). It could be that the Egyptian story of the 'god' who was cast down to Earth (the legend of the Phoenix) was relating the fall of Thoth, Hermes, Baal, Mardum, or Satan.

FIGURE 3

The Three 'emblematical lights' of Freemasons:
The Bible, The Square and The Compass
Also called 'The Jewels of the Lodge'

FIGURE 4

Codes in the angles on both sides?

FIGURE 5

The Winged Solar Light of Wisdom

The New Order in the Sight of the Old Order

If so, this would further indicate that the 'fallen-one' is attempting to rise up from the ashes of his past destruction as a mighty eagle or phoenix. It could be his secret seal. As the reader may wish to pursue this in greater detail, the author has included an extract from page 103 of F.C. Payne's book, **Seal of God** (see Section III reading list). It follows: *"1 = Unity or unit; 2 = separation or witness; 3 = divine perfection or Godhead; 4 = creative work or the world; 5 = grace, free gift of God; 6 = man or man of sin; 7 = God's Seal, often called the perfect number; 8 = resurrection or new beginning; 9 = finality; 10 = ordinal perfection; 11 = disorganization; 12 = perfect government' 13 = sin or rebellion and 37 = living Word of God."*

The nine tail feathers point to the Roman numerals 'MDCCLXXVI.' The clue to understanding the message within these numbers lies in the application of Weishaupt's 'reason' to the gematrical interpretation of these nine **roman** numerals.

Remembering that Weishaupt's objective as well as that of the power behind him was the establishment of a final, pagan order in the physical and spiritual affairs of man, one must look, objectively, at these nine roman numerals for a clue. There are 9 numerals which represents the finality of the new order. The ultimate of spiritual existence is presented as a trinity or 3 items. Within these 9 numerals are 3 groups of 3. Within each of these groups are 2 sub-groups: 1 of 1 by 1 of 2 which act, respectively, as *unity by witness.* Hence, 'MDCCLXXVI' becomes: 'MDC,' 'CLX,' 'XVI,' breaking this into sub-groups yields: 'M.DC,' 'C.LX,' 'X.VI.' Taking the two sub-group patterns into separate numbers yields: 'M.C.X.' and 'DCLXVI' which, respectively, represent: '1110' and '666' (see Fig.1 footnote).

Drawing upon the clues given in the 13th chapter of the Book of Revelation as to the number of the man' who is the antichrist, one finds that the Greek word for 'number' is 'arithmos' which quite frequently is used in the sense of 'a coded number.' Knowing, also, that the 'beast' is referred to elsewhere as 'Mystery Babylon,' one finds that '**666**' is quite possibly a modern mystery code for a Babylonian number. This means that '**666**' could represent another number based on the Babylonian monetary numbering system which used 'dollars' comprised of 60 'cents' as opposed to the modern or Arabic system of 'dollars' comprised of 100 'cents.' The 'sixty-based' Babylonian system was known as a sexigesimal system. The 'one hundred-based' modern system is called a centesimal system. To convert from Babylonian to Arabic, divide the Babylonian number by 100 and then multiply the result by 60. *Hence, '666' Arabic is equal to '1110' Babylonian! Are the Roman numerals on America's Great Seal those of the antichrist and his new world order in both the Babylonian and modern systems of monetary exchange?* The man who put the Great Seal on the back of the United States one-dollar note in 1933 was Franklin D. Roosevelt, an 'illuminist.' This was a symbolic act, because the Seal is supposed to represent 'unity' when it is joined. What better place than on a one-dollar bill?

Any coded system has two parts: the key and the message. The key is always an ordered structure. The message, however, always has the appearance of absolute chaos or randomness. To decipher a chaotic message, the 'random' pattern is passed through the ordered key to yield an ordered message. Both sides of the Great Seal appear to have keys as well as messages encoded into them.

The keys may be those portions which are symmetric to the central vertical axis of the Seal or are otherwise of a logical pattern either symbolically or numerically.

By superimposing the reverse image of each side of the Seal upon itself, many of the asymmetries are quite easily located. On the pyramid side the asymmetries are: the upper words in Latin, the lower scroll words in Latin, the date, the whiteness of the upper letters versus the darkness of the lower letters, the small shrubs or disorder in the land around the pyramid, the tilt of the triangle around the eye and the light rays around the triangle.

On the bird side, the asymmetries are: the light rays around the upper stars, the head of the eagle, the olive branch, the bundle of arrows, the left and right claws, the words on the banner, and the banner, itself. It may be significant that the triangle masks the heads of the arrows when the point is put into the claw with the base over the heads. In this position the arrows are pointing at various letters and numbers on the pyramid side. The clue to the order to read the letters is most probably a numeric progression based on the length of the shafts from the fist.

The three five-pointed stars that appeared when the triangle covered the other ten stars can also represent: 'JAH-BUL-ON' and 'JE-HO-VAH' in the Masonic schools. The five-pointed stars can represent: 'the five Masonic points of fellowship,' 'the five human senses,' 'the five celestial Buddhas,' 'the five great evils of Buddhism: Ignorance, Wrath, Desire, Malignity, and Envy,' and 'the five precepts of Buddhism'.

The first nationally recognized god of Babylon was 'Marduk.' Later his name was changed to: 'Bel,' 'Bul,' and 'Baal,' which meant 'Lord.' His sacred animals were the horse, dog, and **fork-tongued dragon.** One of his favorite instruments was a **triangular** spade. He was depicted in many carvings **wearing a tunic adorned with stars.** He either carried **a bow, spear, thunderbolt**, sceptre, or a net as his weapons in most of the Babylonian records. Recent evidence suggests that his legendary activities may even have pre-dated the times of Thoth, Ammon, Ra, Osiris, and Horus in Egypt.

What Now America?

It would appear that America is either a pawn of the modern Babylonian mystery school or the seat of the 'new Babylon' mentioned in Christian prophecy. It must be significant that America was sealed with the Seal of Babylon. See what a confused jumble its legal system has become. Look at her woeful economy. Notice the unprecedented moral decay and the chaos that shakes the very foundation of American unity. Even the original guarantees of the Constitution and the Bill of Rights have been bent and misconstrued to the point of utter uselessness. Will America be the sacrificial goat in some pagan, global ceremony to 'free mankind of all its ills?' If so, **"Israelites"** are forewarned: "come out of Babylon" (see Section III for more detail). Do the 'Illuminati' really exist in modern affairs? If they do, would their actions be detectable? To find the answers, **let us** next **pretend to be** those hypothetical, 'illuminated' villains.

CHAPTER 4: Model For 'Peace'

Let us pretend that you and I were the Illuminati of the whole world in the early 1950's. In our hypothetical role as those 'obscure men behind the scenes,' we would have been confronted with a real dilemma: the objective of generations of our Illuminati-instigated revolutions, wars, conspiracies, and social experiments was in danger of being forever removed from our grasp. The "Novus Ordo Seclorum" of the Jeffersonian Era was under attack from the modern European branch of the "Great Enlightenment" of that same era. Chaos loomed on the horizon. The civilized world could not have survived another war; so we would have resolved to settle the philosophical differences between the East and the West - between our 'vehicles:' the American Council on Foreign Relations and the English Royal Institute of International Affairs against the European Bilderberg Group - in order that our objectives of world peace, love, and harmony might be reached in a jointly-controlled, "benevolent" dictatorship. That process, though difficult, would prove to be more easily accomplished than obtaining both the support and the dependence of the masses of the world.

Peace By Placebo?

As illuminists we would have supported the philosophy known as 'deistic republicanism.' Such a philosophy would have convinced us that although man had been put on Earth by some unknown process of a "natural" order of things, it was up to man to solve his own crises without any help from his "nature God.' As deistic republicans, we would have considered our chances of negotiating a world peace through the various organizations of our making.

We would have known that the failings of our United Nations and our League of Nations efforts were due not so much to the failings of the leaders of the various countries as to the great cultural barriers that had existed in the minds of the masses of those same countries. We would have seen no possibility of negotiating a peaceful world unity between so many vastly differing cultural prejudices in the short time left to find the solution. 'What is the option?' we would have asked.

Peace By Force?

Seeing that the option of peaceful negotiation would default within the time limit, we would have then considered mobilizing a global army either as some multinational, corporate executive or as an armed form of martial law. Almost before we had suggested such an option we would have realized that the peaceful unity we were seeking could not have been attained in that manner.

That percentage of the masses who would have seen our objectives and would have challenged us would have created a resistance far worse than the French Underground was to Nazi Germany. The ensuing guerilla warfare would have crippled our industrial complexes beyond any hope of technological salvage. What could we have done? Peace by discussion, peace by force - nothing seemed feasible in time to save man from himself.

Peace Alternative Three

In the midst of frantic search to find a third alternative, a stroke of good fortune would have appeared from the research labs of the world. A technological breakthrough of immense value to us would have leaped out of the news headlines. Headlines like: "Anti-gravity: New Spaceship Marvel Seen," "New Energy Source Heralds Technological Revolution," and "Household Flying Saucers: Future Reality?" would have given us *the key to our third alternative: Peace by Pretence.*

The Plan Forms

Using our massive influence, we would have begun to suppress media coverage of any other developments in the 'new technology.' We would have simultaneously begun numerous secret research projects funded by our private, corporate funds as opposed to our puppet governments' budgets. We would have realized - quite apart from its necessity to our plan that the new technology could have collapsed our economic structures, if it were not properly infused to the existing consumer/demand process. The near possibility of cheap and revolutionary products and processes would have immediately changed the masses' consuming habits. We would have known that the masses generally act as individuals without consideration for the well-being of the whole herd. Because of that, they would have stopped consuming the same volumes of old technology in greedy anticipation of the new, thereby, collapsing those industries with huge stocks of old technology. The masses would have killed themselves, their life style, and our plans by their collective act of anticipatory restraint.

Knowing the great majority of the masses were 'blind' and would not 'see' what we would be doing, we would have begun to organize our key scientists, engineers, politicians, industrialists, economists, and bankers into an overnight collection of global problem solvers. We would, also, have made guarded media appeals to the 'watchers' in the masses to support our plan in silence.

We would have **suppressed public awareness of our plans and the new technology.** Those 'unenlightened' inventors and researchers that did make the serendipitous discovery of our new medicine, physics, chemistry, and philosophy would have been made to understand the importance of their future work and discoveries being kept secret for the 'benefit of mankind.' **Unhappily, if we could not have convinced them to join us, they would have had to be eliminated by death or discredit (for the good of the whole, the end would justify the means).** We would have had to suppress the engines that ran on water, the natural cure for cancer and numerous other developments, all in the name of global survival.

We would have secretly financed much of our research through government financed research and defence programs in the corporate sectors.

The people would have never seen where their tax dollars really went. Even the people who would have manufactured our new devices and processes would not have known what they were really constructing. One contractor would have built a secret part numbered 'AX31-423', and would have delivered it to a pre-specified location. At some point way down the line, all the proper pieces of our secret technologies would have been combined by 'our people' into a finished product. That finished product could have been a 'flying saucer' and the manufacturer would never have known that he had done it.

We would have developed our own security organization to maintain the secrecy of our plan. We would have put our 'own people' into the leadership of the various media mechanisms, civilian UFO research societies, national intelligence agencies, foreign policy schools, and economic institutions. This would have given us the necessary controls to even cover-up the accidental discovery of, for instance, one of our own 'flying saucers' caught in a compromising situation or downed with equipment failure. Our plan, our grand design, would have been ready for birth.

"They're Not Ours"

We would have 'conditioned' the masses to believe that 'anti-gravity,' electric 'flying saucers,' limitless energy, and the attendant technologies were just too advanced for any human mind to have developed at that stage of history. We would have easily accomplished their conditioning by using science fiction movies dated several hundred years into the future showing the super technologies as future products, not secret realities.

We would have developed a linear growth pattern in the educational facilities to reduce the chances of too many new discoveries exposing our deception. While the universities, funded by our foundations, plodded along on low-budget, linear research and discovery programs, our own secret research and development facilities would have surged ahead by exponential leaps. We would have recruited those young minds whose talents would have accelerated our own progress towards 'peace.'

First With Fear

To further discourage unnecessary interference from the 'curious' of the masses we would have used the media to implant fear into the general public. Their fear would have been of the *"little green men,"* of *"unfriendly UFOs"* of *"close encounters,"* of *"the unknown."*

During the first fifteen years of our twenty-five year plan for 'peace,' we would have programmed the masses to fear contact with the new and the unknown. For those who reported UFO sightings we would have developed such convincing explanations as: 'marsh gas,' 'weather balloons,' and 'the planet, Venus.' In some cases where such panaceas failed to placate the mob, we would have given several false versions of the incident in a manner that would have frightened the masses into complete apathy. We would have had to drive their heads into the sand to prevent them 'seeing' our machinations.

In the latter ten years of our great plan we would have, presumably, been ready to let the masses pull their heads out. By that time, our secret sociological model and the attendant technology would have been developed to a state of readiness. All we would have had to do was to concoct spurious global crises, break down their cultural boundaries, embarrass their established orders, and threaten them with World War III, or nuclear annihilation. We would have frightened the hell out of them on one hand while, on the other hand, we would have been holding our new social order in reserve.

While they were in a state of shock over the demise of their own social orders, we would have offered them hope in the form of a media-presented escape. We would have presented movies, books, and television which would have portrayed UFO people as friendly, as superior, as 'gods' of ancient legends.

I can see it now, frozen in an energy crisis, saddened by the Watergates of the world, dying from environmental pollution, starving from food shortages, frightened of a global nuclear war, sick of the moral decay, afraid of the daily news, bankrupted by global monetary fluctuations, unemployed from economic depressions, crowded by the ever present birth rate, frightened by the suspicion that a global weather catastrophe was about to happen, mankind that great sluggish mass of which you and I, in reality, are a part, would have been ready for 'Alternative Three.'

In our hypothetical role of the *illuminated global planners* we would have then reversed the earlier fear of future contact with new societies and technologies which were not of human origin. We would have financed television series like 'Star Trek' to replace the effects of 'The Invaders' and "UFO.' We would have financed movies like "Close Encounters of the Third Kind" to counter the conditioning in "War of the Worlds," "It Came From Outer Space," "Invasion of the Body Snatchers," and many others. We would have initiated and funded dozens of new philosophical cults similar to that of the "Urantia Foundation," the "Aetherius Society," "The Uri Geller Cults," the "Heralds of the New Age," the "One World Family," and a variety of other fronts for "The Great White Brotherhood" to convince the young and the foolish of the masses that "their elder brothers from space" were real and were coming to save Earth and its "people of good vibrations.'

Finally, with our teams of highly-trained actors to back us in our bid for peace, love, and harmony through war, hate, chaos and deceit, we would have begun the greatest attempted deception of all history. We would have begun a peaceful invasion of Earth by humans posing as 'elder brothers and sisters' from space, supposedly, from a superhuman culture which would have shown no signs of war, hate, chaos, deceit, and all the negative parts of human endeavour.

Our spacemen messiahs would have revealed themselves from amongst the people and from above the people in their 'light craft.' Demonstrations of the power that our fabricated heroes possessed would have been given in the most visible ways possible. Then, over all operating radios, televisions, and hosts of household electrical devices would have come our message to the people of Earth in a multitude of languages, dialects, and sub-tongues:

"People of Earth... we mean you no harm. We are your friends, your brothers from another star system which is far away. Your race is in danger of destroying itself in the very near future.

It is your right to do so, if you so choose; however, if you wish to accept our help, super-technology, and social order, then we will help you to establish a 'NEW WORLD ORDER' or what you might call, 'a NOVUS ORDO SECLORUM.' We have a corner on global disaster control. Give us your allegiance, your resourcefulness, and your planet; and we will solve Earth for you. The choice is yours: Life our way, or death.'

'WE AWAIT YOUR ANSWER............'

CHAPTER 5: "It's All Happening"

Sitting lazily in the midst of a sunlit field with flowers and trees, listening to the hypnotic 'ch-ch-ch' of the tree crickets' warm afternoon serenade subtly signalling 'siesta-time' and perceiving a variety of little chirping birds that soar and sail endlessly in blue skies filled with fairy floss all seem to wrap the mind in a false sense of peace and safety. In such a state, it is difficult to ponder tales of ancient mystery schools and the hypothetical machinations of their modern descendants, the Illuminati. Yet, one is forced to do so, eventually; for the sun will set; and the elusive warmth of the day will be replaced by the chilling terror of the night.

Roman Empire Revives

There are many signs in current affairs which not only suggest but also confirm the existence of the 'hypothetical Illuminati' of the preceding chapters. One major arm of these 'real Illuminati' recently exposed itself with the official announcement of its formation. Prior to 1968, this 'arm' was quietly known as the 'Top Ten Club' throughout Europe. With the official or public announcement of the groups' formation in 1968, it was dubbed, 'The Club of Rome.'

The Club of Rome was initially formed with a membership of seventy-five select men, from amongst the most prominent scientists, industrialists, and economists of the world. Their president, Dr. Aurelio Peccei, is an economist who was previously an international vice-president for Olivetti and is still the head of the Italian management firm, Italconsult, in Rome. Other men like Kogoro Uemura, president of the Japan Federation of Economic Organizations, and Dr. Alexander King, director-general of scientific affairs for the O.E.C.D., were among the original founders.

The Australasian issue of Time dated 24 January, 1972 (p.30, 31) had some extremely interesting observations to make concerning the Club of Rome. Some of those observations are paraphrased below:

The Club met with favorable support from the Volkswagen Foundation which granted the Club $250,000 in 1970 to finance, *"an international team of scientists led by M.I.T. computer expert, Dennis Meadows, to study the most basic issue of all - survival."* The Club later published their first report entitled, 'Limits to Growth' - which projected disaster for the world. Their computer model was designed to *"simulate the major ecological forces at work in the world today."* Their model showed the urgent necessity to curb population, pollution, non-renewable resource drains, food distribution inequities, and energy wastes. One bright fellow once observed about computer modelling, 'Garbage in gives garbage out.' In this case, the Club eventually admitted to certain 'anomalies in the data they had used.'

A follow-up article in the Australasian issue of Time dated 14 August, 1972 (p.50,51) allowed a few more clues to slip out - further identifying the Club of Rome as an 'Illuminati arm.'

"There is only one way out (of the current world crisis), *says the report* **(Limits to Growth)**: *economic as well as population growth must be stopped cold sometime* **between 1975 and 1990 by** *holding world investment in new plant and machinery equal to the rate at which* **physical** *capital wears out."* The essay went on to say, *"this status quo prescription - the report calls it 'global equilibrium' is* **as chilling as the doomsday prophecy.** *Halting economic growth is not merely a matter of the already affluent giving up such frills as electric toothbrushes or power windows. Sacrifices would be made by the poor, who have not yet collected the benefits of the industrial revolution..."*

"Redistribution of existing wealth is no solution, because the **rich and the middle classes would not give up their wealth unless it was forcibly taken from them** *(remember 'peace by force ?').* *Thus the redistribution* **would imply a series of violent revolutions and wars over the ownership of oil wells, ore mines, and fertile farm land.** *At best, even those could produce only an equality of misery."*

"More than that, a no-growth world **would have extreme difficulty providing either social justice or freedom. It is hard to see how growth could be halted or even substantially slowed without a world dictatorship.**

Only one week after the preceding statements were published, an interview between William Irwin Thompson, former M.I.T. educator, and Time correspondent John Wilhelm was published in the Australian Time issue dated 21 August, 1972 (p.34-36). Excerpts follow:

"In **'At the Edge of History'** *(Thompson) said that there would be an* **invisible college** *surfacing in the '70s or '80s after the exhaustion of the protest movement, and I was surprised to see it come up faster than I expected. I was, however, thinking more in terms of a Cromwellian protectorate than a bunch of behavioural engineers (the Club of Rome) round the world* **who would be trying to consolidate their power.** *The intriguing idea about the Club of Rome is its incredible sophistication as a prestige structure.* **They finesse the whole power situation by not even trying to go for power,** *but they say:* "We're going to show you in our computers that disaster is ahead of us. **However, we happen to be just sitting here cornering the market on disasters and so we're ready when you want to buy disaster control. We'll solve the planet for you...'** "

"Some of our problems stem from the fact that authority today pretty much comes from those who have power. What we need is a clear distinction between authority and power, as in the days of Christ and Caesar before the papacy..."

"So the attempt to create a Club of Rome is useful, but it is such an imperial model. First it's a club, and **it's also the idea of Rome again:** *The old Roman imperial model of the centre of civilization sending its structures out into the provinces."*

"Many of the intellectuals now are so hungry for order that they would be willing to see the end of democracy and some new kind of Napoleonic order coming in. *Arnold Toynbee, in his recent book, 'Surviving the Future,' says that as far as he can see* **we have a choice between a world federal state with an Alexander at the helm or nothing - annihilation."**

*"I think that the intellectuals will be the first people to make accommodation with the new power structure. As long as they can still have their elitist sense as professors and computer scientists, they will be quite happy in an aristocratic pro-management system. They don't stand to lose that much. **Thus the ones who cry the loudest for freedom might not be all that much in favor of it.**"*

Their World System

In 1973, the Club of Rome printed an intra-organizational progress report on their special project: *'The Strategy for Survival Project.'* This interim report entitled, *'Regionalized and Adaptive Model of the Global World System,'* (by Dr. Mihajlo Mesarovic and Dr. Eduard Pestel, directors of the International Club of Rome) was not originally meant for public distribution; so this author had to 'borrow' a copy from the original document. The source of the original was the man who instructed Dr. Mesarovic and Dr. Pestel in computerized systems dynamics. His name was Dr. Donald Drew, an American who was both Professor and Chairman of the Systems Engineering and Management Division of the Asian Institute of Technology in Bangkok, Thailand. This document is paraphrased in the following statements; however, the full document is reproduced in Appendix 7 in the public interest.

According to the report, the Club of Rome have designed their *'Survival Project'* with two specific objectives. They are: *"1) To enable the implementation of scenarios for the future development of the world system which represent visions of the world future stemming from different cultures and value systems and reflecting hopes and fears in different regions of the world; and 2) To develop a planning and options-assessment tool for long-range issues, and thereby **to provide a basis for conflict resolution by cooperation rather than confrontation."***

They devised and tested, even at that time, a computer model of the world and its inhabitants. They would be the first to admit that their model was not perfect; however, that has not deterred them in their purpose: the establishment of a new world order.

Their *"...**world system is represented** in terms of interacting regions with provisions made to investigate any individual country or subregion in the context of regional and global development. Presently the world system is represented by **ten regions: North America, Western Europe, Eastern Europe, Japan, Rest of Developed World, Latin America, Middle East, Rest of Africa, South and South East Asia, and China.'***

Their 'computerized world system' is *"...defined by a given set of laws and principles. Specifically, the levels involved are: geophysical, ecological, technological (man-made energy and mass transfers), economic, institutional, socio-political, value-cultural, and human biological."*

Their *"...model* (had been developed) *up to the stage where it* (could) *be used for policy analysis related to a number of critical issues, such as: **energy resources utilization** and technology assessment; food demand and production; population growth and the effect of timing of birth control programs; reduction of inequities in regional economic developments; depletion dynamics of certain resources, particularly oil reserves; phosphorus use as fertilizer; regional unemployment; constraints on growth due to labour, energy or export limitation, etc."*

By 1973 their computer model had already been used to develop a new world economic system which on its basic or 'micro' level was so detailed as to consider eight production sectors: agriculture, manufacturing, food processing, energy, mining, services, banking and trade, and residential construction. Their computer had also been programmed to assess **the consequences of timing and magnitude of natural disasters such as drought, crop failure due to disease, etc...** (one wonders about earthquakes, blizzards, floods, solar flares, and sunspots).

Their plans for the immediate future at that time were threefold: 1) To assess the changes of the various crises solving options available to them; 2) To implement a global, satellite communications network connecting various test 'regions' to build up a 'joint assessment' tool for the long term value of their model; and 3) To use their model to develop an 'underdeveloped' region in order to assess means of **removing existing obstacles to their proposed new order.**

But, the Club of Rome were not finished. By 1978, five major books had been published by their membership which detailed their format for re-shaping the international order as well as their goals for mankind. In Australia, one member of the Club of Rome, Professor Charles Birch, wrote **Confronting the Future** which listed Australia as the test case for their proposed new order. Professor Birch suggested Australia had all the inherent characteristics necessary to yield the best test of their operational model. He also stated: **"Absolute freedom is not possible. There must be restraints on freedom. Not restraints to dissent, but restraints on activities of individuals or industries that impede agreed social goals. It is utopian and unrealistic to hope that we can simply persuade men to be responsible. Not only must there be some consensus on what constitutes the social good, there must be agreement on sanctions being imposed on those who do not abide by mutually agreed goals."**

Professor Birch is too close to the problem. Although his efforts are noble, his proposed solution is extremely naive. One wonders, *"What makes his vaguely defined model for the perfect society any more likely to work than the countless attempts made by the past generations?"* He is correct in saying absolute freedom among human beings under current situations is impossible; However, his scenario has not allowed for circumstance or process which could change the basic character of mankind. The mystics know this is a distinct probability. Professor Birch has taken a position that is a page from history. Alexander, Napoleon, and Hitler all supported that same position. Even if the world could obtain a truly benevolent Alexander to run things, it would find itself in dire strife upon his death or disability, or even insanity. It is well known that absolute power corrupts absolutely.

Yet, Professor Birch and his ilk still maintain some form of authoritarian democracy (benevolent dictatorship) is the solution to the world's problems. Couched in his general terms like 'restraints', 'that impede', 'simply persuade', and 'consensus of social good' are the horrors of the new Nazism with the same platform Hitler used. Hitler's propaganda appealed to both men of reason and men of mysticism. His candlelit mystical parades which formed the sign of the Illuminati (a burning flame encircled by a ring of light) and the use of the winged eagle over the encircled swastika for his national seal proved him to be an illuminist of the Bavarian school. He was the prototype of one who is yet to come from within a revived Roman Empire which the pseudo-intellectuals of this planet are rapidly bringing to power. In their self-inflicted blindness, they create a 'beast' of unprecedented oppression in human affairs.

The Cosmic Conspiracy Final Edition 2010 © Stan Deyo 2010

In their second report to the Club of Rome, entitled, **Mankind at the Turning Point**, Mesarovic and Pestel gave full details of the systems diagram for their operating computer model of the world economy. Appendix 7 gives a list of every country in the world and where each fits into the 'ten region (or kingdom)' format of their new world order.

It is of some interest to note that Bankamericard, Barclaycard, and Bankcard all use the same single digit 'regional code' that the Club of Rome has 'suggested.' Why have a 'regional code' at all if there is not going to be an international numbering system? If they hold to their time schedule, the new world order will have replaced all existing legal tender with a computer numbered account long before 1982.

Most of their choices for co-regional countries were quite obviously geographic like Region No. 1: Canada and the U.S.A.; However, Region No. 4 represents a region that was obviously born out of future inter-dependencies other than those dictated by geographic proximity. Region No. 4 is: Australia, South Africa (white), Tasmania, Oceania, New Zealand, and **Israel**.

Their model has already divided South Africa into black and white giving black South Africa to the Main African Unity along with Rhodesia. **Their "proposed" model also shows Yugoslavia as a Western European country in Region No. 2 instead of its present position as an Eastern European or communist bloc country.**

Other Signs Manifest

In that hypothetical plan of action in the preceding chapter, various 'conditioning' or propaganda campaigns were suggested as a means to drive the masses into the mold of the Illuminati. Signs of that same technique are manifesting at this very moment. In August of 1978, Carlos Romulo, the Philippines Foreign Minister called for the formation of an 'international economic law which would contribute to world economic progress and enhance world peace prospects.' This is extremely interesting, as it was in the Philippines that Dr. Donald Drew (previously mentioned) used a million selected people to test one of his computer-controlled micro-economies. President Marcos was reportedly so pleased that he wanted the whole of the Philippine economy to go onto Dr. Drew's model.

The author of this book received a great deal of information from Dr. Drew in 1974 while the latter was lecturing in Perth, Australia. Just before he was to leave Australia for the White House to deliver plans for 'hypothetically' inducing massive artificial famine, plague, and riot conditions in South and South East Asia to curb population growths, Dr. Drew told this author that the Shah of Iran had also expressed a desire to use his computerized economy. Please note: Dr. Drew really felt his solution - no matter how bizarre, was the only way to curb the main problem the world has: Over-crowding. Dr. Drew said as he left, *"I wish there were some other way; but as there is not, I intend to return to the area and accept the same sentence my plan will put upon them."*

Before he left, however, this author did introduce a new factor into Dr. Drew's computer model for the area. It was a technological one which, hopefully, has bought a few more years of time for the estimated 100,000,000 victims that the original plan called for. Let us hope so...

In the United States, Mr. Henry Ford warned the U.S. of economic ruin over the growing lack of public confidence. That was in February of 1975. In Berne, the Bank of International Settlements which coordinates the world's central banks warned that the international economy was in danger of collapsing from industrial rundown and mass unemployment. That was in June of 1978. Sweden withdrew from "Snake," the joint float of Western European currencies which was established to create 'stable economic conditions.' That was in August of 1977. In Guatemala City, Kurt Waldheim, Secretary General of the United Nations, predicted a major global financial catastrophe because of excessive borrowing by the developing countries. That was at the biennial conference of the Economic Commission for Latin America in May of 1977. If the Club of Rome really 'advise' as many governments as they say, how is it that there is any real chance of an economic collapse - except by design?

In October of 1977, the Energy Secretary of the U.S., Mr. Schlesinger, announced that the world was living on borrowed time in respect of available energy resources. His concern was primarily with known gas and oil reserves versus projected usage figures of these 'commodities.'

As a result, the Carter administration has put even more pressure on Congress to pass the energy reform bills. In May of 1978, a special report was tabled by the Economic Commission for Europe. In it was the warning that all proved oil resources not under OPEC control would run out by 1996. It spelled-out doom for North America, the Soviet Union, and both Eastern and Western Europe if the OPEC counties were not brought into line; but, is it really true that there is an energy crisis?... No.

A recent study in the U.S. easily answers the question with undeniable facts. The study entitled, The Defence and Foreign Affairs Daily was published on the tenth of December, 1977. Its circulation would have been a bit limited as a subscription costs $350; however, should one wish to pay the price it is still available from The American Institute of Technology in Suite 602 of 2030 M Street at Washington D.C. 20036.

The study stated, *"the commercializing of synthetic fuels could produce up to 9.8 million barrels a day of clean and gaseous liquid fuels from coal and oil shale by the year 2000."* Furthermore, *"...the United States could become totally self-sufficient in hydro-carbons by that time."*

Consider the incident with Dr. Vincent E. McKelvey who has been Director of the United States Geological Survey since 1971. Here is a man whose appointment was approved by not only the President of the United States, but also by the National Academy of Sciences, which nominated him for the post. It would seem that he was well-qualified for his position. In July of 1977 during the time the Carter Energy Bill was being bulldozed through Congress, Dr. McKelvey gave a speech in Boston. During this speech, he noted that *"There may be as much as 60,000 to 80,000 trillion cubic feet of gas sealed in the geo-pressurized zones off the Gulf coast region!"*

Relating those figures to more simple terms would give an amount of gas equal to somewhere between 3,000 and 4,000 times the amount of gas the United States consumed in 1977. This amount is also equal to about **ten times the energy value of all the oil, gas and coal reserves of the United States excluding that in the Gulf coast region.** Also, as part of the reserves classification criteria is that the reserve must be economically viable, many U.S. oil reserves of vast quantities are not listed.

One estimate shows over a trillion barrels of oil could be extracted form the Rocky Mountain shale deposits. These reserves have not been counted as economic - even though economic Japanese ultrasonic extraction processes have been designed and tested since 1968! Also not listed in useful reserves are the billions of barrels of oil left in wells because traditional secondary recovery techniques have made their extraction 'uneconomic.' This author personally knows of at least one newly-patented, U.S. secondary oil recovery process which makes these undeclared reserves a viable source. An energy crisis, you say? What absolute drivel!

Other, non-standard energy resources should have also been added to the reserves; but they were not. Why is it that the electric 'generator' invented by Dr. Oleg Jefimenko of West Virginia University in 1971 has never been released? His 'generator' is a simple, electro-static motor which runs off the differential electric fields of the Earth at either different altitudes or different hemispherical latitudes. His first model only produced a continuous output of eighty watts; however, subsequent models have reportedly done much better using newly-discovered, high-altitude and subterranean conductivity processes. Such a process would also make cheap electricity available anywhere on the Earth, without transportation or cable-laying worries; but one must understand... *"there is an energy crisis.'*

The Failing Establishments

Another part of the previously discussed hypothetical takeover bid was the destruction of the masses' faith in their established orders. What has reality offered? the Watergate scandal, the Brandt spy scandal, the Lockheed scandal, the Whitlam affair, the Gandhi scandals etc,... It is so bad now,... no one can claim he lives in a utopia. In Israel, radical groups cry out for a dictatorship. Recently in England, in a television interview on the steps of Parliament House, a British parliamentarian called for a dictatorship to solve England's woes. Lake sheep running from one loud noise to another, the masses leap from one scandal to another. How short their memory must be. **Hitler did not take Germany; the Germans gave it to him under conditions very similar to the political and economic conditions which today face the entire world.**

Lastly, the spectre of nuclear war looms in the immediate future. Sino-Soviet and Soviet American relationships seem very strained for the moment. The Middle East puzzle is not as close to a lasting peace as Carter, Sadat, and Begin would like the world to believe. In America, some fifty-odd armed groups advocate the violent overthrow of the U.S. Government. In Russia, the civil rights issues are fragmenting internal unity. In China, one intrigue after another is manifesting in the post-Mao power grabs. The Russians are afraid the Chinese will eventually cut-off its Arab oil supplies when the Chinese complete their 'great road to the West.' In such an event Russia would find itself hemmed-in. Its direct access to Africa and the Indian Ocean would be gone.

The Palestinian or West Bank conflict between Israel and Jordan could very well experience a sudden flare-up which might see both sides using new super-technology. Russia could also be visualized moving into Israel and Egypt to try to maintain a military foothold in the Middle East. For that matter, the conflict between Iran and Iraq could suddenly escalate - perhaps simultaneously to an American and Iranian conflict. Yes, this threat of nuclear war, it is real. It needed very little encouragement from the Illuminati at this late date.

The pollution scares are real; however, the intensity of many of the stated situations is far above the reality of the issue. It is obvious: to obtain the attention of the masses one must present a far more disastrous case for ecological failure than really exists to be sure the slow-moving masses react in time to avert disaster in the years to come. Anyone who travels from country to country with a regular frequency can verify this. How many times has one such event been 'played up' in the news of a country far-removed from the event while the country of origin hardly takes notice of the event?

The Alien Encounter

The new Star Trek series will not have the blessing of one of its key characters: 'Mr. Spock,' played by Leonard Nimoy. According to Nimoy, *"The new Star Trek will use science fiction not only as a vehicle for entertainment but also for making political and social statements as well."* This comment of Nimoy's, from November 1977, must surely speak for itself!

In March of 1978, at the annual meeting of the American Association for the Advancement of Science in New York some other, rather interesting, suggestions were presented to the 5,000 assembled scientists. Dr. Robert Jastrow, director of NASA's Goddard Institute for Space Studies, told them he believed this world will be **contacted by aliens within 15 years.** Furthermore, he said, *"Scientific evidence now seems to support the Bible - there was a first cause, a force from which everything came - but I don't know how or why."*

Carl Sagan, astronomer, author, and head of the Laboratory for Planetary Studies at Cornell University in New York had some equally interesting points to make in the Newsweek interview of September, 1977.

Sagan - author of **The Cosmic Connection** and **The Dragons of Eden** is *"... fired by a faith that in the reaches of space there are older planetary civilizations, capable of bestowing on humans the bounty of an unimaginably advanced technology.'* -a point which he made clear in **The Cosmic Connection** when he wrote, **"There must be other starfolk.'**

The Newsweek interview continued, *"He (Sagan) is persuaded that older civilizations, whirling around second - or third - generation stars, would possess technologies so advanced that to earth eyes they would appear 'indistinguishable from magic.' Having survived such technology, which inevitably would include weapons for global destruction,* **they would also have solved their social problems.** *They would be 'benign,' and probably eager to share their secret of survival. Such a possibility awes Sagan. 'The scientific, logical, cultural, and ethical knowledge to be gained by tuning into galactic transmissions may be, in the long run, the most profound single event in the history of our civilization.'"*

Newsweek further observed, *"His faith in salvation through science remains unshakable. He deplores the reversion of young people to mysticism and magic but* **he still hopes to see, in his lifetime, that magical intelligent signal from the cosmos.'**

The evidence is plain for all who wish to see. There is a great conspiracy afoot in the affairs of man. Even Captain Bruce Cathie of New Zealand who wrote, **Harmonic 33**, **Harmonic 695**, **and Harmonic 288**, has realized the conspiracy.

In the September 6th issue of the <u>Western Leader</u> in 1973, Captain Cathie stated that he believed much of the world's knowledge is kept in the hands of a select group of people who would be capable of taking control of world affairs if they continue to keep it secret. He stated, *"Unless this knowledge is given to the world freely, they could form a world government.'*

CHAPTER 6: Circles of Illuminati

There are many ways to view the 'invisible government,' 'the unseen hand,' or as they are more commonly known, 'the Illuminati.' As their organizational charts are not open to the public, one must deduce from other clues what their organization looks like. It is best to start with the foundation of those groups which now comprise the elite of the modern mystery schools.

Chart (6-A) has been prepared to show the major progressions which generated today's mystery schools. As mysticism, symbolism, allegory, and reason all play key roles in shaping Illuminati objectives and administration, it is with mystical symbolism that one can deduce a starting point in the organizational chart of the Illuminati. As their major symbol is a circle with a 'Mogen David' star or shield inside it, this can be a logical start. In the Bavarian branch of the Illuminati, it was decided that a 'circle of best influence' would contain only 100 members. Furthermore, it was thought that six such circles would eventually be sufficient to administrate the entire world. Each group would have an executive of ten people. In the occult mystery schools, (1), (10), and (100) represent 'unity' or 'ordinal perfection'... while (6) represents 'the union of all the apparent duality in reality '... or 'the union of the infinite with the finite'... etc.

A diligent student of recent international affairs could easily find (6) major groups to fill the circles of Chart (6-B). There are three 'foreign policy' bodies which immediately spring to mind: 1) The American Council on Foreign Relations, 2) The English Royal Institute for International Affairs, and 3) The European Bilderberg Group.

A bit of probing into who directs these organizations points out three other groups which are linked by co-directorships:

1) The Tri-lateral Commission
2) The United Nations
3) The Club of Rome

All of these are probably directed by an inner circle of nine members, if the modern Illuminati follow the Bavarian pattern. In fact, when this author was associated with the American F.B.I. under Director J. Edgar Hoover, word filtered down the ranks from Hoover that the 'inner circle of nine' in the Illuminati were so difficult to identify that he thought they bordered on the supernatural. If the modern Illuminati do hold to Dr. Weishaupt's original concepts, then one can expect to find 6 circles with 60 directors and 600 members interwoven into the current fabric of human endeavor.

The Century Association

One possible group of one-hundred initiate Illuminati is the Century Association of America. Its membership is presumably limited to one hundred key members. It has its major membership in New York where the Rockefeller Empire resides... (Refer to Chart 6-C).

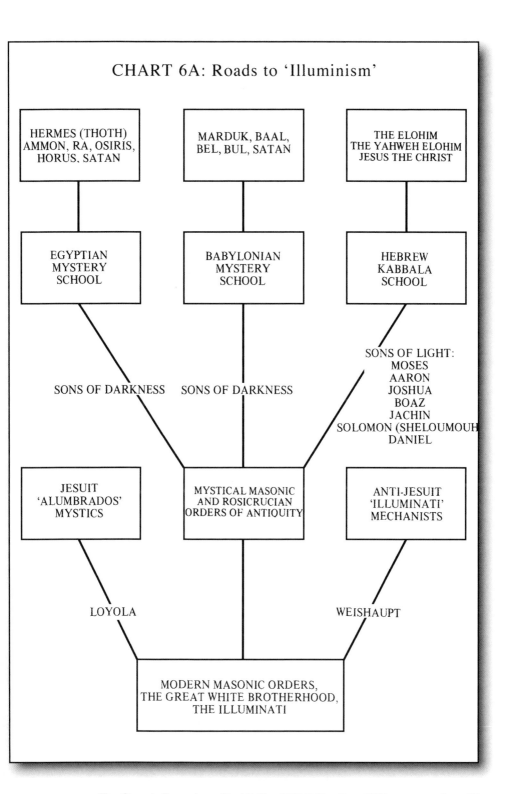

CHART 6A: Roads to 'Illuminism'

HERMES (THOTH) AMMON, RA, OSIRIS, HORUS, SATAN

MARDUK, BAAL, BEL, BUL, SATAN

THE ELOHIM THE YAHWEH ELOHIM JESUS THE CHRIST

EGYPTIAN MYSTERY SCHOOL

BABYLONIAN MYSTERY SCHOOL

HEBREW KABBALA SCHOOL

SONS OF DARKNESS

SONS OF DARKNESS

SONS OF LIGHT:
MOSES
AARON
JOSHUA
BOAZ
JACHIN
SOLOMON (SHELOUMOUH
DANIEL

JESUIT 'ALUMBRADOS' MYSTICS

MYSTICAL MASONIC AND ROSICRUCIAN ORDERS OF ANTIQUITY

ANTI-JESUIT 'ILLUMINATI' MECHANISTS

LOYOLA

WEISHAUPT

MODERN MASONIC ORDERS, THE GREAT WHITE BROTHERHOOD, THE ILLUMINATI

CHART 6-B: INTERLOCKED ILLUMINATI FRONTS

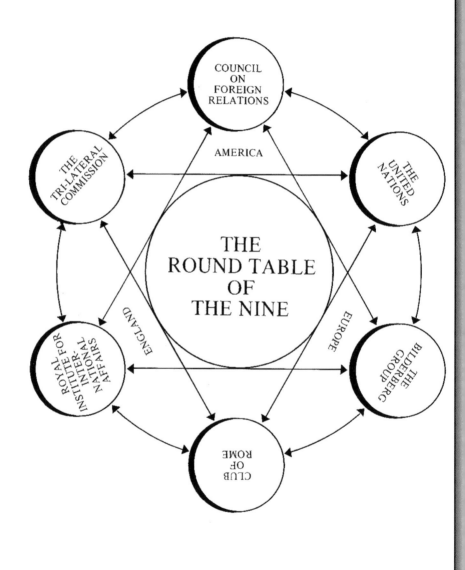

The Cosmic Conspiracy Final Edition 2010 © Stan Deyo 2010

CHART 6-C: INTERLOCKING ORGANIZATIONS

NAME	C.F.R.	T.L.C.	CENT.	BILD.	RAND	COSMOS	C. of R.	NOTES
Zbigniew Brzezinski	Dir.	Dir.	Mem.	Mem.	Mem.			U.S. Nat. Security Council
Carroll L. Wilson	Dir.	Mem.	Mem.			Mem.	Exec. Comm	
Robert V. Roosa	Dir.	Mem.	Mem.		Past Trustee	Mem.		
David Rockefeller	Chrm.	Mem.	Mem.	Mem.				Chrm. Chase Manhatten Bank
Cyrus R. Vance	Dir.	Mem.	Mem.	Mem.				U.S. Sec. of State
Caryl P. Haskins	Dir.		Mem.		Trustee	Mem.		
Joseph E. Johnson	Dir.		Mem.	Adv. Comm.		Mem.		
James A. Perkins	Dir.		Mem.		Past Trustee	Mem.		
Nelson Rockefeller			Mem.	Mem.		Mem.		Former V.P. of U.S.A.
Arthur F. Burns	Mem.		Mem.			Mem.		Chrm. U.S. Fed. Reserve
William A.M. Burden	Dir.		Mem.			Mem.		
Arthur H. Dean	Dir. Em.		Mem.	Adv. Comm.	Mem.			
C. Douglas Dillon	Dir.		Mem.	Mem.				Chrm. Rock. Foundation
John J. McCloy	Dir.		Mem.					Chrm. Ford Foundation
Dr. Robert Scalapino	Dir.		Mem.		Mem.			
Hedley Donovan	Dir.	Mem.	Mem.					Ed. in-Chief: Time, Inc.
George W. Ball	Mem.	Mem.		Adv. Comm.				
August Hecksher	Mem.		Mem.					
Hans J. Morgenthau	Mem.		Mem.					
Henry R. Labouisse	Dir.		Mem.					
Lucian W. Pye	Dir.					Mem.		
Emanuel R. Piore			Mem.			Mem.		Former V.P. at IBM
H.S. Rowen	Mem.				Past. Pres.			
Herman Kahn	Mem.				Mem.			Dir. Hudson Institute Think Tank
Bayless Manning	Pres.		Mem.					
Charles M. Spofford	Dir.							Dir. C.I.A.
Henry Kissinger	Mem.		Mem.					Form. Sec. of State
Andrew Goodpaster	Mem.		Mem.					Supt. West Point
Dean Rusk	Mem.			Adv. Comm.				Former U.S. Sec. of State
George Nebolsine	Mem.			Adv. Comm.				
John S. Coleman	Mem.			Adv. Comm.				
Gen. Walter Bedell Smith	Mem.			Adv. Comm.				Former C.I.A. director
Henry J. Heinz II	Mem.			Adv. Comm.				
Jacob Javits	Mem.			Mem.				U.S. Senator
Edmond de Rothschild		Mem.		Mem.				Pres. Com. Fin. Holding
Dr. Aurelio Peccei				Mem.			Pres.	Italconsult
Gerhard Schroeder			Mem.	Mem.				German Bundestag
Giovanni Agnelli			Mem.	Mem.				Pres. Fiat, Int. Chase M. Bank
Guido Colonna di Paliano			Mem.	Mem.				Pres. La Rinascente
Robert Triffin	Mem.						?	Participant: RIO & GOALS FOR MANKIND
Dr. Alexander King							Ex. Comm.	Participant: RIO Project
David C. Gompert	Mem.							"Goals for Mankind Project"
Sir Zelman Cowen							Dir	Gov. Gen.: Australia
Prof. Charles Birch							Mem.	Aust. Club of Rome
John Stokes								Pres. Aust. Club of Rome
George Boniecki								Aust. Club of Rome
Dr. Mihajlo Mesarovic							Dir.	"Mankind at Turning Point"
Dr. Eduard Pestel							Dir.	"Mankind at Turning Point"
McGeorge Bundy				Mem.				Pres. Ford Foundation
Gerald Ford				Mem.				Former U.S. Pres.
Allen Dulles				Mem.				
Emilio Collado				S. Comm.				V.P. Standard Oil
Georges Pompidou				Mem.				Form. Mgr. Bank of Rothschild
The Duke of Edinburg				Mem.				
Edward Heath				Mem.				Former British Prime Minister
Harold Wilson				Mem.				Former British Prime Minister
Dennis Healey				Mem.				U.K. Chanc. of Exchequer
Giscard d'Estaing				Mem.				Former Pres. of France
Franz Joseph Strauss				Mem.				Former W. Ger. Min. of Finance
Helmut Schmidt				Mem.				W. Ger. Chancellor
Willy Brandt				Mem.				Form. W. Germ. Chanc.
Olaf Palme				Mem.				Swedish Prime Minister
Donald McDonald				Mem.				Canadian Min. of Def.
Prince Bernhard				Chrm.				Former SS Corps, IG Farben
Per Jacobbson				Mem.				Former IMF Mgr.
Henri Spaak				Mem.				Form. Belgium P.M. (Treaty of Rome)
Ludwig Erhard				Mem.				Form. W. Germ. Chanc.
J. Paul Austin		Mem.						Chrm.: Coca-Cola
I.W. Abel		Ex. Comm.						Pres. U.S. United Steelworkers
Robert W. Bonner Q.C.		Ex. Comm.						Bonner & Fouks: Vancouver
Gerard C. Smith		Chrm.						North Amer. Side
Max Kohnstamm		Chrm.						European Side
Takeshi Watanabe		Chrm.						Japanese Side
Francoise Duchene		D. Chrm.						European Side
Tadashi Yamamoto		Sec.						Japanese Side
George S. Franklin		Sec.						North Amer. Side
Harold Brown		Ex. Comm.						Pres. Cal. Inst. Tech.
James E. Carter		Mem.						Pres. of U.S.

NAME	C.F.R.	T.L.C.	CENT.	BILD.	RAND	COSMOS	C. of R.	NOTES
Patrick E. Haggerty		Ex. Comm.						Chrm. Texas Instruments
J.K. Jamieson		Mem.						Chrm. Exxon Corp.
Lane Kirkland		Mem.						Sec./Treas. AFL.-CIO
Walter F. Mondale		Mem.						V.P. of U.S.A.
David Packard		Mem.						Chrm. Hewlett-Packard Corp
Jean-Luc Pepin, P.C.		Ex. Comm.						Pres.: Interimco Ltd.
Edwin O. Reischauer		Ex. Comm.						Former U.S. Amb. to Japan.
Elliot L. Richardson		Mem.						U.S. Am. to U.K.
William M. Roth		Ex. Mem.						Roth Properties
William W. Scranton		Ex. Comm.						Former Gov. of Penn.
Arthur R. Taylor		Mem.						Pres. C.B.S.
Arthur M. Wood		Mem.						Chrm. Sears & Roebuck
Leonard Woodcock		Mem.						Pres. U.A.W.
Alden W. Clausen		Mem.						Pres. Bank of America
Daniel J. Evans		Mem.						Gov. of Washington
Lee L. Morgan		Mem.						Pres. Caterpillar Tractor
Andrew Schonfield		Mem.						Dir. R.I.I.A.: England
Reginald Maulding		Mem.						British Parliament

ABBREVIATIONS:

C.F.R.:	U.S. Council on Foreign Relations	(Past and Present Members)
T.L.C.:	The Tri-Lateral Commission	(Past and Present Members)
CENT.:	The New York Century Association	(Past and Present Members)
BILD.:	The European Bilderberg Meetings	(Former Delegates)
RAND:	RAND Corporation	(Past and Present Members)
COSMOS:	Cosmos Club	(Past and Present Members)
C. of R.:	Internation Club of Rome	(Past and Present Members)
A. Comm.:	Advisory Committee	(Past and Present Members)
S. Comm.:	Steering Committee	(Past and Present Members)
	Past Trustee	(Past and Present Members)
Mem.:	Member	(Past or Present)
Ex Comm.:	Executive Committee	(Past or Present)
Dir.:	Director	(Past or Present)

The Rand Corporation

The RAND Corporation is a privately-owned company. It is primarily involved in research and development of methods to improve: American security, public welfare, computer sciences, economics, bio-sciences, engineering sciences, mathematics, physics, resource analysis, social science and system sciences. It also analyses political, economic, and security situations in most regions of the world.

Furthermore, it provides support and operations of military forces, as well as providing studies of strategic and tactical forces. It is a non-profit organization which is financially supported by the U.S. Government, state and local governments, and certain foundations. Its annual research budget now exceeds $30,000,000. From 1967 to 1972 the president of RAND was H.S. Rowen - a member of the C.F.R. - formerly Deputy Assistant Secretary of Defence for plans and counsel for national security affairs from 1961 - 1965. For some of the other members of the RAND 'school' refer to chart (6-C).

C.F.R.

The Council on Foreign Relations (C.F.R.) was officially founded in 1921 giving their primary objective as: "The study of the international aspects of American political, economic, and strategic problems.' As of 1978 it had 1700 members with 95 'local groups' on the staff record. Some of the past and present members are in Chart (6-C).

The Bilderberg Group

The Bilderberg Group is a post-war, European group formed in 1952 by Joseph Retinger, a Polish refugee in Britain. The organization was, ostensibly, formed as an anti-communist alliance between American and European 'show governments.' Each year since 1954, the Bilderberg Group has invited between 60 and 80 Americans and Europeans together to discuss the world's most pressing problems. It, like the Club of Rome, and the Tri-Lateral Commission is comprised of politicians, economists, businessmen, financiers, and diplomats. Until the Lockheed Scandal removed him, Prince Bernhard of Holland was the principal spokesman and organizer for the Group. He has been replaced by England's Lord Home. Chart (6-C) shows some of those who attended the 1975 Bilderberg Conference in Turkey plus a few of the regular 'guests.'

The Tri-Lateral Commission

The Tri-Lateral Commission is only a few years old. Its 'three sides' are: North America, Europe, and Japan. Most of its members are also members of either the CFR, the Royal Institute for International Affairs, the United Nations, the Bilderberg Group, or the Club of Rome. It, too, is another "club" of illuminists with the major objective of solving global problems peculiar to its three member regions: Region (1) North America. (see Club of Rome model in appendix 7), Region (2) Western Europe (ibid.) and Region (3) Japan (ibid.). A partial list of the executive and general membership is in Chart (6-C).

Club Of Rome

The membership of the Club of Rome is a bit more difficult to obtain than most of the others. It is reasonably certain that the International Club of Rome does not exceed (100) members. It could be that the executive committee is limited to (40), thus giving (4) men to each of the Club's (10) new global regions. The members known to this author are listed in Chart (6-C).

In Australia, a local branch of the Club of Rome has been founded. It is called, simply, "The Australian Club of Rome.' Four of its members are:

1) Sir Zelman Cowen (see Chart 6-C)
2) Professor Charles Birch (see Chart 6-C)
3) Mr. George Boniecki - a Sydney industrialist
4) Mr. John Stokes - President of Australian Club of Rome

According to Professor Birch in his book, 'Confronting the Future,' fifteen other sympathetic people gave him a great deal of help editing his book. These people were all, presumably, Australian residents. It is not known, positively, by this author whether or not any of them are members of the International or Australian Club of Rome; however, the fact that they have helped Professor Birch to produce his book does indicate a close association with his ideals. Their names follow:

1) Mr. John R. Siddons - President - Australian Democrats
2) Professor C.N. Watson-Munro
3) Sir Otto Frankel
4) Sir Willis Connolly
5) Dr. L.C. Noakes
6) Professor E.J. Underwood
7) Dr. F.H.W. Morley
8) Dr. P.G. Law
9) Dr. N. Fisher
10) Dr. R. Mendelsohn
11) Dr. M.F. Day
12) Dr. I.E. Newnham
13) Mr. Christopher Stephen
14) Associate Professor G.W. Ford
15) Mr. M. Llewellyn-Smith

Twentieth Century Fund

In 1919, the Twentieth Century Fund, Inc. was established in Massachusetts. Its donor is listed as Edward A. Filene. Its stated purpose is: 'Research and education on economic and social problems.' It is primarily a research institution which conducts its own studies and distributes the findings on economic, political, and social problems. Its assets at 1965 were $27,936,115. Its annual expenditure was listed as $883,400 which included $407,965 in funds granted for various research projects. Some of the officers and trustees as at 1965 were:

1) Arthur F. Burns (see Chart 6-C)
2) August Hecksher (see Chart 6-C)

3) Erwin Dain Canham - Clerk, newspaper editor, 33rd degree Mason, Christian Scientists, Rhodes scholar

4) Benjamin V. Cohen - Lawyer, counsel for American Zionists and Peace conferences 1919-1921 in London and Paris; senior adviser American delegation to the U.N. General Assembly 1946

5) Arthur Schlesinger

6) Charles P. Raft

7) Robert Oppenheimer

8) Adolf A. Berle - Chairman

9) James Rowe - Vice-Chairman

10) Evans Clark - Secretary

11) H. Sonne - Treasurer

Left Or Right?

From amongst the chaos of mysticism, mechanical reason, and a multitude of groups, clubs, and orders emerges one obvious question: "Is the left wing or the right wing correct?" The answer is: "Neither.." Both are wings of the same bird.

The oldest and most proven strategic rule is: Divide your enemy from within to best conquer him from without. The Illuminati are past-masters of such strategy. The communistic revolutions have all employed "fifth columnists" whose primary functions have been to take positions on both sides of public issues and to use these public issues to create unnecessary division in the target community. By having these 'fifth columnists' or subversives on both sides of the main issues, it mattered little which side was victorious in the normal political process or the revolutionary conflict. Having their subversives on both sides guaranteed victory no matter which side won.

Even the terms 'left' and 'right' are misnomers in most modern political philosophies. Supposedly, 'left' implies 'communist' while 'right' implies 'capitalist;' or 'left' implies 'socialist' while 'tight' implies 'fascist.' These concepts are wrong. They were implanted in the mind of Man through heavy, media conditioning.

The true political spectrum is one which sees all authoritarian or totalitarian philosophies grouped to the left while the complete lack of government or anarchy is the extreme right. These two polarities do not 'meet each other on the circumference of the circle.' Somewhere between total, dictatorial government and anarchy lies a form of limited government which still allows degrees of individual responsibility and, hence, authority to be held by every member of the society. This type of government would be a democratic republic of some description; however, it would still be a government subject to the human failings: conceit and greed. There is no **humanly** feasible way to have a perfect world government without effecting a fundamental change in the human psyche. And that would take an act of God.

The Same Source

Further proof of conspiracy is in the history of both the communist and the national socialist (NAZI) movements. When Adam Weishaupt died in 1830 Karl Marx was only twelve years old. Karl's father, Heinrich Marx, was an outspoken member of the 'European Enlightenment' that followed from Weishaupt's era.

Heinrich was born a Jew; but he was later 'baptized' into a deistic 'church' called the Evangelical Established Church. His son, Karl was, later, also baptized into the same movement. Both Heinrich and his son, Karl, were avid students of the writings of both Kant and Voltaire. Karl Marx was eventually to develop the basis for modern communism. In 1844, when Karl Marx was twenty-six years old, another philosopher, Friedrich Nietzche, was born in Germany. It was Nietzche's development of the post-'Enlightenment,' German-illumination school that laid the foundation for the German NAZI Party. In 1870, when Nietzche, himself, was twenty-six years old, a man named Vladimir Illich Ulyanov was born in Simbirsk, Russia. Ulyanov was later to become that scholar of Latin and Greek whom the world today remembers as Lenin - the architect of the Bolshevik Revolution of 1916-18. Only a year after Lenin was born, Rosa Luxemborg - a Polish Jewess - was born. It was her ideology which in 1916 gave birth to the 'Spartakusband' or 'Spartacus League' (remember Weishaupt was "Spartacus ") which was to become the nucleus of the German Communist Party which supported Lenin's philosophy.

Lenin's early philosophy came under fire from the pure Marxists Plekhanov, Martov, and Trotsky. they claimed that Lenin's hybrid Marxism was leaning toward 'Jacobinism' or the suppression of intraparty discussion which could form a dictatorship **over** the people **not of** the people. This difference of opinion led to the formation of two wings within the Russian Social-Democratic **Worker's Party**. Those wings were Lenin's 'Bolsheviks' (the majority of the party) and Martov's 'Mensheviks' (the minority of the party). Lenin's Bolshevik's eventually gained permanent superiority over the Mensheviks to form the Russian Communist Party. However, Lenin's party still had 'left and right wings' within it. Those of the left favored violent revolution while those of the right favored gradual change within the established order. Lenin took the left path; and in so doing opened the door for Joseph Stalin who was later to form the dictatorship that Martov had predicted.

In 1920, the German National Socialist **Worker's Party**, was formed by Anton Drexler. In 1921, Adolph Hitler took over the Party and played-down the 'Socialist' part while enhancing the **National Worker's Party** portion.

One must remember that the main objective of the real Illuminati is the establishment of a dictatorial world government. Look at the word communism. It is formed from the Latin root communis and means 'common' or 'group' living. On the other hand, look at the word fascism. It is from the Italian word fascio which means 'bundle' or 'group.' Fascio was derived from the Latin or Roman word fasces meaning 'the authority of Rome symbolized by a bundle of rods having among them an axe with the blade projecting.' Both communism and fascism in the modern world are **dictatorial** forms of 'group' living. They are not as different as the Illuminati would have the people believe.

Under communism, the government owns everything; hence, one is told how much of and what items of the government assets he can use. Under fascism, the individual can own property - but he can only use it as the government dictates.

Where is the difference to the people? Both forms are authoritarian and definitely not favoring free thought and action. In no humanly-conceived system does anyone really own property. Anyone who claims to 'own' something is being naive. Man is only a custodian of that which comes under his responsibility from a much higher authority.

If a nation says it 'owns' its land, it is really saying, "We claim this land by right of might and/or occupancy." Think about it; how many of today's nations were formed by one group dispossessing another of its assets? Buying real estate in these times as an investment is profoundly unenlightened. One swift change of the government by revolution or by legislation, and suddenly, no one 'owns' his property any longer. Think about it. The man with the strongest source of power and, hence, authority holds the land - until he, too, is displaced by a bigger, meaner authoritarian.

It has been the objective of the Illuminati from its inception to establish order from the chaos of human affairs. The various sects within the Illuminati have, however, made it painfully obvious that democratic illuminism is not possible. For years, one division of illuminism would back one social ideology while another division would back the opposing ideology. This created war. By having wars, an eventual union of the combatants took lace. Time would pass; and the new union would be set against a new, opposing ideology - controlled by yet another division of the Illuminati. The result of the ensuing war would be a still larger union formed out of the former combatants.

The final limit to such machinations is, hypothetically,. reached when the last conflict unites all the peoples of Earth; however, one wonders if there will be any survivors to that last conflict. If, indeed, there are survivors; and if some sort of an authoritarian structure prevails in the name of order, what common strife will remain to force mankind to maintain that height of human achievement? With no common enemy or opposing nationstate what will give the surviving masses the challenge of conquest? A thesis without the antithesis is illogical to an Illuminatus.

The Illuminati will be faced with division even in their own somewhat-human ranks between at least the Jacobins and the Bavarian Illuminists - not to mention the various oriental illuminist orders. To maintain that crisis state which offers the masses as well as the illuminist orders a reason for uniting, a foe must be invented. It could be an artificial weather problem; it could be an artificial, extraterrestrial landing on Earth by a supposedly advanced civilization; or it could be some wort of contrived - yet unseen and mystical order of villains right here on Earth. Whatever antithesis is to be employed currently remains the secret of the inner circle of those nine, master Illuminati: **"The Round Table of the Nine.'**

Suggested Reading List For Section II

1) Allen, Gary; None Dare Call It Conspiracy; Concord Press, P.O. Box 2686, Seal Beach, California - 1971

2) Birch, Charles; Confronting The Future; Penguin Books Ltd., Middlesex, England - 1976

3) Cantelon, Willard; The Day the Dollar Dies; Logos International,   N.J. 07060

4) Cantelon, Willard; Money Masters of the World; Logos International, Plainfield N.J. 07060

5) Hannah, Walton; Darkness Visible; Augustine Press, 46-48 Princedale Road, London - 1952

6) Laszlo, Ervin; Goals For Mankind; Hutchinson of London, 3 Fitzroy Square, London - 1977

7) Mesarovic, Mihajlo and Pestel, Eduard; Mankind At The Turning Point; Hutchinson and Company, Publishers, 3 Fitzroy Square, London - 1974

8) Robison, John; Proofs of a Conspiracy; Western Islands, Belmont, Massachusetts (02178) 1967 (reprint of original in 1798)

9) Tinbergen, Jan; RIO - Reshaping The International Order; E.P. Dutton & Co. Inc., New York - 1976

10) Webster, Nesta H.; World Revolution... The Plot Against Civilization; Britons Publishing Co. - 1971

SECTION III: Not Of Earth

FORWARD: SECTION III

The following section has been written in a manner that should be explained before it is read. As the preceding sections of this book have painted a very dark picture of the future, the author felt it was necessary to give the reader a choice other than doom.

To be able to do so, the author sought the aid of several learned historians and scholars. In the discussions which followed it became quite clear that the best ideas were being offered by the Jewish and Christian Biblical scholars. To resolve the discrepancies which arose between them concerning the origin of Earth, man, and evil required that the author familiarize himself with the Hebrew language to make his own assessment of the translations.

Early in his analysis of the Hebrew version of the Book of Genesis he managed to chance upon a most enlightening way to resolve the apparent translational discrepancies.

By using morphology or the process of studying word derivatives and origins, the author found that the names, themselves, of key characters and places in Genesis yielded previously obscured detail which enhanced the clarity of many parts of the book.

It was discovered, for example, that the name of a particularly obscure class of beings called the **Nephilim** in Genesis 6:4-5 broke down to reveal the following:

a) They or them (masculine) who were thrown down, descended to a lower level, fell, or fell down...

b) They or them (masculine) of the night, of the time of adversity, or of the divided...

c) They or them (masculine) from the beginning or the foundation.

Reconstructing the name from the preceding parts creates an image of beings who were involved in a great challenge for supremacy from the very beginning of time on Earth. They were beings who lost their challenge and consequently were put down into a lower state of existence. In essence, they were the "bad guys" who threatened an ancient civil war in the universe.

Analyzing two of the Biblical names for **God** yielded an even more fascinating aspect of the real nature of **God**. The name for **God** in the universal state is **Elohim**. When a projection of the omnipresent **Elohim** appeared to mankind it was called the **Yahweh Elohim**... or **Lord God**. The morphology of **Elohim** yielded:

a) They who are bound by an oath...

b) They who are enchanting and fascinating things...

c) They who are of an unknown place...

d) Pause, hush, be silent...

e) The great sea, the great river, or the greatness westward...

By merging these pieces God or **Elohim** becomes They who are a oneness - bound together by an agreed common bond... **They** whose origin is not known... **They** who are fascinating things to be pondered in silence... in awe... like the great river which has no beginning and no end.

Yahweh represents '**the descended existence or form of**'... When this word is coupled with **Elohim** it denoted the **Elohim** in a lower-state projection of themselves in a personal form capable of being touched, communicated with, and... seen.

In, perhaps, more familiar terms, 'God' or '**Elohim**' represents the entire **Creator** and '**Their**' creation while '**the Lord God**' or '**Yahweh Elohim**' represents only the projected portion of **God**. The **Lord God** could be viewed as a three-dimensional communications device used by the totality of **God** to personally communicate with the life forms of Earth - especially mankind, himself.

As morphology had its limitations, the author also used Hebraic numerology and symbolic analysis to further probe the ancient Biblical secrets which are veiled in allegory. The author has preserved the ancient literary form of giving multi-level messages within a single composition. As a result, three levels or degrees of reason have been incorporated into this last section. Let he who has ears, hear... he who has eyes, see... and he who has wisdom, reason these things which are to follow...

For all who read these things there will be some benefit... even for those readers who are of other doctrines such as those:

a) who greet one another on all points of the triangle...

b) who revere the ankh of Horus and bear the initiatory scar on their wrist...

c) who wear the clear number disk...

d) who seek the lavender ray of the Great White Brotherhood, or...

e) who think the only way to the Grand Geometrician of the Universe is by the path of the Sun.

And now, the author invites the reader to let his mind drift backward in time.. back to the time before Earth existed... back to the time which saw the beginning of the only real 'star war' in the Universe...

CHAPTER 1: In The Beginning

They 'has' always been what **They** 'is' and will become. Atoms form worlds within stellar systems which in turn form galaxies comprising a finite link in **Their** great continuum. By **Their** command alone, brilliant galaxies of light tumble and spin through time and space like counterpoint melodies - some throbbing, some singing - yet all locked in **Their** symphonic embrace...

Once upon a time in a far away place on a singular occasion, thousands, even millions of years ago **They** created a new thing of great interest to all who watched... Silently at first and ever so smoothly, it came from within the timeless expanses of the continuum: cruising at first, then racing - a formless mass in the void.

Deep thunderings accompanied by great rushings sounds suddenly burst forth from the compact bundle of vibrating energy. Electric rings flashed behind the hot, gaseous ball as it drilled into the thick darkness of the void. Multi-spectral displays radiated like circular rainbows from the increasingly bright and slowly flattening ball of light... the complex wave form, the beginning, the Word.. had been issued...

And **They** said, 'Let there be Light... and it was so...'

Pulsating power throbbed at regular intervals as the blazing rings of illuminated gases and dust whizzed round the newborn star, the Sun. High-pitched, whining sounds merging with the deep throbbings vibrated in phase with ripples of rainbow light now forming spherical canopies over each spinning ring.

Time passed, and a large foreign mass approached the spinning star system... It began to accelerate quite rapidly as it approached the outer orbital ring of the solar system.

Then it happened. The approaching mass hit an outer ring, and a gigantic shock wave curled backwards into the broken ring as the ring continued its spin. Turbulence formed round the impact region causing the dust and gases to curl into a vortex of energy creating a ball-shaped vortex in the broken ring's orbit.

The hurtling mass - a precipitous seed - continued its rapid journey toward the Sun, smashing into ring after ring until all had been precipitated to spinning orbital balls.

Moments passed in **Their** continuum as **They** watched *the tiny blue ball in the third ring* complete thousands of solar orbits. Large remnants of the seeding mass continued in long, elliptical orbits through the solar system, trailing their cometary plumes behind them as they looped the Sun, having become only faint reminders of the seeding season that had passed.

A canopy of water vapor surrounded the delicate third planet. **They** 'was' pleased...

In the times to come, **They** could see that the atmosphere and its water canopy would diminish to over one thirteenth its current volume.

They knew, also, that a time and season would come in which that beautiful blue orbital water shell would collapse into the surface below. **Their** Consciousness returned to the task at hand...

Before **Them** hung *a sphere of land and water encased in a sphere of gases which in turn were surrounded by a shimmering sphere of water vapor.* As **They** observed the gaseous shell **They** named it, 'the great expanse (or heaven) that divides the orbital water from that water on the surface (seas)'. The land amongst the seas of the surface **They** called 'Earth'.... - a place of great beauty... Earth... an embryonic mystery... **They** mused...

And **They** created vegetation in both the seas and the Earth. Level after level of the new ecology integrated under **Their** flawless command.

Atmospheric temperatures rose to finite limits on the Earth as the heat of the Sun became trapped between the 'waters above' and 'the waters below'... forming a gigantic greenhouse.

The water vapor expanded from the additional heat, forming a huge orbital lens, causing the stars to appear larger and much more numerous than they would appear to people on Earth in the distant years to come. They mused on the things yet to be revealed...

'Let the seas and the great expanse above the seas bring forth every manner of swimming and flying life form.' And it was so...

A great chain of inter-dependent life forms had filled the seas and the air leaving only one level to come: land-dwelling animals which included *man without 'consciousness'* or 'ape-man.'

As **Their** signature the last level came to be and was integrated to the whole.

They, the Elohim, God... the Timeless Consciousness of all existence was pleased; and the **Elohim** rested... contemplating **Their** new creation...

CHALLENGED BY LUCIFER

Shortly after the completion of Earth and the "heavens"[1] around it, the **Elohim** gave the Earth system to the administration of Lucifer - who was the highest ranked of all the free-will beings created by the **Elohim**. Many parts of the infinity of the Universe of the **Elohim** were administered by beings of free will, who - like Lucifer - were called 'the sons of the **Elohim**.' Many of them had countless numbers of sub-administrators under their authority, as appointed by the **Elohim**.

Lucifer administered many other dominions in the Universe in addition to the earth system. His position was so honored that he was in charge of the protocol of the 'seat of consciousness' for the **Elohim**.

---

1   'heavens' here refers to the concentric spherical domains which form the Earth from its core to its outer and somewhat intangible dimensions... not the whole Universe.

It would seem in retrospect that such a being would never even in his wildest dreams have wished to challenge the authority of the **Elohim** - his **Creator**... yet, he did just that.

Details of the exact manner in which he began his rebellion are sketchy to say the least; however, it is known that he had come to believe that his knowledge of the Universe was such that he could overthrow the **Established Order of the Ages**... the **Elohim**.

His basic challenge was that his 'free will' gave him the right to replace the **Elohim** if he could maintain order in the process. The sad fact is, however, that such a transition is impossible *within* an existing system. Furthermore, when there is no other place to set up another system *one must live by the existing system - as long as it is fair.*

There can be no fairer order than one which gives absolute freedom to all conscious life governed only by one law: '**No conscious being may interfere with another being's exercise of free will**'.

Lucifer made the sad mistake of accepting all the praise for his works as personal merit badges. His pride took him farther away from the complete communication he had formerly enjoyed with the **Elohim**. And, since an established order of 'free-will' must (by definition) maintain a central communication to prevent even accidental violation of free-will endeavors, Lucifer immediately began to risk violations of the law by not maintaining complete communication with the **Elohim**.

As time passed Lucifer's agents (referred to as the 'Nephilim' in the Book of Genesis) who were on the paradise planet - Earth - conducted numerous genetic experiments in an attempt to clone beings to help Lucifer in his planned rebellion against the **Elohim**. The clones were to be formed from the cells of the free-will beings under Lucifer. These cells were to be merged with those of the higher life forms on the more dense dimensions of the Earth system where the bipeds and other life forms lived.

The whole conspiracy had depended on their ability to breed a 'super race' *without the consent of the **Elohim**.* Such a deed would be closely akin to modern man trying to 'create[' conscious life forms with human intellect in, say, the physical body of a horse... (something like a centaur).

The **Elohim** knew from the beginning of time that the rebellion *would* eventually arise. Such is the nature of free will. To resolve such a conflict the established order must validate its actions to the faithful beings before removing the free-will of those who dissent. To do any less would be dictatorial.

Then came the rejection of the offers. It was done. Those sons of the **Elohim** who had been responsible for the cloning on Earth were stripped of their powers and put into an inner Earth dimension (referred to in the New Testament as 'Tartarus'). There they would stay until the rebellion was either victorious or was permanently stopped. And so it was that the *'Nephilim'* were put *into* the Earth in that ancient age to await their judgement.

Their dwellings and their clones were then destroyed by the **Elohim** in several swift actions, giving rise to the legends of Atlantis.

It was during those policing actions that many life forms were utterly removed from the ecology of the Earth. A great variety of dinosaurs and now-extinct vegetation forms and many other species of legend and fact were most probably destroyed at this time when the **Elohim** removed all living things connected to the rebellion. Tribes of cloned, man-like apes perished also.

Lucifer protested, hoping to win time to try again from another sector. The **Elohim** listened to the protest. It was decided... Earth wold become the court-room for the resolution of Lucifer's challenge... '**The Cosmic Conspiracy**.'

To preserve the **Established Order of the Ages**, Lucifer's challenge would have to be tested on a scaled-down model of the Universe. Earth had been chosen as the site. The **Elohim** projected a portion of **Their** consciousness to Earth, and **Yahweh Elohim** (The Lord God) then appeared on the Earth... looking like a man... yet being much more than a man. He *was the* **Elohim** *in a single projected person.*

In an area which today lies submerged under the Persian Gulf, the **Yahweh Elohim** created a paradise garden within the existing Earth system. Four rivers had their headwaters in this garden called Eden. (Today, they are known as the Nile, the Tigris, the Euphrates, and the Pishon or 'Persian'.).

The approximate date (within about 10 years) of the creation of this second Garden of Eden was 4000 B.C. The **Yahweh Elohim** had created a miniature model of the **Order of the Ages**. All the Earth was once again operating in harmony. Within the garden complex dwelt the perfect specimens of all life forms including the human (called 'A-dam' in the Hebrew). The **Yahweh Elohim** gave Adam dominion over all the other life forms of the Earth; hence, Adam was the administrator over that which the **Elohim** and the **Yahweh Elohim** had created and renewed as the model Universe to test Lucifer's challenge.

To help Adam in his work the **Yahweh Elohim** cloned a mate for Adam from Adam himself using Adam's rib as the cell source. This mate was called 'Eve'.

In the beginning of this test of Lucifer's challenge for authority there was no need for the Adamic pair to procreate. Their work load was not so demanding as to necessitate any more help. In fact, 'work' was fun. They had free-will within the confines of their garden dwelling. Their free-will gave them the right to challenge the authority of **Yahweh Elohim** - if they so desired.

Now, the model was ready. The Adamic pair were the highest ranking mortal beings on the Earth. The **Yahweh Elohim**, Lucifer, the captive Nephilim and the whole hierarchy of command within the perfection of the **Elohim** was ready for Lucifer's challenge...

Lucifer's name changed to Satan (meaning 'adversary' or 'challenger') as soon as he entered the Earthly Eden complex to try to convince the free-will pair to challenge the command of the **Yahweh Elohim** and, hence the **Law of Elohim**.

The **Yahweh Elohim** had told the Adamic pair to refrain from eating the fruit of only one tree in the Garden. This fruit, they were told, would cause instant death if they ate it for it was from the 'Tree of Knowledge of both good and evil'. To that point their free-will had known only good or harmony. The eating of the fruit would give them a concept of how the harmony could be changed in non-harmonic or evil ways.

To this end, Satan convinced the woman to eat the forbidden fruit by telling her she would not really die by doing so. She finally gave in and ate the fruit, finding much to her amazement that she was not dead. *Yet, when she and Adam had both eaten the fruit they knew something had changed.* Their constant 'spiritual' communication link to the **Elohim** and the **Yahweh Elohim** was *dead*.

By choosing the way that Satan had given, they too, had rebelled against the **Elohim**. The **Yahweh Elohim** was apparently disappointed in their performance.. (or so it appeared to Satan, who figured he had won the challenge).

But then came the surprise to all. The **Yahweh Elohim** told the Adamic pair that they would be able to get back in the 'good books' and later have eternal life in the *real Universe* if they repented (or were sorry for) their deeds against the **Law of the Elohim**; and would pledge allegiance to the coming **Deliverer**. The Adamic pair agreed and the **Yahweh Elohim** told Satan that through the direct descendants of Adam and Eve would come the physical and spiritual son of the **Elohim** who would defeat Satan in the test as well as the real Universe. The '**Deliverer**' *would be more than a trans-dimensional projection.* He would be born into the Earth in a flesh body.

Satan's challenge was far from being won. He subsequently tried to pollute the offspring of the Adamic pair by having Cain kill Abel. Even this did not stop the **Elohim**, for it was decreed that the **Yahweh Elohim**, himself, would incarnate to defeat Satan, through the pure genetic line of Adam.

Satan did not give up. The next six thousand earth-years would tell the tale of a horrific challenge, a challenge that is soon to be resolved in favor of the **Elohim**.

To those earthmen who would live in those six thousand years a **special offer is given**. [The same offer that gave Adam the right to rebirth in the real Universe of the **Elohim** after the conflict was resolved.] **The special offer is** that all who have submitted to the **Elohim** by giving their allegiance to the **Yahweh Elohim** (in the form of Jesus the Christ) will have rebirth into the eternal **Kingdom of the Elohim** (God).

Mankind has been created as a proxy for the administrators (or 'angelic hosts' as he Bible calls them) of the real Universe. Satan's challenge will be resolved by proxy to avoid the disruption of the entire real Universe. Those portions of mankind that repent and voluntarily accept the promise and the **Law of the Elohim** will be given eternal life as elite members of the real Universe by virtue of their oneness with the **Yahweh Elohim** himself. Their position will be one which will let them sit in the final judgement of all those angelics (sons of God) who revolted under Lucifer's will.

Thus, the apparent illogic of chaos or evil existing within order or good is resolved by the realization that the Earth condition is only a mock-up of the reality of the heavenlies. **Chaos does not coexist in reality with order**, *only within this test case, this conflict resolution, this* '**cosmic conspiracy**'.

Our struggle against the principalities of evil and the powers of darkness have commanded the undivided attention of all the heavenly hosts for these last six thousand years. Every Earthly event, no matter how small, is closely monitored by those whom we represent in the conflict. The outcome is already known to be:

-UNITY-

CHAPTER 2: Conflicts Bear Witness

The Old Testament records numerous conflicts between Satan and the **Yahweh Elohim** over various points of order in this great conflict resolution. These conflicts, themselves, bear witness to Satan's repeated violations of even the terms of the test. To list all such conflicts in this book would be too lengthy an exercise. A few of the more outstanding conflicts will suffice to illustrate the purpose behind the predicament of mankind today.

ANGELS AND DAUGHTERS

In a little over seven hundred years from the expulsion of Adam and Eve from the Garden of Eden, the population of mankind grew to vast numbers (some estimates exceed 60,000,000!)

*"And it came to pass when man had begun to multiply on the face of the ground and daughters had been born to them that the sons of **Elohim** (angels) saw the daughters of men that they were fair, so they took themselves wives of whomsoever they chose." (Genesis 6:1-2).*

*'The Nephilim were in the Earth in those days, and also after that when the sons of **Elohim** began to go in unto the daughters of men, and sons were born to them, the same were the heroes that were from age-past times the men of renown.' (Genesis 6:4-5).*

These passages refer to a rather flagrant violation of the terms of the trial. As mankind increased its numbers the 'sons of **Elohim**' (angels, in this case under Satan) took wives from the tribes of Adam's descendants. The ensuing "illegal" unions produced offspring that were the 'heroes', in some cases the villainous heroes of ancient legend about 'gods' and their incessant struggles for supremacy... How could they do otherwise with their sires suffering from the same rebelliousness?

Satan had attempted to completely corrupt the 'pure genetic line' from Adam. In so doing he had hoped to remove all possible physical means of allowing the **Yahweh Elohim** to incarnate on Earth and thereby win the trial against Satan. Some rather unusual hybrids came out of those matings. Some had extreme physical strength; some great beauty; others intellect; yet, others had all three plus a few superhuman powers. Even bestiality occurred giving rise to legends of centaurs and other half-human half-animal beings.

The hybrid breeding continued until about 1646 years after the expulsion of Adam and Eve. *In this time the number of uncorrupted humans remaining had dwindled to only eight out of the millions.*

The **Yahweh Elohim** saw that even the hybrid beings would not follow the Law of the Established Order of the Ages; so with the exception of the eight law-abiding humans and the still pure animal life forms necessary for ecological stability, the **Yahweh Elohim** resolved to destroy all life-bearing flesh on the *Earth* not in the seas.

Satan had polluted the model Universe by allowing members of the real Universe to breed with the model beings. The decision of the **Yahweh Elohim** was just, in that it allowed still one more chance for Satan to prove his challenge - in spite of Satan's treachery.

'Then **Yahweh (the Lord God)** *saw that great was the wickedness (violation of the Law) of man in the Earth, and that every purpose of the devices of his heart was only evil continually; and it grieved* **Yahweh** *that* **He** *had made man on the Earth, and* **He** *took sorrow unto his heart.*'

'And **Yahweh** *said,* '*I must wipe off man whom I created from all the face of the ground, from man unto beast unto creeping thing and unto the bird of the heavens, for I am grieved that I made them.*'

'*But Noah had found favor in the eyes of* **Yahweh**.*' (Genesis 6:5-8).*

As Noah was a righteous man, the 'Lord God' decided to use him as his family and a selection of untainted animal life forms to re-seed the earth after the 'cleansing' took place.

'*And the Earth corrupted itself before* **God (the Elohim)** *and the Earth was filled with violence. And* **God** *beheld the Earth and lo! it had corrupted itself, surely all flesh had corrupted its way on the Earth.*'

'*So* **God** *said to Noah, "The end of all flesh hath come in before* **Me***, for filled is the Earth with violence because of them, behold me than destroying them with the Earth.*' (Genesis 7:11-13).*

The **Yahweh Elohim** told Noah that He would leave Earth at the end of 120 years. Within that time Noah was to construct and stock the Great Ark to survive the Great Flood that would come when the **Yahweh Elohim** left the planet.

In that last days before the rains and destruction started, the **Yahweh Elohim** (God in personal form) left Earth to rejoin the **Elohim** (God in universal form) after Noah and his passengers had all been sealed in the Ark.

The stars appeared to move as the light of the Sun suddenly began to dwindle. Massive grey shapes stretched and swirled across the twilight skies of Earth while great flashes of light forked from one end of the horizon to the other, followed by deafening bombardments of thunder.

The Moon seemed to shake like jelly as the stars began to chart new paths around the shifting planetary axis. It had never rained before - there had only been the vaporous mists that watered the Earth at night. How could they know those grey shapes as 'clouds?' 'What were clouds?... rain?... lightning?'

Fat, splattering drops of water began to fall from the darkened skies as the ground began to move under their feet. Buildings toppled; fear gripped the judged, it continued, unrelenting, just, final.

Gurgling screams, chilling shrieks, stampeding animal sounds and driving water noise spread across *the Earth as that beautiful, shimmering shell of water vapor* condensed and collapsed into the spinning Earth. Billions of tons of water smashed into the Earth, driving oceans from their resting places, breaking subterranean streams and water domes apart.. (They had seen this time... long ago).

Days passed as the deluge continued. The surface glistened in the patchy sunlight as the clouds began to thin. It had rained forty days and nights before it finished. All surface creatures were lost save those inside one small brown spot that gracefully bobbed on the water's surface in the warm sea air. Great clouds of water vapor began to freeze and precipitate at the new poles of Earth. Surging electric currents intensified the polar cooling with their thermoelectric effects. Time passed, and Noah waited for ten months until his Ark came to rest high upon Mount Ararat in Turkey. As he left the Ark through the upper decks the consciousness of the **Elohim** surrounded him saying many things in a new covenant. Among these things of the new covenant was a sign in the clouds. It was the rainbow, a new sign (as there had never been clouds before)... The **Elohim** promised that this sign would be a perpetual reminder of the covenant whereby the world would not be destroyed by flood again. It was a new start... would Satan follow the rules of the trial?

SATAN'S MAN, NIMROD

Nimrod was born about three or four hundred years after the Great Flood. It was Noah's son Ham who sired Nimrod's father, Cush. Nimrod had become a mighty hunter and a strong leader of the entire human tribe before he decided with Satan's help to challenge the Law of the **Elohim**. At the time Nimrod issued his challenge there was only one language spoken by all the humans on Earth. In fact, even though the tribes were settled in various regions they were united in purpose by their common culture and the common language.

It was Nimrod who suggested to humanity that the covenant made by the **Yahweh Elohim** with Noah could not be trusted. Nimrod maintained that humanity had to reach up into the regions (heavens) from which the 'pouring waters' had come to assert itself with the hosts of the **Yahweh Elohim** to prevent any future scattering of man who, by that time, had been illegally given the code to read mankind's future as had been recorded by the **Elohim** in the twelve constellations of the Zodiac. To this end, Nimrod convinced the people to begin construction of the city of Babel. Within that city was to be built a gigantic tower reaching the regions from whence the great rains had come. On the sides of the great tower were to be the keys to interpreting the future of mankind as recorded in the Zodiac. The **Yahweh Elohim** and a host of the Sons of the **Elohim** recognized this expected *breach of due process* by Satan in the Earth system. For, in essence, by convincing Nimrod to organize the humans into a dissident union against the 'heavenlies' Satan had again subverted the *'due process'* in the trial case by coercing mankind to seek-out and challenge the real Universe instead of progressing on Earth as though Earth were their entire real Universe. This ploy, too, was doomed to failure from the start.

The **Yahweh Elohim** and His hosts came down to the Earth and split mankind into diverse regions. The common language of mankind was confused into many different tongues. The idea was to dissuade mankind from further probes into the real Universe until the time of the trial was over. Otherwise, mankind's progress would create a problem in the heavens instead of *solving* the existing one on Earth.

DEMONS AND AMALAKITES

Satan would not give up his illegal attempts to 'rig' his trial on Earth. Firstly, thwarted in his attempt to genetically interfere and then secondly, in his coercive attempt with Nimrod, Satan tried yet another crafty scheme.

He found a unique way to use the influence of the disembodied spirits of his supporters who were killed in the Great Flood. By letting them affix themselves to the minds of living humans, (possess them), he found he could indirectly influence the behavior of even those humans who tried to live the Law of the **Elohim**. The plan was to so pervert or kill even the faithful that the **Yahweh Elohim** would destroy the system and drop *His* case against Satan.

The third ploy was even more subtle, yet just as ineffective as the prior attempts. Satan had used the Amalakites as his human 'hosts.' When the children of Israel became a nation, one of their major directives from the **Elohim** was to destroy the Amalakites. Between Israel and other nations the brutal and cruel Amalakites (who were also known in Egypt as the Hyksos) were utterly destroyed.

'ENOUGH IS ENOUGH'

And so it came to pass, after allowing Satan more than enough chances to have a fair trial on Earth, that the **Elohim** declared an end to the Satanic rebellion. It was decreed that the **Yahweh Elohim**, *Himself*, would incarnate into a flesh body to explain to the humans why they had been suffering for so long; furthermore, He was to officially declare the victory over Satan on Earth; and lastly, He was to initiate the mystical process whereby all human spirits from Adam's time *to the end of this age* who had accepted the Law of the **Elohim** would be rescued from the Earth before Satan's final removal from the Universe, thereupon to be imprisoned on Earth till the final judgement takes place.

KILL THE CHILD

When Satan heard the pronouncement, he immediately despatched his 'angels' to find the time and place of the proposed arrival or birth of the **Yahweh Elohim** in the human form. Their efforts revealed that the prophets of Israel expected the birth of the **Yahweh Elohim** to be in the town of Bethlehem of the Judean province. As it was Herod the Great who ruled Judea, Satan's 'angels' influenced Herod to seek the death of the coming child... who was to be the 'Christ' - *or anointed one*, spoken of from the beginning of the trial (see Genesis 3:15).

Herod had to find where the child was in order to kill Him. So, when a group of astronomers from the East, *'wise men of the East'* - arrived in Jerusalem seeking the 'King of the Jews' whose 'star' they had seen in the East, Herod hatched a plot. He demanded that the priests and scribes of the Jews tell him where the 'Christ' was to be born. When they had told Herod it was to be in Bethlehem of Judea, he then asked the astronomers for the exact date on which they had seen the 'star.' Upon learning the date, Herod secretly issued instructions to the astronomers to go to Bethlehem to find the **Christ** so that he could also come and give allegiance to **Him**. what Herod really meant was that he wanted to kill **Him** (for Satan's cause).

However, an angel, *'messenger,'* of the **Yahweh Elohim** told Joseph and Mary, *the earthly parents of the child called* **Jesus**, to escape to Egypt to prevent Herod's attempt to kill **Jesus**.

Herod, subsequently, had all the male children in Bethlehem under the age of two years put to death. Satan had failed again; yet, he had still another idea...

"Revolt, Follow Me!"

'If,' he reasoned, *'I can convince the* **Yahweh Elohim** *in His descended* (or mortal) *state while He is not so nearly omniscient as in the ascended* (or immortal) *state to swear allegiance to me, I can still win my challenge to the* **Elohim.**.'

Years passed on Earth as **Jesus** grew to manhood. When **Jesus** (the incarnate **Yahweh Elohim**) came to the river where John 'baptized' **Him**, the spirit of the **Elohim** descended as a dove upon **Jesus** endorsing **Him** as the only **Christ** (Luke 3:22).

Satan waited. He knew the next forty days of fasting the human body of **Jesus** would be difficult for **His** human part. When the end of the forty days had come, Satan made his move. Appearing suddenly before **Jesus,** Satan taunted Him saying,

'If you really are the descended **Yahweh Elohim,** *turn these stones to bread* (to feed yourself)'.

Jesus realized Satan's objective and answered,

'Man shall not live by bread alone, but by every word that proceeds out of the mouth of the **Elohim'** (Matt. 4:4).

Satan had hoped to break **Jesus'** reliance on the **Elohim**... but he had failed again, so he tried another approach to get **Jesus** to act for **Himself** without relying on the **Elohim.** He took **Jesus** into Jerusalem and set **Him** upon a pinnacle of the temple, saying to **Him,**

'If you are **the Son of the Elohim** (the descended **Yahweh Elohim**) *throw yourself down; for it is written, the* **Elohim** *shall give His angels charge concerning you, and in their hands they shall bear you up, lest at any time you* (should) *dash your foot against* (even) *a stone'.*

Jesus responded, 'You *shall not put the* **Yahweh Elohim** *to the test'* (Matt. 4:7). Obviously... it was Satan who had been put to the test from the first moment when he originally rebelled against the **Elohim.** Still hoping to find a flaw in the newly formed **son of the Elohim,** Satan made one last temptation.

As **Jesus** stood upon the highest of mountains on the Earth, Satan caused visions of all the kingdoms of the world to appear before Him in their great splendor. Satan's voice ever so smoothly and warmly entreated **Jesus,** saying, *'All those things will I give you, if you will* (but) *fall down and give me your allegiance'.* It wasn't good enough.

Jesus (the descended **Yahweh Elohim**) then turned and with the eyes of the **Elohim** spoke firmly to Satan, *'Begone, Satan; for it is written, you shall give allegiance to the* **Yahweh Elohim,** *your* **Elohim,** *and* **Him** *only shall you serve'* (Matt. 4: 10).

With that, Satan removed himself to plan his next move while the angels of the **Yahweh Elohim** came and attended to the needs of **Jesus... the Christ.**

'You Will Die'

Satan is a bad loser... (and a nasty adversary)! When he realized that the **Yahweh Elohim** had successfully incarnated into a human body and could not be persuaded to rebel against the **Elohim,** Satan decided to kill **the Christ.** This was not, however, as illogical as one might think.

Satan had decided to trap the **consciousness of the Yahweh Elohim** in the lower Earth levels where the spirits (or consciousness) of all dead humans were stored under lock and key. Some sources refer to these levels as 'Hades' and 'Paradise.' The spirits of those people who had *not* responded to the offer of the **Elohim** from the beginning were stored in Hades - *an uncomfortable place* - while those who had supported the **Elohim** were in Paradise - *a peaceful place* - waiting for their **Deliverer** to release them.

It was Satan's guess that once he had killed **Jesus the Christ, His Spirit** would not be able to open the trans-dimensional 'gates' to get **Himself** and the spirits of Paradise out of the lower dimensions... Satan was wrong again... He did kill the **Christ** - but only for three days.

In that time, not only did the **spirit of Jesus the Christ** descend into the underworld of the dead, it also opened the trans-dimensional gates of Paradise to free **His** faithful who waited there. By a special technique which is not yet fully understood, the spirits of the redeemed from Paradise and the spirit of the **Yahweh Elohim** were joined in a mystical union... I suppose I could say they were united by the same fundamental 'frequency' or 'vibration' and removed by **Christ** to the heavens.

When the arisen **Yahweh Elohim** revealed **Himself** to **His** faithful "surface" dwellers of Earth, **He** told them they would be included in a very special future event as a reward for their allegiance. Also, **He** extended a promise of the same reward to all those humans of future generations who would join Him in His cause. In the following chapters I shall discuss this 'reward' in more detail; but, for now, as this snapshot of the great conflict is complete, a few points of additional interest should be noted to help convince those who may still be a bit doubtful that any of this tale really happened.

About That Flood

According to the Bible, the flood of Noah changed the average lifespan of man. Before the Flood man lived to be 600 to 900 years on the average. Within sixty-seven years after the Flood, the average age dropped to 440 years. In 160 years after the Flood the average age had dropped again to 240 years. After only 300 years the average had dropped to 175 years. Now, in modern times a person is lucky to even see seventy-five! Why?

Science tells us that ultra-violet radiation from the Sun is greatly responsible for 'aging' human tissue. The water shell (or canopy) that surrounded Earth prior to the Flood acted as a shield to protect the life forms from these 'aging rays' from the Sun. Not only that, the canopy also maintained a stable 'greenhouse effect' on the Earth's surface. As the water was in orbit, there was no rain as such until the Flood. Prior to that it was surface condensate that watered the fields and rivers at night.

Why else would the **Yahweh Elohim** have told Noah that the rainbow was a 'new sign' in the clouds? If the rainbow had been a phenomenon previously, it could not have been a 'new' sign. Since direct sunlight and rain droplets of a specific size and density are necessary to produce a rainbow, it is quite safe to assume that no clouds graced the skies before the Flood.

The Great Clock

Many of the Biblical events of the past were anticipated by wise men who were astronomers as well as supporters of the **Yahweh Elohim.** They had been taught by the **Yahweh Elohim** to watch for certain star patterns in the heavens as *'event markers'* in human affairs.

These patterns were easily memorised as figures of people and animals formed by the stars. It was from these times that the twelve signs of the zodiac were obtained.

In later centuries these 'star signs' would be misinterpreted by many as *causes* of various events and character traits in human affairs... giving birth to the pseudoscience' of astrology.

Nothing could, of course, have been farther from the truth. The **Yahweh Elohim** used the star patterns and their changing relationships as a mechanical means of signalling the coming of various events in **His** great war with Satan on Earth. Nothing Satan could do could wipe out the 'star clock'. The only thing he could have done was try to hide the secret of the 'star clock's' meaning from mankind or so pollute the pursuit of their true function that no serious scientist would try to find the secret for fear of ridicule. Hasn't astrology filled this role?

How do you think the 'wise men (astronomers) of the East' who 'followed the star' knew what that star meant? They had been told to watch for certain stars to be in certain formations. When such events occurred, the astronomers knew a specific event was due. The beginning of the so-called 'Age of Pisces' began at about the time of the **Christ's** birth... and is ending now as we enter the so-called 'Age of Aquarius'. In the Old Testament an ancient prophecy stated 'and the virgin shall conceive'. Perhaps this had a dual meaning. The constellation of Virgo (the virgin) may have been the origin of the star that led the wise men to Bethlehem where the earthly virgin had conceived. When **Christ** had formed **His** church, they adopted the sign of the fish -Pisces... (or 'ichthus' in the Greek). When **Christ** returns to Earth, **He** will give freely of the 'water of everlasting **Life** that whosoever shall drink of it shall have eternal Life'. Isn't it interesting that we are now entering the 'age of the water bearer Aquarius?'...

-EVIDENCE WITNESSES-

CHAPTER 3: How To Survive

Here we sit caught in the middle of one 'hell' of a big war between Good and Evil... It would seem that our positions as mere mortals in this whole affair is too small to be noticed by the combatants - yet nothing could be farther from the truth!

As the proxies of the heavenly hosts (both good and bad) we are being watched more closely than the world watched Watergate and the Kennedy assassination! We have an incredible *opportunity* - not obligation - laid upon us. Since the beginning of the 'age' Earth, all human beings have had the right to accept the greatest gift ever conceived: eternal life as free-will beings roaming the entire Universe of the **Elohim!**...

When the total of human souls who have accepted this gift reaches a predetermined number, time for this 'age' will cease and the offer will change. **Those who have accepted the terms within the time allowed will be swiftly removed from the Earth in a new, indestructible body as 'reborn sons of the Elohim'.** Those who wait beyond the time limit before deciding to accept will find a different offer exists. It will offer them eternal life - but at a level with less responsibility than that to be held by **the new sons of Elohim** through their Deliverer, Jesus the Christ.

In either situation, those who choose the **Law of the Elohim** will survive forever in the happiest of conceivable circumstances by comparison to life anywhere on Earth now. If you would like to accept the offer at this time, there are just **seven points** to consider. The **fifth** of these points gives the **three** very simple terms of accepting your free gift from the **Elohim**... from the **GOD**... **from They** who love us beyond imagination...

The 1st Point

Since we are all inhabitants of this world under Satan's dominion, and we are all direct descendants of the first man to break the **Law of the Elohim,** we are all not acceptable in the real Universe. The Bible says it this way: *'For all have sinned and fall short of the glory of the* **Elohim'** (Romans III:23); and... *'If we say, sin have we none!,... we are deceiving ourselves and the truth is not in us'* (John 1:8).

The 2nd Point

Those who break the **Law of the Elohim** have to pay a penalty. Until this is paid they [we] are separated from the **Elohim**. The Bible says it this way: *'For the wages of sin is death; but the* **Elohim's** *favor is life age-abiding in* **Jesus Christ** *our* **Lord** [the **Yahweh Elohim**]*'* (Romans VI:23); and... *'He that believeth on **Him** is not to be judged: He that believeth not already hath been judged'* (John III:18).

The 3rd Point

A law breaker cannot redeem himself. It takes an 'intercessor' -which in this case has been provided by the mystical action of **the triune Elohim.** The Bible says it this way: *'For a man who shall keep the whole* **Law** [of the **Elohim**], *but shall stumble in one thing hath become for all things liable'* (James II:10); and... *'Not by works* (good deeds) *which we had done in righteousness, but according to* **His** *mercy He saved us - through means of the bathing of a new birth* (being reborn in the heavens as a son of the **Elohim**)' (Titus III:5).

The 4th Point

A part of that 'mystical action' of **the triune Elohim** was the act of giving the Gift of salvation (or being saved from permanent death). The Bible says it this way: *'For by* **His** *favor have ye been saved through means of faith, and this hath come to pass - not from you,* [but from] *the* **Elohim** *the free gift! Not from works* [good deeds] *should anyone boast'* (Ephesians II:8-9); and... *'For the* **Elohim** *so loved the world that* **His** [**Their**] *only Begotten Son He* (**They**) *gave - that whosoever believeth on* **Him** *might not perish but have life age-abiding'* (John III:16); also... *'***Jesus** *saith unto him, "I am the way and the truth and the life: No one cometh unto the* **Father** [the Elohim] *but through me'* (John XIV:6).

He who has accepted **Jesus the Christ** has been accepted by the **Father (the Elohim)** for the two are one even in their individuality. *'Believe me, that I* [**Jesus**] *am in the* **Father** [the Elohim] *and the* **Father** [the Elohim] *in me - or else on account of the works themselves believe ye. Verily, verily I say unto you - he that believeth on me the works which I am doing he also shall do'* (John XIII:11-12)... Is **He** talking to you?... then...

The 5th Point

Under the *grace* of the **Elohim, three** terms of agreement can be met by those who wish to accept 'salvation' (or, their free ticket out of this world) which **Jesus** paid for by his death on the Cross.

I) You must repent of your violations of the **Law of the Elohim...** (i.e. you must really regret the many transgressions you have and will commit on the Earth and you must agree to let the **Elohim** change you into a new being who avoids evil to the best of his ability). The Bible says it this way: *'Nay! I tell you, but except ye repent ye all in like manner shall perish'* (Luke XIII:3).

2) You must make a verbal declaration that you give allegiance to **Jesus** who is the **Yahweh Elohim** in **His** ascended state of being as **the Son of man.**

The Bible says it this way: *'That if thou shalt confess* (accept the validity of) *the declaration with thy mouth - that* **Jesus is Lord** [Yahweh Elohim], *and shalt believe with thy heart - that* **GOD** [Elohim] *raised Him from among the dead thou shalt be saved'* (Romans X:9).

3) It goes without saying that if you declare **Jesus is Lord** you also believe it. The Bible says it this way: *'And they said: Believe in the* **Lord Jesus,** *and thou shalt be saved, thou and thy house'* (Acts XVI:31).

The 6th Point

Now is the time to decide, before time runs out. This is no threat; it is an entreaty. The Bible says it this way: *'But if it be a vexation in your eyes to serve* **Yahweh** *choose ye for yourselves today whom ye will serve...'* (Joshua XXIV: 15). *'For He saith, I have heard thee in a time accepted, and in the day of salvation have I helped thee; behold, now is the accepted time; behold, now is the day of salvation'* (Corinthians VI:2). Don't delay too long if you haven't accepted the gift yet.

The 7th Point

Having accepted 'the Gift', share with your friends, neighbors, enemies, relatives, and any and all others who would ask you about life and such matters of importance. Witness to others only when you have their attention of their own free will. **Never force the issue. Life under the Elohim** is true *freedom*; do not forget it.

Get together with others and study the entire Bible - but especially the prophetical works such as: Daniel, Ezekiel, Isaiah, Jeremiah, the Psalms, Matthew and the Revelation of John; which hold yet many secrets for us to decode and understand and share. The Torah and the Book of Revelation are revealing more and more to the probing minds of Judeo-Christians new and old alike. Join them. Open your eyes... and ears... raise your understanding of the very essence of Life itself... Seek and you shall find...

- THE TRIUNE: ELOHIM -

CHAPTER 4: Taken To Safety

The first two sections of this book paint a grim picture of today's world; and the worst is yet to come; however, if you have accepted **Christ** as your **Mashiach**, you will not be afflicted by what is to come. You may be in the real Universe 'wearing' a new 'eternal body'. Soon, a series of great mysteries will manifest on Earth and in the surrounding space. One of these mysterious events will be the instantaneous transferring of the entire believing Church (both living and dead believers in **Christ)**, *up from the Earth.* The Church generation living at the start of the 'end times' - sometimes called the 'period of the Great Tribulation on Earth' will not have to experience death. They *will not die.* They will simply put on over the outside of their existing body an immortal body, and the old body will be dissolved and pass away. They will then be caught up inside of shimmering, vibrating, white clouds to a meeting with the **Christ** in the air above the Earth; and shall ever remain with **Him** thereafter. (see 1Thessalonians 4:17). This is called the 'rapture of the Church' by some commentators. It could be looked upon as a transformation from this world to another state of being in another world... the heavens... the real Universe... with the **Elohim.** There are critics who say the rapture is not mentioned in prophecy; however, their failing is only a lack of proper study of the **Word of GOD.** Come as a child... to be taught...

Having read a most comprehensive comparison[2] of the different viewpoints on the 'rapture' question, any doubts I may have personally held on the subject have been completely removed by Dr Walvoord, the author of that comparison. I trust that he will not be offended by my paraphrasing some of his material in an attempt to teach even more people the promise of deliverance from that which is upon us shortly.

Although Dr Walvoord systematically discusses the pros and cons of many viewpoints to the rapture question, I shall limit my discussion here to those points of the positive view... (I do, however, strongly encourage you to get a copy of his book and study it).

There is no doubting that the scriptures predict the Church will be 'raptured' or 'caught up' inside clouds to be with the **Lord** when He comes for them. The Latin translation of the word for 'rapture' in 1Thessalonians 4:17 gives us 'rapere' or 'caught up'. The Interlinear Greek Translation of the whole phrase gives us... *'**shall be seized** inside clouds to a meeting of the **Lord** in air; and so always with* [the] ***Lord** shall be'.* If this is a literal (as opposed to symbolic), future event, then it is a most important part of what the scriptures call the 'hope of the Church'.

---

2 *The Rapture Question* by John F. Walvoord (pres. of the Dallas Theological Seminary in Texas), published by Zondervan Publishing.

Why So Important?

Christ told His disciples in Matthew 24 that there would be certain signs to herald **His** coming at the end of this world system. These signs are spoken of as the 'great tribulation' or as 'Jacob's trouble'.

In the following chapter I shall be looking at the 'great tribulation' in more detail; however, for this moment suffice it to say the events of the tribulation are going to be so horrible that no one in his right mind would want to dwell on Earth during such a time. Obviously, it is of primary concern to believing Christians to know whether or not they (the Church) will have to dwell on Earth during the tribulation.

Dr Walvoord has commented on the situation...

*'The nature of the tribulation is also one of practical importance. If the Church is destined to endure the persecutions of the tribulation, it is futile to claim the coming of the **Lord** as an imminent hope. Instead, it should be recognized that **Christ** cannot come until these predicted sorrows have been accomplished. On the other hand, if **Christ** will come for His church before the predicted time of trouble, Christians can regard His coming as an imminent daily expectation. From a practical stand point, the doctrine has tremendous implications.*[3]*'*

*... 'As illustrated in the writings of Charles Hodge, the post-millennial point of view considers the tribulation a final state of trouble just preceding the grand climax of the triumph of the gospel. The national conversion of Israel and the general conversion of Gentiles is viewed as containing in its last stages a final conflict with antichrist which is equated with Romanism. (Charles Hodge, **Systematic Theology, III,** 812-36.)*[4]*'*

*'The third view, which is popular with pre-millenarians who have specialised in prophetic study, is the pre-tribulational position, which holds that **Christ** will come for His church before the entire seven-year period predicted by Daniel. The church in this point of view does not enter at all into the final tribulation period. This teaching was espoused by Darby and the Plymouth Brethren and popularized by the famous **Scofield Reference Bible.** Generally speaking, the pre-tribulational view is followed by those who consider premillennialism a system of Bible interpretation, while the post-tribulational and mid-tribulational divisions characterise those who limit the area of premillennialism to eschatology.*[5]*'*

... 'Any answer to the rapture question must therefore be based upon a careful study of the doctrine of the church as it is revealed in the New Testament. To a large extent, premillennialism is dependent upon the definition of the church, and premillenarians who fail to distinguish between Israel and the church erect their structure of premillennial doctrine on a weak foundation.[6]*'*

... 'If the term "church" includes saints of all ages, then it is self evident that the "church" will go through the tribulation, as all agree that there will be saints in this time of trouble. If, however, the term church applies only to a certain body of saints, namely, the saints of this present dispensation, then the possibility of the translation of the church before the tribulation is possible and even probable.[7]*'*

---

3   ibid. p.11-12
4   ibid. p.12
5   ibid. p.15
6   ibid. p.16
7   ibid. p.19

Dr. Walvoord goes on to say:

*'All agree the "ecclesia" in its first meaning indicated above is used of Israel in the Old Testament. The issue is whether "ecclesia" is ever used of Israel in the sense of the second, third, and fourth meaning. A study of every use of "ecclesia" in the New Testament shows that all references where "ecclesia" is used in the New Testament in reference to people in the Old Testament can be classified under the first meaning. Of **particular importance is the fact that "ecclesia" is never used of an assembly or body of "saints" except in reference to saints of the present age.**[8] '*

*'The teaching that the body of **Christ** in the New Testament is a separate entity supported by the predictive statement of **Christ** in Matthew 16:18; "Upon this rock **I will build my church."** The figure of speech rests upon a concept of a future undertaking. **Christ** did not say, "I am building" but, "I will build." It is significant that this is the first reference to the church in the New Testament and is here regarded as a future undertaking of **Christ Himself**.[9]"*

*'The classic passage on the baptism of the **Holy Spirit**, I Corinthians 12:13, declares: "For in (by) one spirit were we all baptized into one body, whether Jews or Greeks, whether bond or free; and were all made to drink of one spirit." **The baptism of the spirit is the act of GOD by which the individual believer of Christ is placed into the body of Christ.** The Greek preposition (en) translated (in) in the American Standard Version, is properly rendered (by) in both the Authorised and the Revised Standard Version in recognition of its instrumental use. The Spirit is the agent by whom the work of **GOD** is accomplished. '*

*'In virtue of these significant truths, **it becomes apparent that a new thing has been formed - the body of Christ.** If did not exist before Pentecost, as there was no work of the baptism of the Spirit to form it....[10]'*

This following information will be of great interest to the Seventh Day Adventists: *'Of major importance is the relationship of this to the interpretation of **Daniel's** seventieth week (Daniel 9:27). Those who believe that the present age is a parenthesis regard it as the extended period of time between the close of the sixty-ninth week of Daniel and the beginning of the seventieth week. This would support the teaching of pre-tribulationalists that **the future fulfilment of Daniel's seventieth week has to do with Israel and not the church,** and thereby strengthens the pre-tribulation position. The study of Daniel's seventieth week will sustain the teaching that the church of the present age is a distinct body from those who live in the seventieth week.[11]'*

*'It should be obvious to careful students of the Bible that Mauro [Phillip Mauro, an amillenarian] is not only begging the question but is overlooking abundant evidence to the contrary. Nothing should be plainer to one reading the Old Testament than that the foreview therein provided did not describe the period of time between the two advents. This very fact confused even the prophets. (cf. I Peter 1:10-12). At best such a time interval was only implied and this may be observed in the very passage involved, Daniel 9:24-27. **The anointed one, or the Messiah, is cut-off after the sixty-ninth week, but not in the seventieth week. Such a circumstance could be true only if there were a time interval between these two periods.**[12]'*

---

8   ibid. p.21
9   ibid. p.22
10   ibid. p.22-23
11   ibid. p.24
12   ibid. p.25

Translation of Believers

The scriptures have given us the believers an incredible hope. Not only will we be given eternal life; but the last generation of the Church on Earth will not have to die to get their eternal bodies... (and we are that generation).

'Behold, a mystery I tell to you: We all shall not fall asleep [die in the physical sense], *but we all shall be changed in a moment, in a glance of an eye, at the last trumpet; for a trumpet will sound and the dead will be raised incorruptible, and we all shall be changed. It is required for this corruptible to put on incorruption and this mortal to put on immortality'* (ICorinthians 15: 51-53, Interlinear Greek/English).

Why else would Paul have told us to comfort one another with these words of hope? As Dr Walvoord says, *'The very fact that the hope is presented as a comforting hope is another argument for pre-tribulation rapture for the Church...'* [13]

Hope of Glory

In the scriptures the 'indwelling of **Christ**' in the believers is called the 'hope of glory'. The 'saints' of the tribulation period are never referred to as the 'Church'. Furthermore, there is no mention of the indwelling of **Christ** in those 'saints,' because the **Word of GOD** is to be delivered during the tribulation by prophetic voices as it was in ancient Israel.

The scriptures tell us quite firmly that the days of the last generation of this dispensation age will be *'as it was in the days of Noah'*... and *'as it was in the days of Lot'*. In Noah's time it had never rained; and people were practicing evil in the forms of greed, lust (homo and heterosexual lust), idol worship, necromancy, murder, occultism and other forms of godlessness. In his time Noah was considered a nut and a crank by his contemporaries. In Lot's time fire and burning stone had never rained from the sky to destroy a city; people were practicing the same forms of godlessness - especially sexual intercourse between man and animal and between like sexes of mankind; in his time Lot was considered a nut and a crank by his contemporaries.

In today's time the whole Earth has never been ravaged by fire from the skies or by mankind... yet, evil-doing abounds as it did in those past epochs, and to those of us who cry, *'time is short... repent and be saved... the end of this age is near!'* is attached the social stigma of *'weirdo'*, *'doom prophet'* or *'crank'*, by our contemporaries... Don't let such abuse ruffle your feathers, for it is just as certain that we believers today will be delivered out of chaos and danger as were Noah and Lot and their families in those ages past.

Noah's Ark Today

Today's 'ark' is the **'body of Christ'**. This may seem hard to grasp - but do not close your mind, yet. There is a process whereby all persons of a certain mental attitude like a common faith can be linked together by a common power. The **Holy Spirit** (or the great **Comforter as He** is also known) is that 'common power'. **His effect** on the whole of those who believe in **Jesus the Christ** as their **Lord and Deliverer** is like a melodic song which bonds them together as a mighty and harmonic force.

---

13   ibid. p.36

Furthermore, through **His** mystical action, we, 'believers,' have been unified; so when the **Lord Jesus** is ready to initiate the 'Rapture', only those people on Earth who are in the harmony of the **Holy Spirit** will be selectively extracted from the Earth's surface to a secret meeting with the **Lord Jesus** in the Celestial Heaven,... somewhere. All else will remain on Earth.

There is no need to be in a certain place at a certain time to be raptured. Where you are, the angelic hosts of the **Lord GOD** and the **Holy Spirit,** himself, will 'seize you up inside shimmering clouds of light' instantly... in your eternal body...

If you have not picked-up your 'deliverance ticket' to the 'New Ark', you can still do so (if the Ark hasn't left yet) by not only saying, but also believing the following:

I believe that **Jesus Christ** is the **Son of GOD.** I believe that **He** was born into a human body to become the unique being of the Universe: **GOD** and man. I believe **He** then shed **His** blood on the Cross and died in my place for all the sin I have or will commit in my life on Earth. I believe **He** rose from the dead having defeated Death and Satan on my behalf. I confess that I am a sinner and do hereby seek the new life of a child of **GOD** as well as the fellowship of those others who have made the same decision,... signed:

1).. .. ..

2).. .. ..

3).. .. ..

4).. .. ..

5).. .. ..

6).. .. ..

7).. .. ..

If you have just signed above, then in the name of **Jesus the Christ,** I welcome you 'aboard'... into life everlasting. We are now brothers and sisters awaiting our departure from this planet of chaos to our true home. AMEN! (or, so let it be done), for...

-TODAY'S WORLD IS DOOMED-

CHAPTER 5: Horror With Grace

This is a tale of horror that even Edgar Allan Poe could not have equalled. So much death and destruction will soon strike this generation of mankind that many will hear the cry *"Will no one help the widow's son?..."* - and yet those who hear and understand will fear to respond.

The Tribulation Approaches

When the nation of Israel finally regained control of <u>all</u> the old city of Jerusalem in the 1967 War, a 40-year (360-day levitical year) or perhaps a 45-year (length of time an unrighteous generation has to rebuild the Temple after they return to the Land) countdown began. A 40-year period is referred to by Biblical scholars as one generation. A Biblical generation is equal to that period of time in which the children of Israel wandered in the desert after the Exodus from Egypt. It was during that forty years of wandering that all of the rebellious generation of Israel died out. Since it took forty years, many assume forty years is equal to the lifetime of a generation.

Therefore, from 1967 forward 40 years is 2007 (or 2012 if we use 45 years). According to the prophets and to **Christ, Himself,** all of the prophecies up to and including the second coming of the **Lord Jesus Christ** will have been fulfilled in one Biblical generation from the re-establishment of the kingdom of Israel. This means that 'the great tribulation' or 'the time of Jacob's trouble' should occur in a seven year period between 1967 and 2007 (or 2012).

The great tribulation - that time of many horrors and destructions on the surface of the Earth - does not have to start exactly seven years before the end of this generation. It could start from today forward. The prophecy would still be valid, as it says all these things including the second coming of the **Lord Jesus Christ** will have come to pass within one Biblical generation from the re-establishment of Israel. It is important to know how close this time is. As I pointed out in the previous chapter, the rapture of the Church will occur before this great tribulation.

This is not an attempt to set an exact date for the second coming of the **Lord**. It is merely an attempt to take advantage of the preview that the **Lord, Himself,** gave us to encourage us during these last days. It was given to both those of us who will be in the rapture and to those who would not become followers of the **Elohim** until after the rapture (i.e. during the tribulation). Both groups need hope to endure... for without hope there is no life.

The cry for a <u>great new world order</u> goes up among the peoples of the Earth. Everywhere men cry, "Peace, and safety!" The very elements of this planet and of the Sun, itself, groan in travail as the moment of judgement approaches. Soon, the world will be in the grip of a global dictator unequalled in human history; and wars will increase in frequency and destruction. Soon, the Earth will experience increased earthquake activity; and its axis will began to wobble noticeably. And, soon, the Sun will undergo a catastrophic change in energy output levels.

The **Lord Jesus Christ**, alone, can give and has given the promise of peace, joy, and eternal security to all who would listen and follow His way.

Many 'religious' - though, not really Christian - sects say that we are approaching the beginning of a new age. *They maintain that God will be born into mankind again;* in fact, they maintain that God in human form now lives among us having been born after 1950.

These religious sects (all based on the tenets the 'Great White Brotherhood') *recognize* that a great distress is soon to come upon the Earth and *that many people will disappear from Earth.* I believe these groups are satanic in their allegiance and direction. They speak of the same sudden disappearance of a lot of people which the Bible speaks of when it speaks of the *'Rapture.'* Yet, these satanic sects *maintain those people caught up in this 'Rapture' will be the undesirables of the new age... i.e.* "Those persons not of the new age vibrations or consciousness." An incredibly great deception is definitely afoot in the affairs of man when both sides concede the same event yet attribute different causes to it.

For what you are about to read I must strongly recommend two very good reference books which discuss the Book of Revelation particularly stressing the period of the great tribulation.[14]

The Tribulation Begins

The next event on the prophetic schedule is the rapture of the church. Such an event is going to be hard to disguise to those who remain on the face of the Earth afterwards. Prophecy states that many, many millions of people will not be fooled by the new world order (the satanic force) when it delivers its grand deception to explain the disappearance of those people who were taken in the rapture... (i.e. it will only convince the foolish).

I have given a great deal of thought to the way the lie might be put forward to the people who enter into the tribulation period. Obviously, if millions of people suddenly disappear, it would be almost too easy to identify them as *'raptured Christians'*. If, however, the *'Rapture'* occurs simultaneously to some apparently natural disaster or some other global catastrophe, *the New World Order could then explain away the disappearance of all these people as a natural function of the destruction which hit the planet at that time.*

It makes no difference whether the disappearance of these people is associated with a natural disaster or the supernatural appearance of flying saucers and their attendant psychology and technology. Either mechanism can be used to blind those who dwell on the Earth (after the rapture) to the real truth of the matter. Only those people who will have enough discernment to realize those who disappeared were all the "born again" Christians will be able to resist the lie given out by the antichrist and his 'novus ordo seclorum.'

Now let us look at chart No. 1 (page 56) concerning the alignment of planets. On this chart we can see that a great period of turbulence in the solar system may start at approximately the middle of 1980 (however such a period will exist every 11.5 years thereafter until the solar sunspot cycles cease or change radically).

---

14 *Revelation Illustrated and Made Plain* by Tim Lahaye, Published by Zondervan Books, 1974. and *the Book of Revelation* by Clarence Larkin from the Larkin estate at: 2802 North Park Avenue Philadelphia, 32, Pennsylvania, U.S.A. These books are essential in your 'survival kit' for the future.

This period is approximately eight years long. According to Biblical prophecy the major portion of the geophysical distress to Earth will occur in the last three and a half years of the seven year period of the great tribulation. If we assume that the major geophysical disturbance to the Earth will occur in the 1982-83 period where the majority of the critical alignments occur, then we can see that three and a half year before that must be the start of the seven year tribulation. If such is the case the great tribulation could start anytime from mid 1978 onward.

Again, if such is the case, then the rapture may occur simultaneously to a world-wide disaster which will appear to be of natural or human causes... (some scholars feel the Russian invasion of Israel could also be at this time; and that it could be this disaster will strike the Russian Army down). This is why it could be very important to get this book and the details that it contains to as many people as possible **before** the crunch comes (while there is yet time to prepare).

It is not hard to visualize what may be about to happen. The world has been experiencing its most erratic weather in over half a century. In addition to this, there is massive instability in the world economies; and the threat of nuclear war hangs in the balance over various Middle East tensions - all centred around the West Bank issue in Israel. Furthermore, there is the distinct possibility Russia will move down into the Middle East to capture Israel both to win the support of the surrounding Arab countries and to solve its own, internal commodity shortages.

As I stated earlier in Section 1, President Carter code-named his own special survival project "**Operation Noah's Ark.**" The Russians have a similar project. Both the United States and Russia have been stocking away commodities, technology, and key people getting ready for a great climatic disaster. They are preparing for a global doomsday that will break down civilization without destroying all the people. Underground cities and food storage areas are sprinkled all over the planet to give key members of our population a base from which they could rebuild possibly a better form of the industrialized civilization we have today. The tribulation might well be about to begin... **The rapture of the church and the great disaster may strike simultaneously; so, watch for the event.**

After the rapture and the disaster occur, a multitude of events will take place. In my opinion these events are best elaborated in the order in which they are listed in the Book of Revelation. The tribulation will be divided into two 21-month periods and one 42-month period.

The 1st Quarter

Following the brief chaos that strikes the Earth at the time of the rapture, order will be required to stabilize the human settlement. Most probably the format of the Club of Rome will be adopted in a worldwide state of emergency. A man, apparently a Roman of Jewish blood (which he hides), will be the "antichrist" who will head up this new world order. It will be a dictatorship far more horrible than any of those previously conceived and initiated by elements of mankind.

In an effort to establish order, a new world monetary system will be introduced using a format similar to that used by the Australian Bankcard, the American Mastercharge Card and the European Barclay Card. Instead of using real currency or cheques a system whereby **each person has a computer money number** will be initiated. (note: it will not use the gold standard as Europe is already gearing for an "energy-based" world economy.)

To have such a number and to use the new international exchange facility, the individual will be required to swear allegiance to the new world order. He must, in effect, become a citizen of the new world. It will not be hard to mistake this event when it occurs. In the beginning, people will be encouraged to join this new world money system, simply because no one will have money of any value to buy goods. The number will eventually be tattooed or laser-scanned on the right hand or on the forehead of the individual... (at some point it will become mandatory and no one whether rich or poor, small or great, bond or free will be able to buy or sell without this number or the symbol of the new world order)... (Rev: 13:16-17).

The reasoning behind the placement of the number in such places is quite simple. If one wishes to make point-of-sale "monetary" exchanges efficient when using a computer number to debit and credit the respective account numbers, the number must be visible where it is most frequently and most easily read.

Most people in the world are right-handed and keep their right hand free to tender money, hold car keys or tender a credit card at the point of sale. To put the number on the outside of the hand would be logical; because it would not be seen as easily on the inside if the hand was grasping something (like car keys).

By waving the hand under an 'infrared reader' or a 'bar-code reader' the number could quite easily be read by a cashier and debited into a computer account with great speed. With the exception of the forehead, no other place on the body is as readily accessible to the 'reader' as is the hand. Considering that some people do not even have a right or a left hand, a forehead placement seems the next most logical. For those people who dwell in very cold climates wearing gloves and wrapped up in furs, the eyes, the nose and part of the forehead are always visible. Most people have to navigate by visual means which usually keeps the eyes and a portion of the forehead exposed. For a 'cash number' to be efficient, it has to be readily accessible... not on the bottom of the foot or in the armpit...

No other part of the face or the neck line would suffice in the majority of cases; because men would have beards which would hide their face. Rather than change these cultural traditions or cosmetic appearances it is easily determined that the number should be placed on the forehead plainly visible by all. There would be no *living* being walking around in our society that did not have a head.

The antichrist will have to impose extreme law and order somewhat like a fascist dictatorship to stabilize the world community in the times of crisis that will occur just as the tribulation begins (Rev. 6:2). Such activity will irritate some groups and governments. Immediately resistance to his new world order (which will be quite restrictive in its practices) will rise up from the masses of the world.

War will break out (Rev. 6:3). Three of the ten nation states of the new world confederation will try to wrest power from the antichrist and his super world dictatorship. They will fail and the antichrist will prevail (Dan. 7:20).

In the wake of the wars between the antichrist and the dissident countries (or 'kings' as the Bible puts it) famines are sure to follow (Rev. 6:5-6). Historically, after massive wars, famine follows; because many of the able food producers have been killed off in the fighting. It is not hard to imagine that death will follow around the globe in great numbers. Due to war, famine and the plagues that will attend the events, twenty-five per cent of the population of the world at that time will die (Rev. 6:7-8).

That will be approximately six to seven hundred million people. During that first twenty-one months of the great tribulation the antichrist and his forces will greatly persecute the newly converted 'believers in **Elohim**', because they will refuse to receive the number or the symbol of his new world order (Rev. 6:9-11).

Simultaneously, the greatest evangelistic revival of all recorded history will occur during that time by 144,000 sealed servants of **GOD** who will be from the twelve tribes of Israel. They will bear a visible seal on their foreheads and will go around evangelizing to Jew and Gentile alike (Rev. 7:4-8). It is thought by many that the number of souls saved in this first twenty-one months of the tribulation will be greater than all of the souls saved before that time in the church age (Rev. 7:9-12). These converts will be protected by the **'Grace** of the **Elohim'** even amidst the terrible persecution.

The sixth major catastrophe group of this first quarter will be a great earthquake which will cause widespread damage and fear of the **Elohim's** wrath.

The resulting volcanic eruptions along the surface faultlines may cause the atmosphere to become clouded giving the Sun the appearance of being blackened to virtual extinguishment. The Moon will be the color of blood from the atmospheric pollutants of the erupting magma vents.

Meteorites will fall in great abundance. In fact, part of the clouding of the Sun could be caused by a gigantic dust cloud passing between the Earth and the Sun. The meteorites may be part of that cloud or part of wandering asteroid groups which may be moved into a collision course with the Earth.

Every mountain and island will be shaken and moved from its place from the intensity of the earthquake. The atmosphere will be splashed about as external forces from the cloud collision and the solar 'wind' bombard the Earth. The lens effect of the normally stable atmosphere will be disrupted by the collision. It will appear as though the sky had suddenly been 'rolled away like a scroll'... (Rev. 6:12-16); and the stars will skip through the sky.

The 2nd Quarter

After the six great judgements of the first twenty-one months, a great and dramatic pause preceding the horrible things which are yet to come occurs in the heavens. This event is not witnessed by the dwellers on Earth but solely by those who at that time reside in the heavens with the **Elohim** and **Mashiach.** As the second quarter of the tribulation begins, brilliant flashes of light and yet another great earthquake notify the Earth that the second quarter is about to start (Rev. 8:1-5). Then, hail and fire mixed with blood are cast upon the Earth and one-third of the trees and the grass are burned-up (Rev. 8:7). Could this be caused by the increased solar wind or by a meteoric shower?... or perhaps by super lightning and thunder storms?... Certainly, as the solar energy output drops there will snap freezes until the outer shell of the changing Sun is thrown off.

Then a *"burning mountain"* is cast into the sea and *"one-third of the sea becomes blood and one-third of the sea life dies and one-third of the ships on the sea are destroyed with all life on them"* (Rev. 8:8-9). A *"burning mountain"* could be a comet with a big burning wake. Then a *"great star burning like a torch"* will fall to the Earth. Its arrival pollutes one-third of the rivers and the *"fountains of waters,"* (lakes and springs?) - turning them bitter... (Rev. 8:10-11) (the bitterness may be due to nickel or other metal salts formed by a hot metal reaction to the water).

Next, another unusual thing happens - one which may be invisible to the Earth dwellers. An angel comes down from the heavens with a key to the "bottomless pit" (the inner hollow sphere of Earth?) to unleash what is also called "hell on Earth" by many Biblical scholars. Smoke comes up out of the pit like from a great furnace and the Sun is darkened and the air is filled by this smoke. A great fissure or hole could appear on the Earth's surface leading to the interior of this ball-shaped Earth... after all where is the 'bottom' of the inside of a ball?... A supernatural locust horde comes out of the smoke to sting men and torment them for a period of five months. In fact, some Biblical scholars believe these "locusts" are supernatural and possibly invisible. They will sting those people who are not believers in the **Elohim** during this time so much that the people who are stung *"will seek death and will not be able to find it"* (Rev. 9:1-11).

The next horrible thing that happens is that the river Euphrates is dried up. The Biblical scripture says that four angels (presumably fallen angels) will be released during this time from the place where they were bound in the great river Euphrates; and, it will be their function to slay one-third of mankind existing at that time. The mechanism by which they will do this is in the form of a two-hundred-million man army on "horseback" which will come with fire and brimstone proceeding from the mouths of the horses and their tails wounding the victims to death (Rev. 9:13-19). This situation sounds a bit bizarre and hard to imagine in current technology. It could be these two-hundred-million horse-men are either the armies of the 'kings of the East '(China, Japan, South and South East Asia) or a mystical army of demons on devices and processes that mankind has yet to see in action.

During this period, as always, the "believers" will be protected because they will have the seal of the **Elohim** in their forehead. By this time about fifty per cent of the unregenerated (unsaved, unrepentant) population of the Earth will have been killed. This will terminate the first half of the tribulation; and, this will only be the beginning of the trauma and the pain for mankind dwelling upon the Earth at that time.

The 2nd Half

During the second half or the last 42 months of the great tribulation, the worst plagues of all will strike mankind. At the beginning of this period Satan and his troops and one-third of the angelic hosts will be cast down to Earth. Satan will realize at this point he has been beaten, and his case in the heavens is lost (Rev. 12:7-17). As a result, he will seek revenge on Israel, the Jew and those who sympathise with them. He will try to destroy them and the nations of Earth in the greatest anti-Semitic purge the world has ever known. For the first half of the tribulation the antichrist who will have been the head of the new world order will not have been indwelt by the spirit of Satan - only influenced by him. It is agreed by many Biblical scholars that at this point the antichrist will receive a head wound by the sword or some instrument of war and will appear to die. Satan will then enter the dead body in a mock duplication of the resurrection of **CHRIST**. This will give the world "a great miracle" (a deception in effect) in that the antichrist who will 'appear' to be dead will be raised from the dead by Satan (Rev. 13:3-4). The antichrist will also turn on the 'holy' Roman 'church' of the period... utterly destroying it. He will replace it with worship of himself.

The antichrist will be easy to spot once things start to happen on a global basis. He will come from the darkness of *the great abyss;* he "will come in his own name" which probably means he will "blow his own horn" as well as that his name may actually mean 'challenger' or 'adversary' in Hebrew; he will exalt himself above all else and all others and eventually over the **Elohim;** everything that he will become, will be a lie; one promise after another will be broken including the peace treaty with Israel at the middle of the tribulation; his law will change as his whims do; he will come to destroy; and he will do just that.

The antichrist after being possessed by Satan will be able to work incredible signs and wonders through super technology... ring a bell? Many people will be deceived - even the 'very elect' of Israel if such were possible. New technology and parapsychological "miracles" will abound (Rev. 13:12-15). Those who do not know better will believe these incredible signs and wonders as simply the trappings of a far older and superior race of beings who have made this man their spokesman on Earth.

However, inexplicably, boils and sores will suddenly break out upon all those who have "the mark of the new world order" or the *"number of it"* on their right hand or in their forehead (Rev. 16:2). These foul sores will be somewhat like those of the Egyptian plagues in the Exodus. Here as in all the other plagues of the tribulation the new 'believers' will be protected because they will not have received the mark or the number of the new world order. These boils and sores will probably begin to manifest in the first to the third month of these last forty-two months of the tribulation perhaps as reactions to the injected dyes or laser imprints used to number the people.

Following the plague of sores, the entire sea will be turned to blood like that of a dead man (Rev. 16:3). Remember the plagues of Egypt during the Exodus?... Every living soul in the sea will die from corrupted water turned to dead blood; and I am reasonably certain it could be connected with the *"bitter waters"* and subsequent biochemical reactions. It must be considered that sea water is chemically and functionally similar to blood without the hemoglobin. Adding hot iron and nickel (as the majority of meteorites are made of) could produce a soup of dead marine life and saltwater with the texture and color of dead blood.

Not only will the seas be turned to blood and pollution, all the rivers and fountains of waters in the world will be turned to dead, polluted blood (Rev. 16:4). Very little fresh water will be available; so, many people will die of thirst. In a living organism, the bloodstream is both the food source and the sewer. The prophecy may symbolize the freshwater sources becoming so polluted they will be like stagnant, dead blood.

Suddenly, after all these things (as if they weren't enough) the Sun will become abnormally hot (Rev. 16:8-9). The phase-state changes may produce a much smaller, brighter star for a short time, hence, the light of the Moon would become *"as the light of a normal day"*... which is seven to ten times brighter than now!... The days will be worse! The Earth will heat up and mankind will be scorched and battered by drought and violent thermal storms. Men will curse the name of the **Elohim.** They will blame **GOD** or his 'nature' for all these plagues.

Following the intense heat, a darkness will fall upon the Earth as the Sun appears to go dark for a few weeks. The apparent darkness may be due solely to the fact that the Sun transitions through an ultraviolet or infrared spectral phase in its progression. This dark light may also cause blindness and skin cancer. People will "gnaw their tongues in pain" during the darkness... yet the unrepentant will not yield to the **Lord GOD.** They will stubbornly and needlessly endure the darkness as did Egypt during the Exodus (Rev. 16:10-11).

The scriptures say three 'frog-like spirits' of evil will go forth to deceive the nations as the river Euphrates dries up to let the 'Kings of the East' gather together with the other kings in the valley of Megiddo (Rev. 16:12-14) there to form the combatants of what will be called the '**Battle of Armageddon**' - the last battle on Earth during the tribulation.

After all of these things a voice will sound in the heavens saying, *"It is done."* Many voices, thunderings and lightning will be seen and heard on the Earth and the greatest earthquake in all time will strike the entire planet which is not too hard to imagine as all the major faultlines on Earth are connected by two 'rings' of fault zones... one of which is the 'Pacific Ring of Fire'. All of the great cities of the Earth will fall. Valleys will rise and mountains will be levelled; nothing will be left unchanged. Heavy stones, weighing 135 pounds on the average will fall everywhere in great amounts, ending the tribulation; this will be that time spoken of by the prophets when no human life would be spared if the judgements were to be allowed to continue. However, the **Lord Jesus Christ - the Mashiach** makes **His** return to save Israel and the remaining tribulation believers. In fact, some scholars say that a second 'rapture' may occur just before **Mashiach** 'cleanses' the Earth for **His** 'Kingdom Age' (which is also called the 'Millennium').

In the 24th chapter of Matthew, the **Lord GOD (Mashiach)** stated: *"Immediately after the tribulation of those days the Sun will be darkened, and the Moon will not give its light, and the stars will fall from heaven, and the power of the heavens will be shaken; then will appear the sign of the **Son of Man** and then all the tribes of the Earth will mourn, and they will see the **Son of Man** coming on the clouds in heaven with power and great glory; and **He** will send out **His** angels with a loud trumpet call and they will gather **His** elect from the four winds, from one end of heaven to the other."*

These, then, are the highlights of the horrors that will befall sinful and unrepentant man during the tribulation.The choice is up to the individual. We all have free will to choose which direction we will go in this life and the next. It is up to you. This is not a scare tactic as some people might say; if it frightens you, it is only because you realize that the Truth is in what has been said and it has made you more insecure. Do not be afraid. **The Lord GOD** has provided the escape mechanism, the salvation mechanism, - the ticket - for you.

After you have read about these events if the rapture has not yet occurred, there may still yet be time for you to ask the **Mashiach's** forgiveness to **initiate those 'seven steps' for your ticket immediately.** If the rapture has already occurred when you have finished this book, your road will be a bit more difficult than mine. You should then realize that keeping the **Word of GOD** at all cost is your most vital survival aid. **Avoid swearing allegiance to the new world order; do not accept its mark or its number; let your name be written in the 'Book of Life' which will give you entry into the eternal Kingdom of GOD.**

Remember always that the free gift of GOD is His grace through the Lord Jesus Christ - the Mashiach. It is your personal gift; all you have to do is ask, and it shall be given you... seek, and you shall find... knock, and the **Door** shall be opened to you...

-AMEN: SO LET IT BE-

CHAPTER 6: Until Sin Dies

The close of the great tribulation does not mean the end of man's "time under sin." Although the thousand year reign of **Christ** and **His** Church over the Earth will begin right after the tribulation, there will still be one type of sin present in those dwelling on Earth.

Satan will be chained-up in that "bottomless pit" for the thousand years of the **Kingdom.** Since Satan's sins were rebellion and pride, they will not manifest for the duration of the thousand year rest. The millennial **Kingdom of the Lord** will replace the world systems of governments that man has conceived with Satan's help over these last 6000 years. Since man's systems were of chaos, they were sin and this type of sin will not be present either. The only type of sin remaining will be the temptation to disobey the **Law of Lord** like naughty children - not like Satan's aggressive disobedience -but simply one of weakness to obey.

1000-Year 'Rest'

In Biblical prophecy we are told that a day of the **Lord's** time is as a thousand years of time on the Earth. By this reasoning, the great tribulation could end the 6000 years of man's time on the sixth day of the **Lord**. After that, the following one thousand years of Earth time, will be the seventh day of the **Lord**. From early scripture we can see that this is traditionally a day of rest for the **Lord**. However before the millennial rest can start the Earth must be cleansed. Old relics of the wars and destruction will still be lying around all over Earth. Biochemical by-products, radiation pollution, defoliated soil, and altered weather patterns will have to be re-combined back into harmonic unity on the planet before it is perfect for the millennial age. For this reason many of us think that a second "rapture' like some sort of protective isolation will be given to those *believers* who survive the great tribulation. The scriptures do say "a gathering of the elect from the four winds, from one end of heaven to the other" will occur. Have a look at II Peter 3:3-14, and Isaiah 65:17-20. These passages discuss the renovation of Earth by fire.

The millennial Kingdom will be a fantastically joyful place to live. The mortal men who dwell in that time will find life a great blessing as long as they follow the **Law of GOD** *of their own Free will.* It appears, however, that some of the mortal men living during the millennium will still break the **Law of GOD**. All the nations of the world will be required by the **Lord** to come to an annual feast and honor **Him** with gifts there in Jerusalem. Those who do not come for the annual honoring and the gift giving to the **Lord** will be punished by denial of rain. Rain will not fall upon the crops of their respective countries until they 'shape-up.' They will be victims of their own failings - not Satan's, and not the System's.

Those who are among the many believers during the millennium will enjoy long lives. For those born in the early part of the millennium ages between 900 and 1000 years will be common. For those people born during the millennium who will not follow the **Mashiach's** way, death will come by the age of 100. The world will have a perfect environment like it had in the Garden of Eden. There will be no taxes; there will be no heavy burdens laid upon it; there will only be the Law, the rule for harmony among all the created things of the Universe.

During this 1000 years of peace on Earth, Satan and his 'cast-out crew' will have been chained in the bottomless pit waiting for their last chance to make war with the **Lord GOD** and **His** Saints, on Earth. At the end of the 1000-year period Satan and his hordes will be released from their imprisonment in their abyss; and will gather support from the millennial nations for their cause. Once again two forms of sin will exist in the world; one will be the sin of disobedience and the other the sin of Satan's temptation to destroy the Law.

Satan will raise up armies from the millennial nations to make war on Jerusalem and **the Mashiach.** Satan's efforts will prove fruitless. As his armies surround Jerusalem wherein the **Lord GOD** and the Church reside, Satan's armies of human, demonic, and disembodied beings alike will all be utterly defeated. They will be thrown off of the planet and cast into the 'burning lake of fire.' At this point all sin will be destroyed forever. In Revelation 21:1 the old renovated Earth of the millennium is destroyed by fire and 'a great rushing wind.' A new Earth and a new Heaven will be formed afterwards... (remember 'heaven' here could mean the atmospheric or etheric heaven around Earth; but there are at least three 'heavens' mentioned in the Bible).

One heaven is in the atmosphere of Earth; another houses the stars in the skies; and the third is the dwelling place of the **Lord GOD Almighty.** Remember most Biblical scholars think the *'heaven'* which will 'cease to exist' is the *'atmospheric heaven'* where Satan and his hosts previously ruled. At this time the destruction of the atmospheric Heaven and the Earth will be the destruction of the 'mini-heaven' or the 'courtroom' which was used as Satan's mini-model of the real Universe to prove or disprove his challenge to the throne of **GOD**...

"Heaven and Earth shall pass away but My words shall not pass away" (Matthew 24:35). Yet a new Heaven (not Heavens) and a new Earth which shall have no sea shall come to be... (read Revelation 21:1).

Even though the old shall pass away a new heaven and a new Earth will be created. We don't know what size it will be, but it will be big enough to house a literal **Kingdom of GOD** 1500 miles on a side and in either a square or pyramid shape. When sin has been removed completely only the harmony of joy, beauty, and perfection will remain. There will be things to see, senses to sense, places to go, and concepts to conceive that at present are completely inaccessible by the mind of mortal man. The concept of sitting on clouds and playing harps is a far cry from what really waits for us in the Heavens. For approximately the last 1900 years (by Earth time) the **Lord GOD** has been in the Heavens preparing a place for us who will join Him. *He did say that in **His Father's** house there are many mansions...* (John 14:2)

The Heavens where spiritual and immortal man will have dwelt with **Christ - the Mashiach** during the great tribulation will *eventually* descend on the new Earth. *"And it shall be that the Holy City descends and abides on the new Earth forever..."* Become *"as a child"* before the **Lord** because...

-MAN'S TIME UNDER SIN WILL PASS...-

CHAPTER 7: Spread The Word

The Lord Jesus Christ told us before **He** left Earth there would be wars and rumors of wars, and that false prophets would spring up before the time that **He** would return for **His** Church. **He** said the final days of this age would be like the birth pangs of a woman.

As I reflect over the last 1900 years since **His** leaving I can see more clearly what **He** meant. When **He** left there was a period of intense evangelization in the real Church. This was early in the process of delivery. Then Rome entered into the Church and the real Church died... (or so it would appear). After a few hundred years had passed there was new light shed on the scriptures showing that the second coming (i. e. the deliverance of the Church) had not failed to come but was yet to come, and that only a lack of fully enlightened understanding of the prophecy had led the early Church fathers to think the second coming was to have been immediate. Then came other lulls and other revivals with the time between them growing shorter and shorter. Now the birth pangs quicken... In modern times the revivals have come and gone with only a few decades and finally only a few years between them. I think we are now on the verge of the last revival - the delivery is imminent. Those of us here in Australia at almost the uttermost reaches of the "civilised" world are probably the last major nation to be evangelized before the process is complete... before the rapture occurs... before the Church is delivered.

The time is so short that **we must dwell constantly upon spreading the Word that the Lord entrusted to us.** Frivolous pursuits must be curtailed. Do not waste time, food, or effort in these last days. Use the talents the **Lord** gave you, and your reward will be great in Heaven. Throw away the talents the **Lord** gave you and your reward will be minimal in the Heavens - if you do get there at all.

Read your Bible; study doctrine as it is written in the scriptures... and get with a Bible teacher. Encourage each other in these times with the news that the Rapture is imminent and our deliverance with it. Read your local newspapers; watch the world news (especially the Middle East events); and keep yourself posted on current events to see how close we draw to the time when we who have a ticket out of this place shall be leaving... going home... returning as sojourners so long absent.

Look around you and see how much like 'the days of Noah and Lot' the times and people around us are. The number of real "born again" Christians on Earth today would probably be less than three per cent of the population. It is the **Lord GOD's Will** that no soul should perish but that all should be saved. Even though all will not accept this, it is offered to all. Do not let an opportunity to witness to one of your brothers pass by. Give up some of your recreation time to witness and to study, -if the **Lord** calls it.

Do not be ashamed to use the name of the **Lord Jesus Christ** to witness in these times, for the **Holy Spirit** does the work for you. 'Witnessing' does not mean cramming the Bible doctrine down somebody's throat. 'Witnessing' means sharing information to those who ask… to those who knock at your door and say, *"can you help me?"* When they ask, you should answer in the name of the **Lord Jesus Christ** -the **Sinless One**- who has already paid for their 'ticket' to deliverance in **Him** -the '**New Ark**'. Their baptism should follow.

Remember, also, there is a special blessing on those who read the Book of Revelation; it is one of the key books of our age. The reading list at the end of this section is a survival kit for both those who will be raptured before the tribulation and those alive on Earth during the tribulation. There will be those friends and relatives close to you who are hard to the **Word of GOD** at this time. Do them a favor and buy them the library of books they will need. This is not a book-sales pitch; it is a plea to set up information in various places all over the world so that the powers of the antichrist cannot completely destroy the written **Word** which will be of great help to those dwelling on Earth during the great tribulation. Many of the 'born again' Christians in Australia have not only hidden such libraries, but have also hidden video and audio tape libraries on the **Word of GOD**. They maintain these libraries and tell their non-Christian friends where the libraries are; so that when the rapture removes these 'born-again' Christians, their friends who remain on Earth can salvage the libraries before the new world order has a 'book burning.'

In these last few hours before the 'enlightened' dupes set-up their dictatorial 'New World Order', one last set of instructions must be given to both Christian and non-Christian. The one single thing that the 'enlightened' **fear** the most is the ability of those they would enslave to remain independent of their 'New World Order.'

Remember the Illuminati plan to collapse the entire world economy and replace it with their own. This will be accomplished by several processes, some of which are:

a)   Collapsing transportation systems in all the industrialized countries of the world,

b)   Producing temporary artificial famines in major cities of industrialized and nonindustrialized nations alike by using strikes and weather control mechanisms,

c)   Breaking-down social order in the major cities by destroying communal utilities and law.

Their objective will be to force the highly-dependent city-dwellers to become totally dependent on the 'New World Order.' Each "born again" Christian and anyone else who wishes to oppose the Illuminati should become independent of the normal industrialized means of obtaining the biological necessities in the major civilization centres. This means that each person in that category should be able to survive in the wilderness both physically and spiritually. Do not waste time.

If you plan to endure at your present location stock the necessary provisions where you live (see my wife's book, *"Dare To Prepare"*). If you wish to prepare for being somewhat mobile while you endure, acquire warm clothing, good walking boots, means of procuring food from the land, a basic medical kit, and a Bible.

Lastly, become as physically fit and independent of medicines and addictive items such as cigarettes, coffee, alcohol and other 'social drugs' so that you are prepared for the certain collapse of the world systems as we know them today.

Remain calm; do not give up hope; and do not forget: The **Author** of all that is good once said, "Remember that **I**, **Jesus,** am the **Way,** the **Truth** and the **Life**…

IT IS FINISHED…

- MAY THE YAHWEH ELOHIM BLESS THIS TESTIMONY -

'Survival Kit' Reading List For Section III

1) *The New American Standard Bible* from the **Lord GOD**... This book is the most important of all the books you will have in your survival kit library. In addition, you should try to find a copy of the *Interlinear Greek/English New Testament* to go with it. A version is available from Samuel Baxter and Sons Ltd., 72 Maryleborne Lane, London WI.

2) *Revelation Illustrated And Made Plain* by Dr. Tim Lahaye, published by Zondervan Publishing House of the Zondervan Corporation, Grand Rapids, Michigan 49506, U. S. A. copyright 1973. Other than the Bible this book is the most important of any of the books that you can have in your survival kit. It is the most definitive work that I have seen on the Book of Revelation. Because of the immensity of the information in the Book of Revelation, and the incredible depth required to properly assimilate it, it is necessary that all of us have the work of a teacher on one hand as we try to read the book on the other hand. Get this book and keep it. I cannot recommend it highly enough.

3) *The Rapture Question* by John F. Walvoord (a comprehensive Biblical study of the translation of the church), published by Zondervan publishing house of the Zondervan Corporation, Grand Rapids, Michigan 49506, U. S. A. This book is a book for now before the rapture of the church. Admittedly, it will have value after the rapture as it will show people what has happened; but it should be read by those who are border-line Christians now. Get it. I recommend it highly.

4) *Armageddon, Oil and the Middle East Crisis* by John F. Walvoord, published by Zondervan publishing house of the Zondervan Corporation, Grand Rapids, Michigan 49506, U.S.A. This books is also essential to your library as it will have function in your kit before and after the rapture.

5) *The Book of Revelation* by Clarence Larkin, for sale by the Reverend Clarence Larkin Estate, 2802 North Park Avenue, Philadelphia, Pennsylvania, U.S.A.

6) *The Spirit World* by Clarence Larkin, for sale by the Reverend Clarence Larkin Estate, 2803 North Park Avenue, Philadelphia, Pennsylvania, U.S.A. This book gives an in-depth study of the nature, origin and distribution of the spirits, demons, and angelic hosts of **GOD**'s Kingdom. This book may be essential during the great tribulation.

7) *Satan is Alive and Well on Planet Earth* by Hal Lindsay with C. C. Carlson; published by Bantam Books under licence from the Zondervan Corporation 1974.

8) *The Late Great Planet Earth* by Hal Lindsay; published by Zondervan Corporation 1974.

9) *Seal of God* by F. C. Payne; copyright *1972* by E. R. Finck, 2 Howard Street, Windsor Gardens, South Australia 5087, Australia...

The *Seal of God* may be hard to get in other countries as it is printed locally in South Australia. It is a book dealing with the numerics of the Bible. It is an excellent book on Biblical numerology. If you can, get a copy of it; because it will help you to decipher part of the mystery that I have written into my own book and definitely the mystery that still resides in the Book of Revelation and the entire Bible itself. The **Word of GOD** is logical Truth.

10) *Revised Standard Version of the Holy Bible* from the **Lord GOD**...

He is the Alpha and the Omega...
He is the Beginning and the End...
He is the First and the Last...

APPENDIX No. 1: New York Herald Tribune Articles on Anti-Gravity

NEW YORK HERALD-TRIBUNE: Sunday, Nov. 20, 1955, pp. I & 36

"CONQUEST OF GRAVITY AIM OF TOP SCIENTISTS IN U. S."

(Photo above): "ANTI-GRAVITY RESEARCH - Dr. **Charles T. Dozier**, left, senior research engineer and guided missiles expert of the **Convair Division of General Dynamics Corp.**, conducting a research experiment toward control of gravity with **Martin Kaplan**, Convair Senior electronics engineer."

(Photo insert): "IN CHARGE - **George S. Trimble jr,** vice-president in charge of advanced design planning of **Martin Aircraft Corp.**, is organizing a new research institute for advanced study to push a program of theoretical research on gravitational effect"

"CHANGES FAR BEYOND THE ATOM ARE THE PRIZE"

(Revolution in Power, Air, Transit Is Seen)

This is the first of a series on new pure and applied research into the mysteries of gravity and efforts to devise ways to counteract it. Written by **Ansel E. Talbert, military and aviation editor, N.Y.H.T.**

The initial steps of an almost incredible program to solve the secret of gravity and universal gravitation are being taken today in many of America's top scientific laboratories and research centres. **A number of major, long-established companies in the United States aircraft and electronics industries also are involved in gravity research.** Scientists, in general, bracket gravity with life itself as the greatest unsolved mystery in the Universe. But there are increasing   numbers who feel that there must be a physical mechanism for its propagation which can be discovered and controlled.

Should this mystery be solved it would bring about a greater revolution in power, transportation and many other fields than even the discovery of atomic power. The influence of such a discovery would be of tremendous import in the field of aircraft design - where the problem of fighting gravity's effects has always been basic.

A FANTASTIC POSSIBILITY

One almost fantastic possibility is that if gravity can be understood scientifically and negated or neutralized in some relatively inexpensive manner, it will be possible to build aircraft, earth satellites, and even space ships that will move swiftly into outer space, without strain, beyond the pull of earth's gravity field. They would not have to wrench themselves away through the brute force of powerful rockets and through expenditure of expensive chemical fuels.

Centres where pure research on gravity now is in progress in some form include **the Institute for Advanced Study at Princeton, N.J.** and also at Princeton University: **the University of Indiana's School of Advanced Mathematical Studies** and **the Purdue University Research Foundation.**

A scientific group from the **Massachusetts Institute of Technology**, which encourages original research in pure and applied science, recently attended a seminar at the **Roger Babson Gravity Research Institute of New Boston, N.H.**, at which **Clarence Birdseye**, inventor and industrialist, also was present. Mr. Birdseye gave the world its first packaged quick-frozen foods and laid the foundation for today's frozen food Industry: more recently he has become interested in gravitational studies.

A proposal to establish at the **University of North Carolina at Chapel Hill, N.C.**, an 'Institute of Pure Physics' primarily to carry on theoretical research on gravity was approved earlier this month by the University's board of trustees. This had the approval of **Dr. Gordon Gray** who has since retired as president of the University. **Dr. Gray** has been **Secretary of the Army, Assistant Secretary of Defence, and special assistant to the President of the United States.**

FUNDS COLLECTED: Funds to make the institute possible were collected by **Agnew H. Bahnson jr.**, an industrialist of **Winston Salem, N.C.** The new **University of North Carolina** administration is now deciding on the institute's scope and personnel. The directorship has been offered to **Dr. Bryce S. Dewitt** of the **Radiation Laboratories** at the **University of California at Berkeley**, who is the author of a **Roger Babson** prize-winning scientific study entitled, '**New Directions for Research in the Theory of Gravity.**'

The same type of scientific disagreement which occurred in connection with the first proposals to build the hydrogen bomb and an artificial earth satellite -now under construction - is in progress over anti-gravity research. Many scientists of repute are sure that gravity can be overcome in comparatively few years if sufficient resources are put behind the project. Others believe it may take a quarter of a century or more.

REFUSE TO PREDICT: Some pure physicists, while backing the general program to try to discover how gravity is propagated, refuse to make predictions of any kind. Aircraft industry firms now participating or actively interested in gravity include **Glenn L. Martin Co. of Baltimore**, builders of the nation's first giant jet-powered flying boat; **Convair of San Diego**, designers and builders of the giant B-36 intercontinental bomber and the world's first successful vertical take-off fighter; **Bell Aircraft of Buffalo**, builders of the first piloted airplane to fly faster than sound and a current jet take-off and landing airplane, and **Sikorsky Division of United Aircraft**, pioneer helicopter builders.

Lear, Inc., of Santa Monica, one of the world's largest builders of automatic pilots for airplanes; **Clarke Electronics of Palm Springs, California**, a pioneer in its field, and the **Sperry Gyroscope Division of Sperry-Rand Corp., of Great Neck, L.I.**, which is doing important work on guided missiles and earth satellites, also have scientists investigating the gravity problem.

USE EUROPEAN EXPERT: **Martin Aircraft** has just put under contract one of Europe's leading theoretical authorities on gravity and electromagnetic fields - **Dr. Burkhard Heim of Goettingen University** where some of the outstanding discoveries of the century in aerodynamics and physics have been made, and **Dr. Pascual Jordan of Hamburg University**, Max Planck Medal winner whose recent work called 'Gravity and the Universe' has excited scientific circles throughout the world.

Dr. Heim, now **professor of theoretical physics at Goettingen**, and who was a member **of Germany's Bureau of Standards during World War II**, is certain that gravity can be overcome. Dr. Heim lost his eyesight and hearing, and had both arms blown off at the elbow in a World War II rocket explosion. He dictates his theories and mathematical calculations to his wife.

Martin Aircraft, at the suggestion of **George S. Trimble**, its vice-president in charge of advanced design planning, is building between **Washington** and **Baltimore** a new laboratory for the **Research Institute for Advanced Study... A theoretical investigation of the implications for future gravity research in the 'United Field Theory' of the late Dr. Albert Einstein is now underway here.**

Although financed by **Martin**, the Institute will have no connection with the day-to-day business of building airplanes. Its general manager is **Welcome Bender.**

Up to now no scientist or engineer - so far as is known in the scientific circles - has produced the slightest alteration in the magnitude or direction of gravitational 'force' although many cranks and crackpots have claimed to be able to do this with 'perpetual motion machines.'

NO ACCEPTED THEORY: There is no scientific knowledge or generally accepted theory about the speed with which it travels across interplanetary space, making any two material particles or bodies - if free to move - accelerate toward each other. But the current efforts to understand gravity and universal gravitation both at sub-atomic level and at the level of the Universe have the positive backing today of many of America's outstanding physicists.

These include **Dr. Edward Teller of the University of California**, who received prime credit for developing the hydrogen bomb; **Dr. J. Robert Oppenheimer**, director of the **Institute for Advanced Study** at Princeton; **Dr. Freeman J. Dyson, theoretical physicist** at the Institute, and **Dr. John A. Wheeler**, professor of physics at **Princeton University** who made important contributions to America's first nuclear fission project.

PURE RESEARCH VIEW: It must be stressed that scientists in this group approach the problem only from the standpoint of pure research. They refuse to predict exactly in what directions the search will lead or whether it will be successful beyond broadening human knowledge generally.

Other top-ranking scientific minds being brought to bear today on the gravity problem are those of **Dr. Vaclav Hlavaty, of the University of Indiana**, who served with **Dr. Einstein** on the faculty of **Charles University in Prague** and later taught advanced mathematics at the **Sorbonne in Paris**; and of **Dr. Stanley Deser and Dr. Richard Arnowitt** of the **Princeton Institute for Advanced Study.**

Dr. Hlavaty believes that gravity simply is one aspect of electro-magnetism - the basis of all cosmic forces - and eventually may be controlled like light and radio waves.

HOPE TO FIND KEY: **Dr. Deser and Dr. Arnowitt** are of the opinion that very recently discovered nuclear and sub-nuclear particles of high energy which are difficult to explain by any present-day theory, may prove to be the key that eventually unlocks the mystery. It is their suggestion that the new particles may prove to be basic gravitational energy which is being converted continually and automatically in an expanding Universe directly into the most useful nuclear and electromagnetic forms.' In a recent scientific paper they point out:

'One of the most hopeful aspects of the problem is that until recently gravitation could be observed but not experimented on in any controlled fashion, while now with the advent in the past two years of the new high-energy accelerators (the **Cosmotron** and the even more recent **Berkeley Bevatron**) the new particles which have been linked with the gravitational field can be examined and worked with at will.'

An important job of encouraging both pure and applied gravity research in the United States through annual prizes and seminars as well as, the summarizing of new research for engineers and scientists in industry looking forward to a real 'hardware solution' to the gravity problem is being performed by the **Gravity Research Foundation of New Boston, N.H.**

This was **founded** and **endorsed** by **Dr. Roger Babson**, economist, who is an alumnus of M.I.T. and a lifelong student of the works of Sir Isaac Newton, discoverer of gravity. Its president is **Dr. George Rideout of Boston.**

(Photo Inset): BLACKBOARD MATH - **Dr. Vaclav Hlavaty**, of the **University of Indiana's** graduate **Institute of Advanced Mathematics**, who has stimulated research on gravity control, working on a problem."

(Photo Inset): "ANTI-GRAVITY AND AVIATION - **George S. Trimble jr.** vice-president in charge of advanced design planning of **Martin Aircraft Corp.**, left discussing the application of **anti-gravitational research to aviation** with two Martin scientists, **J.D. Pierson**, centre, and **William B. Yates.**"

- END ARTICLE ONE -

NEW YORK HERALD-TRIBUNE: Monday, Nov. 21, 1955, pp. 1 & 6

"SPACE-SHIP MARVEL SEEN IF GRAVITY IS OUTWITTED"

(Photo insert): "FLYING SAUCER OF THE FUTURE? - A reproduction of an oil painting by Eugene M. Gluhareff, president of Gluhareff Helicopter & Airplane Corp. of Manhattan Beach, Calif., showing a 'saucer-shaped' aircraft or space ship for exploring far beyond the earth's atmosphere and gravity field. Mr. Gluhareff portrays it operating at 'moderate speed' over the New York - New England area and notes that in the painting a 'propulsive blast of the electron beams from the rear of the saucer is visible, giving the saucer a translational force,'" "SPEEDS OF THOUSANDS OF MILES AN HOUR WITHOUT A JOLT HELD LIKELY"

This is the second in a series on new pure and applied research into the mysteries of gravity and efforts to devise ways to counteract it. written by **Ansel E. Talbert, military and Aviation editor, N.Y.H.T.**

"Scientists today regard the earth as a giant magnet. Many in America's aircraft & electronics industries are excited over the possibility of using its magnetic and gravitational fields as a medium of support for amazing **'flying vehicles'** which will not depend on the air for lift.

Space ships capable of accelerating in a few seconds to speeds many thousand of miles an hour and making sudden changes of course at these speeds without subjecting their passengers to the so-called 'G-forces' caused by gravity's pull also are envisioned. These concepts are part of a new program to solve the secret of gravity and universal gravitation already in progress in many top scientific laboratories and long-established industrial firms of the nation.

NUCLEAR RESEARCH AIDS: Although scientists still know little about gravity and its exact relationship to electromagnetism, recent nuclear research and experiments with 'high energy machines' such as the **Brookhaven Cosmotron** are providing a flood of new evidence believed to have a bearing on this.

William P. Lear, inventor and chairman of the board of **Lear, Inc.**, one of the nation's largest electronics firms specializing in aviation, for months has been going over new developments and theories relating to gravity with his chief scientists and engineers.

Mr. Lear in 1950 received the **Collier Trophy** from the **President of the United States** 'for the greatest achievement in aviation in America' through developing a light-weight automatic pilot and approach control system for jet fighter planes. **He is convinced that it will he possible to create artificial 'electro-gravitational fields' whose huge polarity can be controlled to cancel out gravity.'**

He told this correspondent: 'All the (mass) materials and human beings within these fields will be part of them. They will be adjustable so as to increase or decrease the weight of any object in its surroundings. They won't be affected by the earth's gravity or that of any celestial body.

'This means that if any person was in an anti-gravitational airplane or space ship that carried along its own gravitational field - no matter how fast you accelerated or changed course - your body wouldn't any more feel it than it now feels the speed of the earth.'

Scientists and laymen for centuries have been familiar with the phenomena that 'like' poles of two magnets - the north and the north poles for example - repel each other while 'unlike' poles exert an attraction. In ancient times 'lodestones' possessing natural magnetism were thought to possess magical powers.

FARADAY'S DISCOVERIES: But the nineteenth century discoveries of England's great scientist, **Michael Faraday**, paved the way for construction of artificial 'electro-magnets' - in which magnetism is produced by means of electric currents. They retain it only so long as the current is flowing. An electromagnet can be made by winding around a soft iron 'core' - a coil of insulated wire carrying electric current. Its strength depends primarily on the number of turns in the coil rather than the strength of the current.

Even today, America's rapidly expanding electronics industry is constantly finding new uses for electromagnets. For example, **Jack Fletcher**, a young electronics and aeronautical engineer of Covina, Calif., has just built a 'Twenty-First Century Home' containing an electronic stove functioning by magnetic repulsion.

PAN FLOATS IN AIR: In it seven coils of wire on laminated iron cores are contained inside a plywood cabinet of blond mahogany. The magnetic field from these coils induces 'eddy currents' in an aluminum cooking pan nineteen inches in diameter, which interact and lift the pan into space like a miniature 'flying saucer.'

The cooking pan floats about two inches in the air above the stove in a stabilized condition; 'eddy currents' generate the heat that warms it while the stove top itself remains cold. The aluminum pan will hold additional pots and it can be used as a griddle. It is, of course, a variation of several other more familiar magnetic repulsion gadgets including the 'mysterious floating metal ball' of science hall exhibits.

No type of electromagnet known to science or industry would have any application to the building of a real aircraft or 'flying vehicle'. But one of America's most brilliant young experimental designers, **Eugene M. Gluhareff, president of Gluhareff Helicopter and Airplane Corp. of Manhattan Beach, Calif., has made several theoretical design studies of round or saucer-shaped 'vehicles' for travel into outer space, having atomic generators as their basic 'engines'.**

SON OF COPTER DESIGNER: **Mr. Gluhareff** is the son of Michael E. Gluhareff, chief designer for **Dr. Igor I. Sikorsky,** helicopter and multi-engined aircraft pioneer. **Dr. Sikorsky** and the elder **Mr. Gluhareff,** who has won the **Alexander Klemin award,** one of aviation's highest honours, are themselves deeply concerned in the problem of overcoming gravitation.

The younger **Mr. Gluhareff** already has been responsible for several successful advanced designs of less amazing "terrestrial" aircraft. He envisions the electric power obtained from the atomic generators operating electronic reactors -'that is, obtaining propulsion by the acceleration of electrons to a very high velocity and expelling them into space in the same manner that hot gases are expelled from jet engines.' Such an arrangement would not pollute the atmosphere with radioactive vapors.

COULD CONTROL ACCELERATION: Because of its 'long-lasting fuel', an atomic-electronic flying disk would be able to control its acceleration to any speed desired and there would be no need for being 'shot into space' according to **Mr. Gluhareff. Radial electronic beams around the saucer's rim would be operating constantly and would sustain flight by 'acting against gravity.'**

Mr. Gluhareff thinks that control can be achieved by a slight differentiation of the deflection of electronic beams in either direction: the beams would act in the same way as an orthodox plane's ailerons and elevator.

GRAVITATIONAL CHANGES: **Mr. Gluhareff** agrees with **Dr. Pascual Jordan of Hamburg University,** one of Europe's outstanding authorities on gravitation who proved many parts of the **'Quantum Theory' of Dr. Max Planck, that it will be possible to induce substantial changes in the gravitational fields of rotating masses through electromagnetic research.** Dr. Jordan has just signed a contract to do research for **Martin Aircraft Corp.** of Baltimore.

Norman V. Peterson guided missiles engineer of the **Sperry-Gyroscope Division of Sperry-Rand Corp.** of **Great Neck, L.I.,** who as president of the **American Astronautical Society** attended the recent 'earth satellite' meeting in Copenhagen corroborates the theory that 'nuclear powered - or solar powered - ion electron beam reactors - will give impetus to the conquest of space.'

(Photo insert): "FLOATING COOKING PAN - The 'electronic stove' functioning by magnetic repulsion built by **Jack Fletcher,** a young engineer of West Covina, Calif., The aluminum cooking pan, nineteen inches in diameter, floats two inches above the cabinet like a miniature 'flying saucer'. It is completely stable while 'hovering' and can be used as a griddle or as a holder for additional pots and pans. 'Eddy currents' from a magnetic field created by an electromagnet inside the cabinet have warmed the pan - although the stove top remains completely cold."

END ARTICLE TWO

NEW YORK HERALD-TRIBUNE: Tuesday, Nov. 22, 1955, pp. 6 & 10

"NEW AIR DREAM-PLANES FLYING OUTSIDE GRAVITY"

(Photo insert): **Lawrence D. Bell,** founder and president of **Bell Aircraft Corp., of Buffalo,** using a Japanese ivory ball to illustrate his view that humans before long will operate planes outside the earth's atmosphere, then outside the gravity field of the earth. The pilots with him, three top test pilots of the Air Force, are, left, **Lt. Col. Frank J. Everest;** centre, in light suit, **Maj. Charles Yeager,** and, in uniform next to **Mr. Bell, Maj. Arthur Murray."**

"ABLE TO GO WHERE WE WANT"

This is the third in a series of three articles on new pure and applied research into the mysteries of gravity and the efforts to devise ways to overcome it. Written by **Ansel E. Talbert, Military and Aviation Editor.** N.Y.H.T.

"The current interest in America's aircraft and electronics industries in finding whether gravity can be controlled or 'cancelled-out' is not confined to imaginative young graduates of engineering and scientific schools. Some of the two industries' most experienced and highly regarded leaders today are engaged directly or deeply interested in theoretical research relating to gravity and universal gravitation. Their basic aim is eventually to build 'hardware' in the shape of planes, earth satellites, and space ships 'which can go where we want and do what we want without interference from gravity's mysterious trans-spatial pull.'

BELL IS OPTIMISTIC: **Lawrence D. Bell, whose company in Buffalo** built the first piloted aircraft in history to fly faster than sound, is certain that practical results will come out of current gravity research. He told this correspondent: 'Aviation as we know it is on the threshold of amazing new concepts. **The United States aircraft industry already is working with nuclear fuels and equipment to cancel out gravity instead of fighting it.**

'**The Wright Brothers** proved that man does not have to be earth-bound. Our next step will be to prove that we can operate outside the earth's atmosphere and the third will be to operate outside the gravity of the earth.'

OPTIMISM SHARED: **Mr. Bell's** company during the last few days made the first powered flights with its new Bell X-2 rocket plane designed to penetrate deep into the thermal or heat barrier encountered due to atmospheric fiction at a speed above 2,000 miles per hour. It also is testing a revolutionary new jet vertical-rising-and-landing 'magic carpet' airplane.

Grover Loening, who was the first graduate in aeronautics in an American University and the first engineer hired by the Wright Brothers, holds similar views. Over a period of forty years, Mr. Loening has had a distinguished career as an aircraft designer and builder recently was decorated by the United States Air Force for his work as a special scientific consultant. 'I firmly believe that before long man will acquire the ability to build an electromagnetic contra-gravity mechanism that works,' he says. 'Much the same line of reasoning that enabled scientists to split up atomic structures also will enable them to learn the nature of gravitational attraction and ways to counter it.'

Right now there is considerable difference of opinion among those working to discover the secret of gravity and universal gravitation as to exactly how long the project will take. George S. Trimble, a brilliant young scientist who is head of the new advanced design division of Martin Aircraft in Baltimore and a member of the sub-committee on high-speed aerodynamics of the National Advisory Committee for Aeronautics, believes that it could be done relatively quickly if sufficient resources and momentum were put behind the program.

'I think we could do the job in about the time that it actually required to build the first atom bomb if enough trained scientific brainpower simultaneously began thinking about and working towards a solution,' he said. 'Actually, the biggest deterrent to scientific progress is a refusal of some people, including scientists, to believe that things which seem amazing can really happen.

'I know that if Washington decides that it is vital to our national survival to go where we want and do what we want without having to worry about gravity, we'd find the answer rapidly.'

SIKORSKY CAUTIOUS: Dr. Igor I. Sikorsky, one of the world's outstanding airplane and helicopter designers, is somewhat more conservative but equally interested. He believes that within twenty-five years man will be flying beyond the earth's atmosphere, but he calls gravity, 'real, tangible, and formidable.' It is his considered scientific observation that there must be some physical carrier for this immense trans-spatial force.

Dr. Sikorsky notes that light and electricity, once equally mysterious, now have become 'loyal, obedient servants of man. appearing or disappearing at his command and performing at his will a countless variety of services.' But in the case of gravitation he says the more scientists attempt to visualize the unknown agent which transmits it, 'the more we recognize we are facing a deep and real mystery.'

The situation calls for intensive scientific research, Dr. Sikorsky believes. Up to now all gravity research in the United States has been financed out of the private funds of individuals or corporations. Leaders of the nation's armed forces have been briefed by various scientists about the theoretical chances of conquering gravitation but so far their attitude is 'call us when you get some hardware that works.'

Dudley Clarke, president of Clarke Electronics laboratories of Palm Springs, Calif., one of the nation's oldest firms dedicated to electronic research and experimentation, is one scientist in the hardware stage of building something that he believes will prove gravity can be put to useful purposes.

Mr. Clarke's company has just caused a stir in the electronics industry by developing pressure-sensitive resistors having unusual characteristics for parachute and other aviation use, according to 'Teletech and Electronic Industries' magazine of 480 Lexington Ave. Mr. Clarke who years ago worked under Dr. Charles Steinmetz, General Electric Company's electrical and mathematical 'wizard' of the 1930s, is sure that this successful harnessing of gravitation will take place sooner than some of these 'ivy tower' scientists believe.

Like Sir Frank Whittle, Britain's jet pioneer who was informed in 1935 by the British Air Ministry that it could see no practical use for his jet aircraft engine, Mr. Clarke has a particularly cherished letter. It was written about the same time by the commanding general at Wright Field giving a similar analysis of a jet design proposal by Mr. Clarke.

Mr. Clarke notes that the force of gravity is powerful enough to generate many thousand times more electricity than now is generated at Niagra Falls and every other water-power centre in the world - if it can be harnessed. This impending event, he maintains, will make possible the manufacture of anti-gravity 'power packages' which can be bought for a few hundred dollars. These would provide all the heat and power needed by one family for an indefinite period.

VOLUME XI - No. 1, 1956   INTER✈AVIA

Artist's conception of a vertical rising, disc-shaped aircraft which could result from a project under development for the U.S. Air Force by Avro Ltd., Canada (Official U.S. Air Force photo).

Dr. W.R.G. Baker, vice-president and general manager of **General Electric Co.'s** electronics division, points out that scientists working in many fields actually are beginning to explore the universe, learning new things about the makeup of 'outer space' and formulating new concepts. He says:

'Today we in electronics are deeply interested in what lies beyond the earth's atmosphere and its gravity field. For there we may find the electronics world of what now. Such questions usually have been reserved for the realm of physics and astronomy. But through entirely new applications in radar for example science already is able to measure some of the properties of the world beyond. 'Warm bodies radiate microwaves, and by recording noise signals, we are learning about invisible celestial forces we did not even know existed.'

Dr. Arthur L. Klein, professor of aeronautics at the California Institute of Technology, is certain that 'if extra-terrestrial flight is to be achieved, something will be required to replace chemical fuels.'

Dr. Hermann Oberth, Germany's greatest rocket pioneer, who is now working on guided missiles for the **United States Army,** calculates that 40,000 tons of liquid propellents will be required to lift a payload of only two tons beyond the earth's gravitation. Regarding this chemical fuel problem **Dr. Klein** says, 'there are no other serious obstacles.'

Many thoughtful theoretical scientists and practical engineers see a space vehicle de-gravitized to a neutral weight and following an electronically-controlled route charted by radar as the ultimate answer."

APPENDIX No. 2: Towards Flight - without Stress or Strain... or Weight

BY INTEL, WASHINGTON, D.C.

The following article is by an American journalist who has long taken a keen interest in questions of theoretical physics and has been recommended to the Editors as having close connections with scientific circles in the United States. The subject is one of immediate interest and Interavia would welcome further comment from initiated sources. --Editors.

Washington D. C. — March 23, 1956: Electrogravitics research, seeking the source of gravity and its control, has reached a stage where profound implications for the entire human race begin to emerge. Perhaps the most startling and immediate implications of all involve aircraft, guided missiles -- atmospheric and free space flight of all kinds.

If only one of several lines of research achieve their goal -- and it now seems certain that this must occur -- gravitational acceleration as a structural, aerodynamic and medical problem will simply cease to exist. So will the task of providing combustible fuels in massive volume in order to escape the earth's gravitic pull -- now probably the biggest headache facing today's would-be "space men".

And towards the long-term progress of mankind and man's civilization, a whole new concept of electrophysics is being levered out into the light of human knowledge.

There are gravity research projects in every major country of the world. A few are over 30 years old . Most are much newer. Some are purely theoretical and seek the answer in Quantum, Relativity and Unified Field Theory mathematics -- Institute for Advanced Study at Princeton, New Jersey; University of Indiana's School of Advanced Mathematical Studies; Purdue University Research Foundation; Goettingen and Hamburg Universities in France, Italy, Japan and elsewhere. The list, in fact, runs into the hundreds.

Some projects are mostly empirical, studying gravitic isotopes, electrical phenomena and the statistics of mass. Others combine both approaches in the study of matter in its super-cooled, super-conductive state, of jet electron streams, peculiar magnetic effects or the electrical mechanics of the atom's shell. Some of the companies involved in this phase include Lear Inc., Gluhareff Helicopter and Airplane Corp., The Glenn L. Martin Co., Sperry-Rand Corp., Bell Aircraft, Clarke Electronics Laboratories, the U.S. General Electric Company.

The concept of weightlessness in conventional materials which are normally heavy, like steel, aluminium, barium, etc., is difficult enough, but some theories, so far borne out empirically in the laboratory, postulate that not only can they be made weightless, but they can in fact be given a negative weight. That is: the force of gravity will be repulsive to them and they will -- new sciences breed new words and meanings for old ones -- loft away contra-gravitationally.

In this particular line of research, the weights of some materials have already been cut as much as 30% by "energizing" them. Security prevents disclosure of what precisely is meant by "energizing" or in which country this work is under way.

A localized gravitic field used as a ponderamotive force has been created in the laboratory. Disc airfoils two feet in diameter and incorporating a variation of the simple two-plate electrical condenser charged with fifty kilovolts and a total continuous energy input of fifty watts have achieved a speed of seventeen feet per second in a circular air course twenty feet in diameter. More lately these discs have been increased in diameter to three feet and run in a fifty-foot diameter air course under a charge of a hundred and fifty kilovolts with results so impressive as to be highly classified. Variations of this work done under a vacuum have produced much greater efficiencies that can only described as startling. Work is now under way developing a flame-jet generator to supply power up to fifteen million volts.

Such a force raised exponentially to levels capable of pushing man-carrying vehicles through the air -- or outer space -- at ultrahigh speeds is now the object of concerted effort in several countries. Once achieved it will eliminate most of the structural difficulties now encountered in the construction of high-speed aircraft. Importantly the gravitic field that provides the basic propulsive force simultaneously reacts on all matter within that field's influence. The force is not a physical one acting initially at a specific point in the vehicle that needs then to be translated to all the other parts. It is an electrogravitic field acting on all parts simultaneously.

Subject only to the so-far immutable laws of momentum, the vehicle would be able to change direction, accelerate to thousands of miles per hour, or stop. Changes in direction and speed of flight would be effected by merely altering the intensity, polarity and direction of the charge.

Man now uses the sledge-hammer approach to high-altitude high-speed flight. In the still-short life-span of the turbo-jet airplane he has had to increase power in the form of brute thrust some twenty times in order to achieve just a little more than twice the speed of the original jet plane. The cost in money in reaching this point has been prodigious. The cost in highly-specialized man-hours is even greater. By his present methods man actually fights in direct combat the forces that resist his efforts. in conquering gravity he would be putting one of his most competent adversaries to work for him. Antigravitics is the method of the picklock rather than the sledge-hammer.

The communications possibilities of electrogravitics, as the new science is called, confound the imagination. There are apparently in the ether an entirely new unsuspected family of electrical waves similar to electromagnetic radio waves in basic concept. Electrogravitic waves have been created and transmitted through concentric layers of the most efficient kinds of electromagnetic and electrostatic shielding without any apparent loss of power in any way. There is evidence, but not yet proof, that these waves are not limited by the speed of light. Thus the new science seems to strike at the very foundations of Einsteinian Relativity Theory.

But rather than invalidating current basic concepts such as Relativity, the new knowledge of gravity will probably expand their scope, ramification and general usefulness. It is this expansion of knowledge into the unknown that more emphasizes how little we do know; how vast is the area still awaiting research and discovery.

The most successful line of the electrogravitics research so far reported is that carried on by Townsend. T. Brown, an American who has been researching gravity for over thirty years. He is now conducting research projects in the U.S. and on the Continent. He postulates that there is between electricity and gravity a relationship parallel and/or similar to that which exists between electricity and magnetism. And as the coil is the usable link in the case of electromagnetics, so is the condenser that link in the case of electrogravitics. Years of successful empirical work have lent a great deal of credence to this hypothesis.

The detailed implications of man's conquest of gravity are innumerable. In road cars, trains and boats the headaches of transmission of power from the engine to wheels or propellers would simply cease to exist. Construction of bridges and big buildings would be greatly simplified by temporary induced weightlessness, etc. Other facets of work now under way indicate the possibility of close controls over the growth of plant life; new therapeutic techniques, permanent fuel-less heating units for homes and industrial establishments; new manufacturing techniques; a whole new field of chemistry. The list is endless... and growing.

In the field of international affairs, other than electrogravitics' military significance, what development of the science may do to raw materials values is perhaps most interesting to contemplate. Some materials are more prone to induced weightlessness than others. These are becoming known as gravitic isotopes: Some are already quite hard to find, but others are common and, for the moment, cheap. Since these ultimately may be the vital lofting materials required in the creation of contra-gravitational fields, their value might become extremely high with equivalent rearrangement of the wealth of natural resources, balance of economic power and world geo-strategic concepts.

How soon all this comes about is directly proportional to the effort that is put into it. Surprisingly, those countries normally expected to be leaders in such an advanced field are so far only fooling around. Great Britain, with her Ministry of Supply and the National Physical Laboratory, apparently has never seriously considered that the attempt to overcome and control gravity was worth practical effort and is now scurrying around trying to find out what its all about. The U.S. Department of Defense has consistently considered gravity in the realm of basic theory and has so far only put token amounts of money into research on it. The French, apparently a little more open-minded about such things, have initiated a number of projects, but even these are still on pretty much of a small scale. The same is true throughout most of the world. Most of the work is of a private venture kind, and much is being done in the studies of university professors and in the traditional lofts and basements of badly undercapitalized scientists.

But the word's afoot now. And both Government and private interest is growing and gathering momentum with logarithmic acceleration. The day may not be far off when man again confounds himself with his genius; then wonders why it took him so long to recognize the obvious.

Of course, there is always a possibility that the unexplained 3% of UFOs, "Unidentified Flying Objects", as the U.S. Air Force calls "flying saucers", are in fact vehicles so propelled, developed already and undergoing proving flights - by whom... U.S., Britain... or Russia? However, if this is so, it's the best kept secret since the Manhattan Project, for this reporter has spent over two years trying to chase down work on gravitics and has drawn from Government scientists and military experts the world over only the most blank of stares.

This always the way of exploration into the unknown...

INTERAVIA   Volume XI - No. 5, 1956, pages 373-374

The American scientist Townsend T. Brown has been working on the problems of electrogravitics for more than thirty years. He is seen here demonstrating one of his laboratory instruments, a disc-shaped variant of the two-plate condenser.

Townsend Brown's free-flying condenser. If the two arc-shaped electrodes (on the left and right rims) are placed under electrostatic charge, the disc will move, under the influence of interaction between electrical and gravitational fields, in the direction of the positive electrode. the higher the charge, the more marked will be the electrogravitic field. With a charge of several hundred kilovolts the condenser would reach speeds of several hundred miles per hour.

Author's diagram illustrating the electrogravitic field and the resulting force on a disc-shaped electrostatic condenser. The centre of the disc is of solid aluminium. The solid rimming on the sides is perspex, and in the trailing and leading edges (seen in the direction of motion) are wires separated from the aluminium core chiefly by air pockets. The wires act in a manner similar to the two plates of a simple electrical condenser and, when charged, produce a propulsive force. On reaching full charge, a condenser normally loses its propulsive force; but in this configuration the air between the wires is also charged; so that in principle the charging process can be maintained as long as desired. As the disc also moves - from minus to plus - the charged air is left behind, and the condenser moves into new, uncharged air. Thus, both charging process and propulsive force are continuous.

Author's sketch of a supersonic space ship roughly 50 ft. in diameter, whose lift and propulsion are produced by electrogravitic forces. The vehicle is supported by a "lofting cake" L consisting of "gravitic isotopes" of negative weight, and is moved in the horizontal plane by propulsion elements T_1 and T_2.

Towards Flight
without Stress or Strain... or Weight

BY INTEL, WASHINGTON, D.C.

The following article is by an American journalist who has long taken a keen interest in questions of theoretical physics and has been recommended to the Editors as having close connections with scientific circles in the United States. The subject is one of immediate interest, and Interavia *would welcome further comment from initiated sources.* — Editors.

Washington, D.C. — March 23, 1956 : Electro-gravitics research, seeking the source of gravity and its control, has reached a stage where profound implications for the entire human race begin to emerge. Perhaps the most startling and immediate implications of all involve aircraft, guided missiles — atmospheric and free space flight of all kinds.

If only one of several lines of research achieve their goal — and it now seems certain that this must occur — gravitational acceleration as a structural, aerodynamic and medical problem will simply cease to exist. So will the task of providing combustible fuels in massive volume in order to escape the earth's gravitic pull — now probably the biggest headache facing today's would-be " space men ".

And towards the long-term progress of mankind and man's civilization, a whole new concept of electro-physics is being levered out into the light of human knowledge.

There are gravity research projects in every major country of the world. A few are over 30 years old.[1] Most are much newer. Some are purely theoretical and seek the answer in Quantum, Relativity and Unified Field Theory mathematics — Institute for Advanced Study at Princeton, New Jersey ; University of Indiana's School of Advanced Mathematical Studies ; Purdue University Research Foundation ; Goettingen and Hamburg Universities in Germany ; as well as firms and Universities in France, Italy, Japan and elsewhere. The list, in fact, runs into the hundreds.

Some projects are mostly empirical, studying gravitic isotopes, electrical phenomena and the statistics of mass. Others combine both approaches in the study of matter in its super-cooled, super-conductive state, of jet electron streams, peculiar magnetic effects or the electrical mechanics of the atom's

shell. Some of the companies involved in this phase include Lear Inc., Gluhareff Helicopter and Airplane Corp., The Glenn L. Martin Co., Sperry-Rand Corp., Bell Aircraft, Clarke Electronics Laboratories, the U.S. General Electric Company.

The concept of weightlessness in conventional materials which are normally heavy, like steel, aluminium, barium, etc., is difficult enough, but some theories, so far borne out empirically in the laboratory, postulate that not only can they be made weightless, but they can in fact be given a negative weight. That is : the force of gravity will be repulsive to them and they will—new sciences breed new words and new meanings for old ones— loft away contra-gravitationally.

In this particular line of research, the weights of some materials have already been cut as much as 30 % by " energizing " them. Security prevents disclosure of what precisely is meant by " energizing " or in which country this work is under way.

A localized gravitic field used as a ponderamotive force has been created in the

The American scientist Townsend T. Brown has been working on the problems of electrogravitics for more than thirty years. He is seen here demonstrating one of his laboratory instruments, a disc-shaped variant of the two-plate condenser.

laboratory. Disc airfoils two feet in diameter and incorporating a variation of the simple two-plate electrical condenser charged with fifty kilovolts and a total continuous energy input of fifty watts have achieved a speed of seventeen feet per second in a circular air course twenty feet in diameter. More lately these discs have been increased in diameter to three feet and run in a fifty-foot diameter air course under a charge of a hundred and fifty kilovolts with results so impressive as to be highly classified. Variations of this work done under a vacuum have produced much greater efficiencies that can only be described as startling. Work is now under way developing a flame-jet generator to supply power up to fifteen million volts.

Such a force raised exponentially to levels capable of pushing man-carrying vehicles through the air—or outer space—at ultra-high speeds is now the object of concerted effort in several countries. Once achieved it will eliminate most of the structural difficulties now encountered in the construction of high-speed aircraft. Importantly, the gravitic field that provides the basic propulsive force simultaneously reacts on all matter within that field's influence. The force is not a physical one acting initially at a specific point in the vehicle that needs then to be translated to all the other parts. It is an electro-gravitic field acting on all parts simultaneously.

Subject only to the so-far immutable laws of momentum, the vehicle would be able to change direction, accelerate to thousands of miles per hour, or stop. Changes in direction and speed of flight would be effected by merely altering the intensity, polarity and direction of the charge.

Man now uses the sledge-hammer approach to high-altitude, high-speed flight. In the still-short life-span of the turbo-jet airplane, he has had to increase power in the form of brute thrust some twenty times in order to achieve just a little more than twice the speed of the original jet plane. The cost in money

[1] Ultimately they go back to Einstein's general theory of relativity (1916), in which the law of gravitation was first mathematically formulated as a field theory (in contrast to Newton's " action-at-a-distance " concept).   — Ed.

Townsend Brown's free-flying condenser. If the two arc-shaped electrodes (on the left and right rims) are placed under electrostatic charge, the disc will move, under the influence of interaction between electrical and gravitational fields, in the direction of the positive electrode. The higher the charge, the more marked will be the electrogravitic field. With a charge of several hundred kilovolts the condenser would reach speeds of several hundred miles per hour.

in reaching this point has been prodigious. The cost in highly-specialized man-hours is even greater. By his present methods man actually fights in direct combat the forces that resist his efforts. In conquering gravity he would be putting one of his most competent adversaries to work for him. Anti-gravitics is the method of the picklock rather than the sledge-hammer.

The communications possibilities of electro-gravitics, as the new science is called, confound the imagination. There are apparently in the ether an entirely new unsuspected family of electrical waves similar to electro-magnetic radio waves in basic concept. Electro-gravitic waves have been created and transmitted through concentric layers of the most efficient kinds of electro-magnetic and electro-static shielding without apparent loss of power in any way. There is evidence, but not yet proof, that these waves are not limited by the speed of light. Thus the new science seems to strike at the very foundations of Einsteinian Relativity Theory.

But rather than invalidating current basic concepts such as Relativity, the new knowledge of gravity will probably expand their scope, ramifications and general usefulness. It is this expansion of knowledge into the unknown that more and more emphasizes

how little we do know ; how vast is the area still awaiting research and discovery.

The most successful line of the electro-gravitics research so far reported is that carried on by Townsend T. Brown, an American who has been researching gravity for over thirty years. He is now conducting research projects in the U.S. and on the Continent. He postulates that there is between electricity and gravity a relationship parallel and/or similar to that which exists between electricity and magnetism. And as the coil is the usable link in the case of electro-magnetics, so is the condenser that link in the case of electro-gravitics. Years of successful empirical work have lent a great deal of credence to this hypothesis.

The detailed implications of man's conquest of gravity are innumerable. In road cars, trains and boats the headaches of transmission of power from the engine to wheels or propellers would simply cease to exist. Construction of bridges and big buildings would be greatly simplified by temporary induced weightlessness etc. Other facets of work now under way indicate the possibility of close controls over the growth of plant life ; new therapeutic techniques ; permanent fuel-less heating units for homes and industrial establishments ; new sources of industrial power ; new manufacturing techniques ; a whole new field of chemistry. The list is endless . . . and growing.

In the field of international affairs, other than electro-gravitics' military significance, what development of the science may do to raw materials values is perhaps most interesting to contemplate. Some materials are more prone to induced weightlessness than others. These are becoming known as *gravitic isotopes*. Some are already quite hard to find, but others are common and, for the moment, cheap. Since these ultimately may be the vital lofting materials required in the creation of contra-gravitational fields, their value might become extremely high with equivalent rearrangement of the wealth of national natural resources, balance of economic power and world geo-strategic concepts.

Author's sketch of a supersonic space ship roughly 50 ft. in diameter, whose lift and propulsion are produced by electrogravitic forces. The vehicle is supported by a " lofting cake " L consisting of " gravitic isotopes " of negative weight, and is moved in the horizontal plane by propulsion elements T, and T₂.

How soon all this comes about is directly proportional to the amount of effort that is put into it. Surprisingly, those countries normally expected to be leaders in such an advanced field are so far only fooling around. Great Britain, with her Ministry of Supply and the National Physical Laboratory, apparently has never seriously considered that the attempt to overcome and control gravity was worth practical effort and is now scurrying around trying to find out what it's all about. The U.S. Department of Defense has consistently considered gravity in the realm of basic theory and has so far only put token amounts of money into research on it. The French, apparently a little more open-minded about such things, have initiated a number of projects, but even these are still on pretty much of a small scale. The same is true throughout most of the world. Most of the work is of a private venture kind, and much is being done in the studies of university professors and in the traditional lofts and basements of badly undercapitalized scientists.

But the word's afoot now. And both Government and private interest is growing and gathering momentum with logarithmic acceleration. The day may not be far off when man again confounds himself with his genius ; then wonders why it took him so long to recognize the obvious.

Of course, there is always a possibility that the unexplained 3 % of UFO's, " Unidentified Flying Objects ", as the U.S. Air Force calls " flying saucers ", are in fact vehicles so propelled, developed already and undergoing proving flights –by whom . . . U.S., Britain . . . or Russia ? However, if this is so it's the best kept secret since the Manhattan project, for this reporter has spent over two years trying to chase down work on gravitics and has drawn from Government scientists and military experts the world over only the most blank of stares.

This is always the way of exploration into the unknown.

*

Author's diagram illustrating the electrogravitic field and the resulting propulsive force on a disc-shaped electrostatic condenser. The centre of the disc is of solid aluminium. The solid rimming on the sides is perspex, and in the trailing and leading edges (seen in the direction of motion) are wires separated from the aluminium core chiefly by air pockets. The wires act in a manner similar to the two plates of a simple electrical condenser and, when charged, produce a propulsive force. On reaching full charge a condenser normally loses its propulsive force, but in this configuration the air between the wires is also charged, so that in principle the charging process can be maintained as long as desired. As the disc also moves—from minus to plus—the charged air is left behind, and the condenser moves into new, uncharged air. Thus both charging process and propulsive force are continuous.

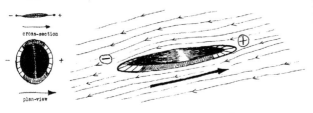

cross-section

plan-view

APPENDIX No. 3: The Gravitics Situation

The Gravitics Situation

Please note: the following paper has six appendices numbered (I - VI). They should not be confused with the appendices of *The Cosmic Conspiracy* which are numbered (No.1 - No.7).

December 1956
Gravity Rand Ltd.
66 Sloane Street
London S.W. 1

Theme of the science for 1956-1970:

SERENDIPITY

Einstein's view:-

"It may not be an unattainable hope that some day a clearer knowledge of the processes of gravitation may be reached; and the extreme generality and detachment of the relativity theory may be illuminated by the particular study of a precise mechanism".

CONTENTS

Thanks to the Gravity Research Foundation for Appendix II - VI

(page 3) I Engineering note on present frontiers of knowledge

Gravitics is likely to follow a number of separate lines of development: the best known short term proposition is Townsend Brown's electrostatic propulsion by gravitators (details of which are to be found in the Appendix I). An extreme extrapolation of Brown's later rigs appears to suggest a Mach 3 interceptor type aircraft. Brown called this basically force and motion, but it does not appear to be the road to a gravitational shield or reflector. His is the brute force approach of concentrating high electrostatic charges along the leading edge of the periphery of a disk which yields propulsive effect. Brown originally maintained that his gravitators operate independently of all frames of reference and it is motion in the absolute sense - relative to the universe as a whole. There is however no evidence to support this. In the absence of any such evidence, it is perhaps more convenient to think of Brown's disks as electrostatic propulsion which has its own niche in aviation. Electrostatic disks can provide lift without speed over a flat surface. This could be an important advance over all forms of airfoil which require induced flow; and lift without air flow is a development that deserves to be followed up in its own right, and one that for military purposes is already envisaged by the users as applicable to all three services. This point has been appreciated in the United States and a program in hand may now ensure that development of large sized disks will be continued. This is backed by the U.S. Government but it is something that will be pursued on a small scale. This acceptance follows Brown's original suggestion embodied in Project Winterhaven. Winterhaven recommended that a major effort be concentrated on electrogravitics based on the principle of his disks. The U.S. Government evaluated the disks wrongly, and misinterpreted the nature of the energy. This incorrect report was filed as an official assessment, and it took some three years to correct the earlier misconception. That brings developments up to the fairly recent past. and by that time it was realized that no effort on the lines of Winterhaven was practical, and that more modest aims should be substituted. These were re-written around a new report which is apparently based on newer thoughts and with some later patents not yet published which form the basis of current U.S. policy. It is a matter of some controversy whether this research could be accelerated by more money but the impression in Gravity Rand is that the base of industry is perhaps more than adequately wide. Already companies are specializing in evolution of particular components of an electrogravitics disk. This implies that the science is in the same state as the ICBM - namely that no new breakthroughs are needed, only intensive development engineering. This may be an optimistic reading of the situation: it is true that materials are now available for the condensers giving higher k figures than were postulated in Winterhaven as necessary, and all the ingredients necessary for the disks appear to be available. But industry is still some way from having an adequate power sources and possessing any practical experience of running such equipment.

The long term development of gravity shields, absorbers, and 'magic metals' appears at the moment however to be a basically different problem, and work on this is not being sponsored (officially, that is) so far as is known. The absorber or shield could be intrinsically a weapon of a great power, the limits of which are difficult to foresee. The power of the device to undermine the electrostatic forces holding the atom together is a destructive by-product of military significance. In unpublished work Gravity Rand has indicated the possible effect of such a device for demolition. The likelihood of such work being sponsored in small countries outside the U.S. is slight, since there is general lack of money and resources and in all such countries quick returns are essential.

Many people hold that little or no progress can be made until the link in the Einstein unified field theory has been found. This is surely a somewhat defeatist view, because although no all embracing explanation of the relationship between the extraordinary variety of high energy particles continually being uncovered is yet available much can be done to pin down the general nature of anti-gravity devices.

There are several promising approaches one of them is the search for negative mass, a second is to find a relationship between gravity and heat, and a third is to find the link between gravitation and the coupled particles. Taking the first of these: negative mass, the initial task is to prove the existence of negative mass, and Appendix II outlines how it might be done. This is Mozer's approach which is based on the Schroedinger time independent equation with the center of mass motion removed. As the paper shows, this requires some 100 bev - which is beyond the power of existing particle accelerators: however the present Russian and American nuclear programs envisage 50 bev bevatrons within a few years and at the present rate of progress in the nuclear sciences it seems possible that the existence of negamass will be proved by this method of a Bragg analysis of the crystal structure - or disproved.

If negamass is established, the precise part played by the subnuclear particles could be quickly determined. Working theories have been built up to explain how negative masses would be repelled by positive masses and pairs would accelerate gaining kinetic energy until they reach the speed of light and then assume the role of the high energy particles. It has been suggested by Ferrell that this might explain the role of neutrino, but this seems unlikely without some explanation of the spin ascribed to the neutrino. Yet the absence of rest mass or charge of the neutrino makes it especially intriguing. Certainly, further study of the neutrino would be relevant to gravitational problems. If, therefore, the aircraft industry regards anti-gravity as part of its responsibilities it cannot escape the necessity of monitoring high energy physics or the neutrino. There are two aircraft companies definitely doing this; but little or no evidence that most of the others know even what a neutrino is.The relationship between electrical charges and gravitational forces however will depend on the right deductions being drawn from excessively small anomalies (See Appendix VI). First clues to such small and hitherto unnoticed effects will come by study of the unified field theory. such effects may be observed in work on the gravithermals, and interacting effect of heat and gravity. Here, at least, there is firmer evidence materials are capable of temperature changes depending on gravity. This, as Beams says, (see Appendix III) is due to results from the alignment of the atoms. Gravity tensions applied across the ends of a tube filled with electrolyte can produce heat or be used to furnish power. The logical extension of this is an absorber of gravity in the form of a flat plate and the gravitative flux acting on it (its atomic and molecular structure, its weight density and form are not, at this stage, clear) would lead to an increase in heat of the mass of its surface and subsurface particles.

The third approach is to aim at discovering a connection between nuclear particles and the gravitational field. This also returns to the need for interpreting macroscopic relativistic phenomena at one extreme in terms of microscopic quantum mechanical phenomena at the other. Beaumont in suggesting a solution recalls how early theory established rough and ready assumptions of the characteristics of electron spin before the whole science of the atomic orbital was worked out. These were based on observation and they were used with some effect at a time when data was needed. Similar assumptions of complex spin might be used to link the microscopic to the macroscopic. At any rate, there are some loose ends in complex spin to be tied up, and these could logically he sponsored with some expectation of results by companies wondering how to make a contribution.

If a real spin or rotation is applied to a planar geoid the gravitational equipotentials can be made less convex, plane or concave. These have the effect of adjusting the intensity of the gravitational field at will which is a requirement for the gravity absorber. Beaumont seemed doubtful whether external power would have to be applied to achieve this. but it seems reasonable to suppose that power could be fed into the system to achieve a beneficial adjustment to the gravitational field, and conventional engineering methods could ensure that the weight of power input services would be more than offset by weightlessness from the spin inducer. The engineering details of this are naturally still in the realms of conjecture; but, at least, it is something that could be worked out with laboratory rigs; and, again, the starting point is to make more accurate observations of small effects. The technique would be to accept any anomalies in nature and from them to establish what would be needed to induce a spin artificially.

* * *

It has been argued that the scientific community faces a seemingly impossible task in attempting to alter gravity when the force is set up by a body as large as this planet and that to change it might demand a comparable force of similar planetary dimensions. It was scarcely surprising therefore that experience had shown that while it has been possible to observe the effects of gravity it resisted any form of control or manipulation. But the time is fast approaching when for the first time it will be within the capability of engineers with bevatrons to work directly with particles that it, is increasingly accepted, contribute to the source of gravitation; and whilst that in itself may not lead to an absorber of gravity, it will at least throw some light on the sources of the power.

Another task is solution (See Appendix IV) of outstanding equations to convert gravitational phenomena to nuclear energy. The problem, still not yet solved may support the Bondi-Hoyle theory that expansion of the universe represents energy continually annihilated instead of being carried to the boundaries of the universe. This energy loss manifests itself in the behaviour of the hyperon and K-particles which would, or might, form the link between the microcosm and macrocosm. Indeed Deser and Arnowitt propose that the new particles are a direct link between gravitationally produced energy and nuclear energy. If this were so it would be the place to begin in the search for practical methods of gravity manipulation. It would be realistic to assume that the K-particles are such a link. Then a possible approach might be to disregard objections which cannot be explained at this juncture until further unified field links are established. As in the case of the spin and orbital theories, which were naive in the beginning, the technique might have to accept the apparent forces and make theory fit observation until more is known.

* * *

Some people feel that the chances of finding such a unified field theory to link gravity and electrodynamics are high; yet think that the finding of a gravity shield is slight because of the size of the energy source, and because the chances of seeing unnoticed effects seem slender. Others feel the opposite and believe that a link between nuclear energy and gravitational energy may precede the link between the Einstein general relativistic and Quantum Theory disciplines. Some hope that both discoveries may come together; while a few believe that a partial explanation of both may come about the same time,, which will afford sufficient knowledge of gravitational fields to perfect an interim type of absorber using field links that are available.

This latter seems the more likely since it is already beginning to happen. There is not likely to be any sudden full explanation of the microcosm and macrocosm; but one strand after another joining them will be fashioned, as progress is made towards quantizing the Einstein theory.

(page 10) II Management note on the Gravitics Situation

The present anti-gravity situation as one of watching and waiting by the large aircraft prime contractors for lofting inventions or technological breakthroughs. Clarence Birdseye in one of his last utterances thought that an insulator might be discovered by accident by someone working on a quite different problem; and in 500 years gravity insulators would be commonplace. One might go further than Birdseye and say that principles of the insulator would, by then, be fundamental to human affairs; it would be as basic to the society as the difference today between the weight of one metal and another. But at the same time it would be wrong to infer from Birdseye's remark that a sudden isolated discovery will be the key to the science. The hardware will come at a time when the industry is ready and waiting for it. It will arrive after a long period of getting accustomed to thinking in terms of weightlessness, and naturally it will appear after the feasibility of achieving it in one form or another has been established in theory (But this does not mean that harnessed forces will be necessarily fully understood at the outset.).

The aim of companies at this stage must therefore surely be to monitor the areas of progress in the world of high energy physics which seem likely to lead to establishment of the foundations of anti-gravity. This means keeping a watchful eye on electrogravitics, magnetogravitics gravitics isotopes; and electrostatics in various forms for propulsion or levitation. This is not at the present stage a very expensive business, and investment in laboratory man-hours is necessary only when a certain line of reasoning which may look promising comes to a dead-end for lack of experimental data, or only when it might be worth running some laboratory tests to bridge a chasm between one part of a theory and another or in connecting two or more theories together. If this is right, anti-gravity is in a state similar to nuclear propulsion after the NEPA findings, yet before the ANP project got under way. It will be remembered that was the period when the Atomic Energy Commission sponsored odd things here and there that needed doing. But it would be misleading to imply that hardware progress on electrostatic disks is presently so far along as nuclear propulsion was in that state represented by ANP. True the NEPA men came to the conclusion that a nuclear-propelled aircraft of a kind could be built, but it would be only a curiosity. Even at the time of the Lexington and Whitman reports it was still some way from fruition: the aircraft would have been more than a curiosity but not competitive enough to be seriously considered.

It is not in doubt that work on anti-gravity is in the realm of the longer term future. One of the tests of virility of an industry is the extent to which it is so self confident of its position that it can afford to sponsor R&D which cannot promise a quick return. A closing of minds to anything except lines of development that will provide a quick return is a sign of either a strait-laced economy or of a pure lack of prescience, (or both).

Another consideration that will play its part in managerial decision is that major turning points in anti-gravity work are likely to prove far removed from the tools of the aircraft engineer. A key instrument for example that may determine the existence of negamass and establish posimass-negamass interaction is the super bevatron. It needs some 100 bev gammas on hydrogen to perform a Bragg analysis of the elementary particle structure by selective reflection to prove the existence of negamass. This value is double as much the new Russian bevatron under construction and it is 15 times as powerful as the highest particle accelerations in the Berkeley bevatron so far attained. Many people think that nothing much can be done until negamass has been observed.

If industry were to adopt this approach it would have a long wait and a quick answer at the end. But the negamass-posimass theory can be further developed; and, in anticipation of its existence, means of using it in a gravitationally neutralized body could be worked out. This, moreover, is certainly not the only possible approach: a breakthrough may well come in the interaction between gravitative action and heat theory at the moment suggests that if gravity could produce heat the effect is limited at the moment to a narrow range (See Appendix V). But the significant thing would be establishment of a principle.

History may repeat itself thirty years ago, and even as recently as the German attempts to produce nuclear energy in the war, nobody would have guessed that power would be unlocked by an accident at the high end of the atomic table. All prophecies of atomic energy were concerned with how quickly means of fusion could be applied at the low end. In anti-gravity work, and this goes back to Birdseye, it may be an unrelated accident that will be the means of getting into the gravitational age. It is a prime responsibility of management to be aware of possible ways of using theory to accelerate such a process. In other words serendipity.

It is a common thought in industry to look upon the nuclear experience as a precedent for gravity, and to argue that gravitics will similarly depend on the use of giant tools, beyond the capabilities of the air industry and that companies will edge into the gravitational age on the coat-tails of the Government as industry has done, or is doing, in nuclear physics. But this over looks the point that the two sciences are likely to be different in their investment. It will not need a place like Hanford or Savannah River to produce a gravity shield or insulator once the knowhow has been established. As a piece of conceptual engineering the project is probably likely to be much more like a repetition of the turbine engine. It will be simple in its essence, but the detailed componentry will become progressively more complex to interpret in the form of a stable flying platform and even more intricate when it comes to applying the underlying principles to a flexibility of operating altitude ranging from low present flight speeds at one extreme to flight in a vacuum at the other. This latter will be the extreme test of its powers. Again the principle itself will function equally in a vacuum - Townsend Brown's saucers could move in a vacuum readily enough - but the supporting parts must also work in a vacuum. In practice, they tend to give trouble, just as gas turbine bits and pieces start giving trouble in proportion to the altitude gained in flight.

But one has to see this rise in complexity with performance and with altitude attainment in perspective: eventually the most advanced capability may be attained with the most extremely simple configurations. As is usual however in physics developments the shortest line of progress is a geodesic, which may in turn lead the propulsion trade into many roundabout paths as being the shortest distance between aims and achievement.

But aviation business is understandably interested in knowing precisely how to recognize early discoveries of significance and this Gravity Rand report is intended to try and outline some of the more promising lines. One suggestion frequently made is that propulsion and levitation may be only the last - though most important - of a series of others, some of which will have varying degrees of gravitic element in their constitution. It may be that the first practical application will be in the greater freedom of communications offered by the change in wave technique that it implies. A second application is to use the wave technique for anti-submarine detection, either airborne or seaborne. This would combine the width of horizon in search radar with the underwater precision of Magnetic Airborne Detection, and indeed it may have the range of scatter transmissions. Chance discoveries in the development of this equipment may lead to the formulation of new laws which would define the relationship of gravity in terms of usable propulsion symbols. Exactly how this would happen nobody yet knows and what industry and government can do at this stage is to explore all the possible applications simultaneously, putting pressure where results seem to warrant it.

In a paper of this kind it is not easy to discuss the details of the wave technique in communications, and the following are some of theories, briefly stated which require no mathematical training to understand, which it would be worth management keeping an eye on. In particular, watch should be made of quantitative tests on lofting, and beneficiation of material. Even quite small beneficiation ratios are likely to be significant. There are some lofting claims being made of 20% and more, and the validity of these will have to be weighed carefully. Needless to say much higher ratios than this will have to be attained. New high-k techniques and extreme-k materials are significant. High speeds in electrostatic propulsion of small discs will be worth keeping track of (by high speed one means hundreds of m.p.h.) and some of these results are beginning to filter through for general evaluation. Weight mass anomalies, new oil-cooled cables, interesting megavolt gimmicks, novel forms of electrostatic augmentation with, hydrocarbon and non-hydrocarbon fuels are indicative, new patents under the broadest headings of force and motion may have value, new electrostatic generator inventions could tip the scales and unusual ways of turning condensers inside-out, new angular propulsion ideas for barycentric control; and generally certain types of saucer configuration are valuable pointers to ways minds are working.

Then there is the personnel reaction to such developments. Managements are in the hands of their technical men, and they should beware of technical teams who are dogmatic at this state. To assert electrogravitics is nonsense; is as unreal as to say it is practically extant. Management should be careful of men in their employ with a closed mind - or even partially closed mind - on the subject.

This is a dangerous age: when not only is anything possible, but it is possible quickly. A wise Frenchman once said you have only to live long enough to see everything 'and the reverse of everything;' and that is true in dealing with very advanced high energy physics of this kind.

Scientists are not politicians: they can reverse themselves once with acclaim - twice even with impunity. They may have to do so in the long road to attainment of this virtually perfect air vehicle. It is so easy to get bogged down with problems of the present; and whilst policy has to be made essentially with the present in mind - and in aviation a conservative policy always pays - it is management's task and duty to itself to look as far ahead as the best of its technicians in assessing the posture of the industry.

(page 17) GLOSSARY

Gravithermals:	alloys which may be heated or cooled by gravity waves. (Lover's definition)
Thermisters:	materials capable of being influenced by gravity.
Electrads:	materials capable of being influenced by gravity.
Gravitator:	a plurality of cell units connected in series: negative and positive electrodes with an interposed insulating member (Townsend Brown's definition).
Lofting:	the action of levitation where gravity's force is more than overcome by electrostatic or other propulsion.
Beneficiation:	the treatment of an alloy or substance to leave it with an improved mass-weight ratio.
Counterbary:	this, apparently, is another name for lofting.
Barycentric control:	the environment for regulation of lofting processes in a vehicle.
Modulation:	the contribution to lofting conferred on a vehicle by, treatment of the substance of its construction as distinct from that added to it by outside forces. Lofting is a synthesis of intrinsic and extrinsic agencies.
Absorber; insulator:	these terms - there is no formal distinction between them as yet - are based on an analogy with electromagnetism. This is a questionable assumption since the similarity between electromagnetic and gravitational fields is valid only in some respects such as both having electric and magnetic elements. But the difference in coupling strengths, noted by many experimenters, is fundamental to the science. Gravity moreover may turn out to be the only non-quantized field in nature, which would make it, basically, unique. The borrowing of terms from the field of electromagnetism is therefore only a temporary convenience. Lack of Cartesian representation makes this a baffling science for many people.
Negamass:	proposed mass that inherently has a negative charge.
Posimass:	mass the observed quantity - positively charged.
Shield:	a device which not only opposes gravity (such as an absorber) but also furnishes an essential path along which or through which, gravity can act. Thus whereas absorbers reflectors and insulators can provide a gravitationally neutralized body, a shield would enable a vehicle or sphere to 'fall away' in proportion to the quantity of shielding material.
Screening:	gravity screening was implied by Lanczos. It is the result of any combination of electric or magnetic fields in which one or both elements are not subject to varying permeability in matter.
Reflector:	a device consisting of material capable of generating buoyant forces

which balance the force of attraction. The denser the material, the greater the buoyancy force. When the density of the material equals the density of the medium the result will be gravitationally neutralized. A greater density of material assumes a lofting role.

Electrogravitics: the application of modulating influences in an electrostatic propulsion system

Magnetogravitics: the influence of electromagnetic and meson fields in a reflector.

Bosun fields: these are defined as gravitational electromagnetic, π and r meson fields (Metric tensor).

Fermion fields: these are electrons neutrinos muons nucleons and V-particles (Spinors).

Gravitator cellular body: two or more gravitator cells connected in series within a body (Townsend Brown's definition)

(page 20) REFERENCES

Mackenzie, Physical Review. 2. pp 321-43.

Eotvos, Pekar and Fekete Annalen der Physik. 68. (1922) pp. 11-16.

Heyl, Paul R. Scientific Monthly, 47, (1938) p. 115.

Austin, Thwing, Physical Review, 5, (1897) pp. 494-500.

Shaw, Nature (April 8,1922), p. 462, Proc. Roy. Soc., 102 (Oct. 6, 1922), p. 46.

Brush, Physical Review, 31, p. 1113 (A).

Wold, Physical Review, 35, p. 296 (abstract).

Majorana, Attidella Reale Academie die Lincei, 28, (1919) pp. 160, 221, 313, 416, 480, 29, (1920), pp. 23, 90, 163, 235 Phil. Mag., 39 (1920) p. 288.

Schneiderov, Science, (May 7, 1943), 97 sup. p. 10.

Brush, Physical Review, 32 p. 633 (abstract).

Lanczos, Science, 74, (Dec 4, 1931), sup. p. 10.

Eddington, Report on the Relativity Theory of Gravitation, (1920), Fleetway Press, London.

W.D. Fowler et al, Phys, Rev. 93, 861, (1954).

R.L. Arnowitt and S. Deser, Phys. Rev. 92, 1061, (1953).

R. L. Arnowitt Bull,,A.P.S. 94 798, (1954) S. Deser, Phys. Rev. 93, 612, (1954).

N. Schein D.M. Haskin and R.G. Glasser, Phys. Rev. 95, 855, (1954).

R.L. Arnowitt & S. Deser unpublished, Univ. of California Radiation Laboratory Report, (1954)

H. Bondi and T. Gold, Mon. Not. R. Astr. Soc., 108, 252, (1948).

F. Hoyle, Mon. Not. R. Astr. Soc., 108, 372, (1948).

B.S. DeWitt, New Directions for Research in the Theory of Gravitation, Essay on Gravity, 1953.

C. H. Bondi, Cosmology, Cambridge University Press, 1952.

F.A.E. Pirani and A. Schild, Physical Review 79, 986 (1950).

Bergman, Penfield, Penfield, Schiller and Zatzkis, Physical Review, 80, 81 (1950).

B.S. DeWitt, Physical Review 85, 653 (1952).

See, for example, D. Bohm, Quantum Theory, New York, Prentice-Hall, Inc. (1951) Chapter 22.

B.S. DeWitt, Physical Review. 90, 357 (1953), and thesis (Harvard, 1950).

A. Pais, Proceedings of the Lorentz Kamerlingh Onnes Conference, Leyden, June 1953.

For the treatment of spinors in a unified field theory see W. Pauli, Annalen der Physik, 18, 337 (1933). See also B.S. DeWitt and C.M. DeWitt, Physical Review, 87, 116 (1952).

The Quantum Mechanical Electromagnetic Approach to Gravity F.L. Carter Essay on Gravity 1953. On Negative mass in the Theory of Gravitation Prof. J.M. Luttinger Essay on Gravity 1951.

(page 22) Appendix I (of the Gravitics Situation)

SUMMARY OF TOWNSEND BROWN'S ORIGINAL PATENT SPECIFICATION

A Method of and an Apparatus or Machine for Producing Force or motion.

This invention relates to a method of controlling gravitation and for deriving power therefrom, and to a method of producing linear force or motion. The method is fundamentally electrical.

The invention also relates to machines or apparatus requiring electrical energy - that control or influence the gravitational field or the energy of gravitation; also to machines or apparatus requiring electrical energy that exhibit a linear force or motion which is believed to be independent of all frames of reference save that which is at rest relative to the universe taken as a whole, and said linear force or motion is furthermore believed to have no equal and opposite reaction that can be observed by any method commonly known and accepted by the physical science to date.

Such a machine has two major parts A and B. These parts may be composed of any material capable of being charged electrically. Mass A and mass B may be termed electrodes A and B respectively. Electrode A is charged negatively with respect to electrode B, or what is substantially the same, electrode B is charged positively with respect to electrode A, or what is usually the case, electrode A has an excess of electrons while electrode B has an excess of protons. While charged in this manner the total force of A toward B is the sum of force g (due to the normal gravitational field), and force e (due to the imposed electrical field) and force x (due to the resultant of the unbalanced gravitational forces caused by the electro-negative charge or by the presence of an excess of electrons of electrode A and by the electro-positive charge or by the presence of an excess of protons on electrode B).

By the cancellation of similar and opposing forces and by the addition of similar and allied forces the two electrodes taken collectively possess a force $2x$ in the direction of B. This force $2x$, shared by both electrodes, exists as a tendency of these electrodes to move or accelerate in the direction of the force, that is, A toward B and B away from A. Moreover any machine or apparatus possessing electrodes A and B will exhibit such a lateral acceleration or motion if free to move.

In this Specification I have used terms as 'gravitator cells' and 'gravitator cellular body' which are words of my own coining in making reference to the particular type of cell I employ in the present invention. Wherever the construction involves the use of a pair of electrodes, separated by an insulating plate or member, such construction complies with the term gravitator cells, and when two or more gravitator cells are connected in series within a body, such will fall within the meaning of gravitator cellular body.

The electrodes A and B are shown as having placed between them an insulating plate or member C of suitable material, such that the minimum number of electrons or ions may successfully penetrate it. This constitutes a cellular gravitator consisting of one gravitator cell.

It will be understood that, the cells being spaced substantial distances apart, the separation of adjacent positive and negative elements of separate cells is greater than the separation of the positive and negative elements of any cell and the materials of which the cells are formed being the more readily affected by the phenomena underlying my invention than the mere space between adjacent cells, any forces existing between positive and negative elements of adjacent cells can never become of sufficient magnitude to neutralize or balance the force created by the respective cells adjoining said spaces. The uses to which such a motor, wheel or rotor may be put are practically limitless, as can be readily understood, without further description. The structure may suitably be called a gravitator motor of cellular type.

In keeping with the purpose of my invention an apparatus may employ the electrodes A and B within a vacuum tube. Electrons, ion, or thermions can migrate readily from A to B. The construction may be appropriately termed an electronic, ionic, or thermionic gravitator as the case may be.

In certain of the last named types of gravitator units it is desirable or necessary to heat to incandescence the whole or a part of electrode A to obtain better emission of negative thermions or electrons or at least to be able to control that emission by variation in the temperature of said electrode A.

Since such variations also influence the magnitude of the longitudinal force or acceleration exhibited by the tube, it proves to be a very convenient method of varying this effect and of electrically controlling the motion of the tube.

The electrode A may be heated to incandescence in any convenient way as by the ordinary methods utilizing electrical resistance or electrical induction.

Moreover, in certain types of the gravitator units, now being considered it is advantageous or necessary also to conduct away from the anode or positive electrode B excessive heat that may be generated during the operation. Such cooling is effected externally by means of air or water cooled flanges that are in thermo connection with the anode, or it is effected internally by passing a stream of water, air, or other fluid through a hollow anode made especially for that purpose.

The gravitator motors may be supplied with the necessary electrical energy for the operation and resultant motion thereof from sources outside and independent of the motor itself. In such instances they constitute external or independently excited motors. On the other hand, the motors when capable of creating sufficient power to generate by any method whatsoever all the electrical energy required therein for the operation of said motors are distinguished by being internal or self-excited. Here it will be understood that the energy created by the operation of the motor may at times be vastly in excess of the energy required to operate the motor. In some instances the ratio may be even as high as a million to one. Inasmuch as any suitable means for supplying the necessary electrical energy, and suitable conducting means for permitting the energy generated by the motor to exert the expected influence on the same may be readily supplied, it is now deemed necessary to illustrate details herein. In said self-excited motors the energy necessary to overcome the friction or other resistance in the physical structure of the apparatus, and even to accelerate the motors against such resistance, is believed to be derived solely from the gravitational field or the energy of gravitation. Furthermore, said acceleration in the self excited gravitator motor can be harnessed mechanically so as to produce usable energy or power, said usable energy or power, as aforesaid, being derived from or transferred by the apparatus solely from the energy of gravitation.

`The gravitator motors function as a result of the mutual and unidirectional forces exerted by their charged electrodes. The direction of these forces and the resultant motion thereby produced are usually toward the positive electrode. This movement is practically linear. It is this primary action with which I deal.

As has already been pointed out herein, there are two ways in which this primary action can accomplish mechanical work. First, by operating in a linear path as it does naturally, or second, by operating in a curved path. Since the circle is the most easily applied of all the geometric figures, it follows that the rotary form is the important.

There are three general rules to follow in the construction of such motors. First, the insulating sheets should be as thin as possible and yet have a relatively high puncture voltage. It is advisable also to use paraffin saturated insulators on account of their high specific resistance. Second, the potential difference between any two metallic plates should be as high as possible and yet be safely under the minimum puncture voltage of the insulator. Third, there should in most cases be as many plates as possible in order that the saturation voltage of the system might be raised well above the highest voltage limit upon which the motor is operated.

Reference has previously been made to the fact that in the preferred embodiment of the invention herein disclosed the movement is towards the positive electrode. However, it will be clear that motion may be had in a reverse direction determined by what I have just termed 'saturation voltage' by which is meant the efficiency peak or maximum of action for that particular type of motor; the theory, as I may describe it, being that as the voltage is increased the force or action increases to a maximum which represents the greatest action in a negative to positive direction. If the voltage were increased beyond that maximum the action would decrease to zero and thence to the positive to negative direction.

The rotary motor comprises, broadly speaking, an assembly of a plurality of linear motors fastened to or bent around the circumference of a wheel. In that case the wheel limits the action of the linear motors to a circle, and the wheel rotates in the manner of a fireworks pin wheel.

I declare that what I claim is

1. A method of producing force or motion, which comprises the step of aggregating the predominating gravitational lateral or linear forces of positive and negative charges which are so co-operatively related as to eliminate or practically eliminate the effect of the similar and opposing forces which said charges exert.

2. A method of producing force or motion, in which a mechanical or structural part is associated with at least two electrodes or the like, of which the adjacent electrodes or the like have charges of differing characteristics, the resultant, predominating, uni-directional gravitational force of said electrodes or the like being utilized to produce linear force or motion of said part.

3. A method according to Claim 1 or 2, in which the predominating force of the charges or electrodes is due to the normal gravitational field and the imposed electrical field.

4. A method according to Claim 1, 2 or 3 in which the electrodes or other elements bearing the charges are mounted, preferably rigidly, on a body or support adapted to move or exert force in the general direction of alignment of the electrodes or other charge-bearing elements.

5. A machine or apparatus for producing force or motion, which includes at least two electrodes or like elements adapted to be differently charged, so relatively arranged that they produce a combined linear force or motion in the general direction of their alignment.

6. A machine according to Claim 5 in which the electrodes or like elements are mounted, preferably rigidly on a mechanical or structural part, whereby the predominating uni-directional force obtained from the electrodes or the like is adapted to move said part or to oppose forces tending to move it counter to the direction in which it would be moved by the action of the electrodes or the like.

7. A machine according to Claim 5 or 6 in which the energy necessary for charging the electrodes or the like is obtained either from the electrodes themselves or from an independent source.

8. A machine according to Claim 5, 6 or 7, whose force action or motive power depends in part on the gravitational field or energy of gravitation which is controlled or influenced by the action of the electrodes or the like.

9. A machine according to any of Claims 3 to 8, in the form of a motor including a gravitator cell or a gravitator cellular body, substantially as described.

10. A machine according to Claim 9, in which the gravitator, cellular body or an assembly of the gravitator cells is mounted on a wheel-like support, whereby rotation of the latter may be effected, said cells being of electronic, ionic or thermionic type.

11. A method of controlling or influencing the gravitational field or the energy of gravitation and for deriving energy or power therefrom comprising the use of at least two masses differently electrically charged, whereby the surrounding gravitational field is affected or distorted by the imposed electrical field surrounding said charged masses, resulting in a unidirectional force being exerted on the system of charged masses in the general direction of the alignment of the masses, which system when permitted to move in response to said force in the above mentioned direction derives and accumulates as the result of said movement usable energy or power from the energy of gravitation or the gravitational field which is so controlled, influenced, or distorted.

12. The method of and the machine or apparatus for producing force or motion by electrically controlling or influencing the gravitational field or energy of gravitation.

(page 30)   Appendix II (of the Gravitics Situation)

A Quantum Mechanical Approach to the Existence of Negative Mass and Its Utilization in the Construction of Gravitationally Neutralized Bodies

Since the overwhelming majority of electrostatic quantum mechanical effects rely for their existence on an interplay of attractive and repulsive forces arising from two types of charge, few if any fruitful results could come from a quantum mechanical investigation of gravity, unless there should be two types of mass. The first type, positive mass; (hereafter denoted as posimass) retains all the properties attributed to ordinary mass, while the second type, negative mass (hereafter denoted as negamass) differs only in that its mass is an inherently negative quantity.

By considering the quantum mechanical effects of the existence of these two types of mass, a fruitful theory of gravity will be developed. Theory will explain why negamass has never been observed, and will offer a theoretical foundation to experimental methods of detecting the existence of negamass and utilizing it in the production of gravitationally neutralized bodies.

To achieve these results, recourse will be made to Schroedinger's time independent equation with the center of mass motion removed. This equation is:

$$ -\hbar^2/2\mu \, \nabla^2 \psi + V\psi = E\psi $$

where all symbols represent the conventional quantum mechanical quantities. Particular attention will be paid to the reduced mass $\mu = (m_1 m_2) \div (m_1+m_2)$ where m_1 and m_2 are the masses of the two interacting bodies.

One can approach the first obstacle that any theory of negamass faces, namely the explanation of why negamass has never been observed, by a consideration of how material bodies would be formed if a region of empty space were suddenly filled with many posimass and negamass quanta. To proceed along these lines, one must first understand the nature of the various possible quantum mechanical interactions of posimass and negamass.

Inserting the conventional gravitational interaction potential into Schroedinger's equation and solving for the wave function ψ , yields the result that the probability of two posimass quanta being close together is greater than the probability of their being separated. Hence, there is said to be an attraction between pairs of posimass quanta. By a similar calculation it can be shown that while the potential form is the same two negamass quanta repel each other. This arises from the fact that the reduced mass term in Schroedinger's equation is negative in this latter case. The type of negamass posimass interaction is found to depend on the relative sizes of the masses of the interacting posimass and negamass quanta, being repulsive if the mass of the negamass quantum is greater in absolute value than the mass of the posimass quantum, and attractive in the opposite case. If the two masses are equal in absolute value the reduced mass is infinite and Schroedinger's equation reduces to $(V-E)\psi - 0$. Since the solution $\psi - 0$ is uninteresting physically, it must be concluded that $V - E$, and, hence, there is no kinetic energy of relative motion. Thus, while there is an interaction potential between the equal mass posimass and negamass quanta, it results in no relative acceleration and thus, no mutual attraction or repulsion while much could be said about the philosophical implications of the contradiction between this result and Newton's Second Law, such discussion is out of the scope of the present paper, and the author shall, instead, return with the above series of derivations to a consideration of the construction of material bodies in a region suddenly filled with many posimass and negamass quanta.

Because of the nature of the posimass-posimass and negamass-negamass interactions, the individual posimass quanta soon combine into small posimass spheres, while nothing has, as yet, united any negamass quanta. Since it is reasonable to assume that a posimass sphere weighs more than a negamass quantum in absolute value, it will attract negamass quanta and begin to absorb them. This absorption continues until the attraction between a sphere and the free negamass quanta becomes zero due to the reduced mass becoming infinite. The reduced mass becomes infinite when the sphere absorbs enough negamass quanta to make the algebraic sum of the masses of its component posimass and negamass quanta equal to the negative of the mass of the next incoming negamass quantum. Thus the theory predicts that all material bodies after absorbing as many negamass quanta as they can hold, weigh the same very small amount, regardless of size.

Since this prediction is in violent disagreement with experimental fact, one must conclude that the equilibrium arising as a result of the reduced mass becoming infinite has not yet been reached. That is, assuming that negamass exists at all, there are not enough negamass quanta present in the universe to allow posimass spheres to absorb all the negamass they can hold. One is thus able to explain the experimental fact that negamass has never been observed by deriving the above mechanism in which the smaller amounts of negamass that may be present in the universe are strongly absorbed by the greater amounts of posimass producing bodies composed of both posimass and negamass, but which have a net positive, variable, total mass.

Having thus explained why negamass has never been observed in the pure state, it is next desirable to derive an experimental test of the existence of negamass through considering the internal quantum mechanical problem of small amounts of negamass in larger posimass spheres. One is able to gain much physical insight into this problem by simplifying it to the qualitatively similar problem of one negamass quantum in the field of two posimass quanta that are fixed distance apart. Further simplification from three dimensions to one dimension and replacement of the posimass quanta potentials by square barriers, yields a solution in which the ground state energy E_0 of the negamass quantum in the field of one posimass quantum, is split into two energy levels in the field of the two posimass quanta. These two levels correspond to even and odd parity solutions of the wave equation where E_{even} lies higher and E_{odd} lower than E_0. The magnitudes of the differences $E_{even}-E_0$ and E_0-E_{odd} depend on the separation distance between the two posimass quanta, being zero for infinite separation and increasing as this separation distance is decreased.

Since the energy of a system involving negamass tends to a maximum in the most stable quantum mechanical configuration, the negamass quantum will normally be in state E_{even}. When the system is excited into state E_{odd}, the negamass quantum will favor the situation in which the two posimass quanta are as far apart as possible, since E_{odd} increases with increasing separation distance between the two posimass quanta, and the system tends toward the highest energy state. Thus independent of and in addition to the attractive posimass posimass gravitational interaction, there is a repulsive quantum mechanical exchange interaction between pairs of posimass quanta when the system is in state E_{odd}. The result of these two oppositely directed interactions is that the two posimass quanta are in stable equilibrium at some separation distance.

Since this equilibrium occurs between all posimass pairs in an elementary particle, a necessary consequence of the existence of negamass is that when in the first excited state elementary particles have a partial crystal structure.

This theoretical conclusion is capable of experimental verification by performing a Bragg analysis of the elementary particle crystal structure through shining high energy gamma rays on hydrogen. Part of the gamma ray energy will be utilized in lowering the system from energy E_{even} to E_{odd}, and if selective reflection is observed, it will constitute a striking verification of the existence of negamass. An order of magnitude calculation shows that, if the equilibrium distance between pairs of posimass quanta is one one-millionth the radius of an electron, 100 bev gamma rays will be required to perform this experiment.

Having discussed why negamass has never been observed, and having derived an experimental test of its existence it is next desirable to develop an experimental method of utilizing negamass in the production of gravitationally neutralized bodies by further consideration of some ideas previously advanced. It has been pointed out that if a source of negamass is present, a posimass sphere continues to absorb negamass quanta until equilibrium is reached as a result of the reduced mass becoming infinite. Because the sphere thus produced is practically massless and because the gravitational interaction between two bodies is proportional to the product of their respective masses, it follows that the sphere is practically unaffected by the presence of other bodies. And thus, the problem of making gravitationally neutralized bodies is reduced to the problem of procuring a source of negamass quanta. This will be the next problem discussed.

The binding energy of a negamass quantum in a posimass sphere may be obtained as one of the eigenvalue solutions to Schroedinger's Equation. If the negamass quanta in a body are excited to energies in excess of this binding energy by shining sufficiently energetic gamma rays on the body these negamass quanta will be emitted and negamass source will thus be obtained.

To estimate the gamma ray energy required to free a negamass quantum from a posimass body, certain assumptions must be made concerning the size and mass of posimass and negamass quanta. Since these quantities are extremely indefinite, and since the whole theory is at best qualitative, attempting to estimate the energy would be a senseless procedure. Suffice it to say that because of the intimate, sub-elementary particle nature of the posimass-negamass interaction, it seems reasonable to assume that quite energetic gamma rays will be required to break this strong bond.

To briefly review what has been shown a quantum mechanical theory of negamass has been developed based on the assumptions that gravitational interactions obey the laws of quantum mechanics and that all possible interactions of negamass and posimass with themselves and each other follow the well known inverse square law. This theory explains the experimental fact that negamass has never been observed, and outlines plausible experimental methods of determining the existence of negamass and utilizing it in the construction of gravitationally neutralized bodies. While these experimental methods may perhaps be out of the realm of practicality at the present, there is every reason to hope that they will be performable in the future. At that time, the plausibility of the existence of negamass and the theory behind the construction of gravitationally neutralized bodies from it, will meet their final tests.

SUMMARY PARAGRAPH

A quantum mechanical theory of negative mass is developed, based on the assumptions that gravitational interactions obey the laws of quantum mechanics, and that all possible interactions of negative and positive mass with themselves and each other follow the well-known inverse square law. This theory explains the experimental fact that negative mass has never been observed, and outlines plausible experimental methods of determining the existence of negative mass and utilizing it in the construction of gravitationally neutralized bodies.

Prof. F. Mozer

(page 37) Appendix III (of the Gravitics Situation)
GRAVITY EFFECTS

The order of magnitude of the heat given off by an alloy as a result of the separation by gravity tension can be reliably estimated. Suppose we assume that an alloy of half tin and half lead completely fills a tube 5 meters long and 100 cm^2 cross section which is maintained accurately at a temperature 277° C. At this temperature the alloy is liquid suppose next that the tube is raised from a horizontal plane into a vertical position, i.e. to a position where its length is parallel to the direction of gravity. If, then, the alloy is free from convection as it would be if it is maintained at uniform temperature and if it is held in this position for several months, the percentage of tin at the bottom of the tube will decrease while the relative amount at the top will increase. A simple calculation shows that the concentration of tin at the top is about one tenth of one percent greater than at the bottom and that approximately one calorie of heat is given off in the separation progress. If after several months the tube is again placed so that its length is in a horizontal plane the tin and lead will remix due to the thermal agitation of the atoms and heat is absorbed by the alloy.

Another interesting effect occurs when an electrolyte is subjected to gravity tension. Suppose a five meter glass tube is filled with a water solution of say barium chloride and the electrical potential between its ends is measured first when the length of the tube is parallel to the horizontal and second when its length is vertical. The difference in potential between the two ends is practically zero when the tube is horizontal and approximately eighty five microvolts when it is vertical.

This effect was discovered by Des Coudres in 1892. If a resistor is attached across the ends when the tube is vertical, heat of course is produced. If the tube is maintained at constant temperature the voltage decreases with time and eventually vanishes. The effect is believed to result from the fact that the positively charged barium ions settle faster than the lighter negatively charged chlorine ions as a result of gravity tension.

In conclusion, we have seen that gravity tension effects an alloy in such a way that it gives off heat. This phenomenon results from the alignment of the atoms and from their separation by the gravitational field, the contribution of the latter being larger than that of the former. Also, the gravity tension sets up a potential across the ends of a tube filled with an electrolyte and this potential when applied across an external circuit may produce heat or drive an electric motor to furnish power. Several other small thermal effects possibly may arise from gravity tension in addition to those discussed above but space is not available to consider them in this essay. Also, studies of the effect of gravitational fields and their equivalent centrifugal fields upon matter will no doubt be of great value in the future.

J.W. Beams

(page 39)   Appendix IV (of the Gravitics Situation)

LINK BETWEEN GRAVITATION AND NUCLEAR ENERGY
by Dr Stanley Deser and Dr Richard Arnowitt

Quantitatively we propose the following field equations:

$$-kT_{\mu\nu} = R_{\mu\nu} + \tfrac{1}{2} Rg_{\mu\nu} + C_{\mu\nu}(\Phi\Psi)$$
$$\left(\tfrac{1}{i}\delta^{\mu}\partial_{j\mu} + m + \lambda\sigma^{\mu\nu}K_{\mu\nu}(x)\right)\Psi = 0$$

with a similar equation for Φ. In the above, Ψ represents the hyperon wave functions and Φ the K-particle quantized field operators. The first three terms in the first equation are the usual structures in the Einstein General Relativity. The last term, $C_{\mu\nu}$ is the "creation" tensor which is to give us our conversion from gravitational to nuclear energy. It is like $T_{\mu\nu}$ in being an energy momentum term. In the second equation $\partial_{j\mu}$ represents the covariant derivative while δ^{μ} is a generalized Dirac matrix arranged so that the second equation is indeed covariant under the general group of coordinate transformations. The $\sigma^{\mu\nu}K_{\mu\nu}$ term will automatically include the higher hyperon levels. $C_{\mu\nu}$ is a functional of the hyperon and K-field variables Ψ and Φ. As can be seen these equations are coupled in two ways first the creation term $C_{\mu\nu}$ depends upon the field variables Ψ and Φ while the gravitational metric tensor $g_{\mu\nu}$ enters through the covariant derivative etc. λ is a new universal constant giving the scale of the level spacings of the hyperons. Rigorously speaking the field equations should be, of course, second quantized. For purposes of obtaining a workable first approximation it is probably adequate to take expectation values and solve the semi classical equations. The creation tensor $C_{\mu\nu}$ must be a bilinear integral of the Φ and Ψ fields and may have cross terms as well of the form $\int \Phi \overline{\Psi} \psi(dx)$. These equations will indeed be difficult to solve; but upon solution will give the distribution of created energy and, hence, lead eventually to the more practical issues desired.

Appendix V (of the Gravitics Situation)

Gravity/Heat Interaction

Let us suppose that we have to investigate the question whether gravitative action alone upon some given substance or alloy can produce heat. We do not specify its texture, density nor atomic structure; we assume simply the flux of gravitative action followed by an increase of heat in the alloy.

If we assume a small circular surface on the alloy, then the gravitative flux on it may be expressed by Gauss' theorem and it is $4\pi M$, where M represents mass of all sub-surface particles; the question is, can this expression be transformed into heat. We will assume it can be. Now recalling the relativity law connecting mass and energy:

$$M = m_0 + T/c^2 \text{ (by Einstein)}$$

where:- T = Kinetic energy
   m = Initial mass
   c = Velocity of Light

we set $4\pi M = m_0 + T/c^2 = m_0 + (m_0 v^2)/2c^2$

But v^2/c^2 is a proper fraction: hence $M = m_0 + m_0/2k$

In the boundary case v = c, $M = m_0(1+1/k)$ for all other cases $4\pi M = m_0((k+1)/k)k \neq 0$. Strictly M should be preceded by a conversion factor $1/k$ but if inserted, it does not alter results. Thus if gravity could produce heat, the effect is limited to a narrow range, as this result shows.

It merits stress that in a gravitational field the flow lines of descent -- are Geodesics.

J.W. Wickenden

Appendix VI (of the Gravitics Situation)

WEIGHT-MASS ANOMALY

There is a great need for a precise experimental determination of the weight to mass ratio of protons or electrons. Since the ratio for a proton plus an electron is known already, the determination of the ratio for either particle is sufficient. The difficulty of a direct determination of the gravitational deflection of a charged particle in an experiment similar to the neutron or neutral atom experiment is due to electrical forces being much greater than gravitational forces. For example, one electron five meters away from a second electron exerts as much force on that second electron as the gravitational field does. Thus stray electrons or ions which are always present on the walls of an apparatus can exert sufficient force to completely mask the gravitational force. Even if the surface charges are neglected, image charges of the electron beam itself and self repulsion in the beam may obscure the gravitational deflection. An additional problem is the Earth's magnetic field. Electrons of even a few volts energy will feel a force due to the Earth's field a thousand billion times larger than the gravitational deflection. This last problem is avoided in a static measurement of the ratio such as a weighing of ionised matter. However, this last method has the additional difficulty of requiring a high proportion of ionized to unionized matter in the sample being weighed. Of course all these problems can be resolved to some extent; but it is questionable if an experiment of either of the above types can be designed in which all the adverse effects can simultaneously be sufficiently minimized. Probably a completely new type of experiment will have to be devised to measure the weight to mass ratio of the proton or electron.

Such a measurement may detect a deviation from the law of constant weight to mass ratio. If such an anomaly can be shown to exist there is the possibility of finding a material which would be acted upon in an unusual manner in a gravitational field.

Martin L. Perl.

APPENDIX No. 4: The Maxfield Letters

Radiological Center
Maxfield Clinic - Hospital
Dallas, Texas 75219
May 10,1972

The Chief Superintendent of the Aeronautical Research Lab
Box 4331, G.P.O.
Melbourne, Australia 3001

Dear Sir:

I had the opportunity about seven months ago of meeting a young man who migrated to Australia at my suggestion. He is a competently trained engineer and he is now working as a Computer Systems Analyst for a tractor company and would like to make a change into the Aeronautical Research Lab.

He has been working in a field that Dr. Edward Teller and I have been interested in. I would hope that you could interview Mr. Deyo and perhaps, find a place for him in your organization.

With best wishes, I am
Sincerely yours,

J. R. Maxfield Jr., MD

JRMjr:e
13704-0

MAXFIELD RADIOLOGICAL CENTER

Maxfield Clinic - Hospital   May 10, 1972
2711 Oak Lawn Avenue
Dallas, Texas 75219

Mr. Stan Deyo, Jr.
1282 Glenhuntly Rd.
Carnegie, Vic 3163
Australia

Dear Stan:

I appreciate your note very much and I am glad that you have gotten into Australia and are going to develop into a fine citizen for them. It was a real privilege to help you.

I have written to my friend, Sir John Williams, Australian New Zealand Steamship Lines, who is the prominent man in the Melbourne area and was one that I contacted about your going down originally. He is interested in helping young people and is interested in Australia.

I would suggest that you go by and have a visit with him. I am writing him and sending him a copy of this letter requesting that after the visit that he also contact the Chief Superintendent of the Aeronautical Research Lab at Melbourne. I am writing to the Chief Superintendent for you myself also.

Drop me a note as to how you are getting along --- after your visit with Sir John and with the Aeronautical Research Lab.

I expect to be in Australia the latter part of October with Dr. Teller; and perhaps, we can all get together at that time.

With best wishes, I am

Sincerely yours,

J. R. Maxfield, jr., MD

JRMjr:e
13704-1

APPENDIX No. 5: Einstein's Relativity Error

The physical sciences in 1873 seemed to once again take on an air of stability as James Clerk Maxwell published his, 'Treatise on Electricity and Magnetism'.

In this paper he discussed electricity, magnetism, and electromagnetism as functions of waves in a fluid space (ether). His theory held popular support until the year 1887 when the two U.S. physicists, A.A. Michelson and Edward W. Morley performed their historic experiment with light. Their experiment (the 'Michelson-Morley Experiment') was designed to use light as a means to determine if space were a 'fluid' as Maxwell's equations had assumed.

The 'M-M' test results, however, appeared to deny the existence of fluid (or ether) space. To explain the 'apparent' failure of the M-M test to detect the ether, Hendrik Lorentz and George Fitzgerald developed their now famous 'transforms' (the Lorentz-Fitzgerald transforms - 1902) in which length contractions, mass increase, and time lag were offered as explanation for the negative test result. Note that the Lorentz-Fitzgerald transforms still treated space as an inertial fluid... one undetectable by known technology.

Einstein, who first *began* the formulation of his Special Theory of Relativity in 1895, published it in 1905. He seized upon the Lorentz-Fitzgerald transforms and the M-M test results as evidence of a universal axiom: the velocity of light is (to the observer) the limit *measurable* velocity in the universe - (this does not mean it is the limit velocity in the universe, however.)

The Discipline Details

Einstein was faced with an apparent paradox as to the nature of space. It behaved like a fluid in many ways - yet in others it behaved like an abstract, ten-component Ricci Tensor from the Reimannian model of the Universe. The failure of the M-M test to detect an ether was the final straw. Yet, hard as he tried, Einstein failed to remove the "ether" from $E = mc^2$. The following discussion should illustrate this point:

Diagram 1 is a schematic of the M-M test. It was conducted on the basis that if an ether existed, the earth would by moving **through** it. Hence, there would be a relative velocity between earth and the fluid of space.

It was reasoned that by splitting a beam of light (F) into two parts; sending one out and back in-line with the direction of earth's orbital path, to mirror (A) from half-silvered mirror (G); sending the other at right angles to the direction of earth's orbital path to mirror (B) through half-silvered mirror (G) and glass plate (D); and re-combining the two beams in the interferometer (E) one should be able to detect a shift in the phases of the two beams relative to one another. This shift could be accurately predicted by knowing the velocity of light (c) and the velocity (v_e) of earth through orbital space. Their reasoning was as follows (refer diag. 1, 2a, 2b):

Assuming:

v_e = velocity of ether *wind* or *drift*
c = velocity of light = velocity from G to B by fixed extra-terrestrial observer
s = distance GA = GB
t_1 = go-return time in-line (GA - AG)
t_2 = go-return time at right angles (GB - BG)
t = .5t_2
v_1 = apparent velocity from G to B by earth observer

Then the time (t_1) is determined by: $[s/(c - v_e)] + [s/(c + v_e)] = t_1$ which reduces to:

(Eq. 1)
$$2sc/(c^2 - v_e^2) = t_1$$

DIAGRAM 1

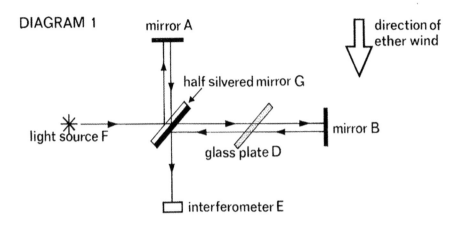

mirror A

half silvered mirror G

light source F

direction of ether wind

mirror B

glass plate D

interferometer E

DIAGRAM 2ᵃ

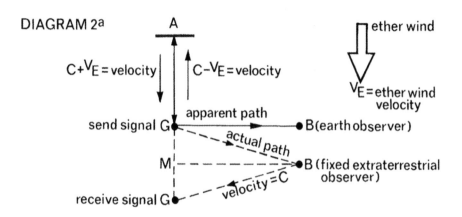

A

$C+V_E$ = velocity

$C-V_E$ = velocity

ether wind

V_E = ether wind velocity

send signal G

apparent path

B (earth observer)

actual path

M

B (fixed extraterrestrial observer)

receive signal G

velocity = C

DIAGRAM 2b

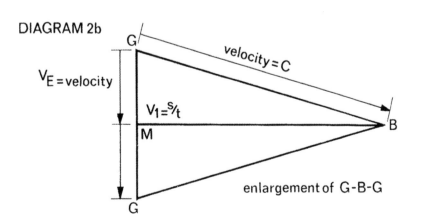

G

V_E = velocity

velocity = C

$V_1 = ^s/_t$

M

B

G

enlargement of G-B-G

Also, the time (t_2) is determined by first solving for (v_1) in terms of (c) and (v_e) using the Pythagorean Theorum ($c^2 = a^2 + b^2$)... or, in this instance: (G to B)^2 = (G to M)^2 + (M to B)^2.

by substitution, $\qquad\qquad c^2 = v_e^2 + v_1^2 \qquad\qquad$ hence:

(Eq. 2) $\qquad\qquad\qquad\qquad v_1 = (c^2 - v_e^2)^{.5}$

Now, solving for the time (t) - which is the same over GM, GB, MB - of the GB trip by substituting $s/t = v_1$ in (Eq. 2), one obtains:

(Eq. 3) $\qquad\qquad\qquad\qquad s/t = (c^2 - v_e^2)^{.5} \qquad$ and, rearranging...

(Eq. 3) $\qquad\qquad\qquad\qquad t = s/(c^2 - v_e^2)^{.5}$

then, substituting: $\qquad\qquad t = .5t_2$

gives: $\qquad\qquad\qquad\qquad t_2/2 = s/(c^2 - v_e^2)^{.5}$ or:

(Eq. 4) $\qquad\qquad\qquad\qquad t_2 = 2s/(c^2 - v_e^2)^{.5}$

By comparing the ratio of the in-line go-return time (t_1) to the right angle go-return time (t_2) one obtains:

(Eq. 5) $\qquad\qquad\qquad t_1/t_2 = [2sc/(c^2 - v_e^2)] [(c^2 - v_e^2)^{.5}/2s]$ or:

(Eq. 5) $\qquad\qquad\qquad t_1/t_2 = (1 - v_e^2/c^2)^{-.5}$

Now then, **if the light source is at rest with respect to the ether,** one sees:

(Eq. 6) $\qquad\qquad\qquad\qquad v_e = 0 \qquad\qquad$ hence:

(Eq. 7) $\qquad\qquad\qquad\qquad t_1/t_2 = 1/(1-0)^{.5} = 1/1 = 1$

Such a ratio as (Eq. 7) shows is exactly what every successive try of the **linear** M-M test has obtained... (notice: *linear* not *angular).* Lorentz and Fitzgerald knew there had to be an ether; so they developed their well-known transforms - an act which was in essence a way of saying, *there has to be an ether... we'll adjust our **observed** results by a factor which will bring our hypothetical expectations and our test results into accord...* **Their whole transform was based on the existence of ether space!** Their transform, in essence, said that length shortened, mass flattened, and time dilated as a body moved through the ether; hence it was possible to detect the ether.

Einstein came along in 1905 saying the Michelson-Morley test showed the velocity of light to be a *universal constant* to the observer. Seizing upon this and the Lorentz-Fitzgerald transforms, Einstein was able to formulate his Special Relativity which resulted in the now famous $E = Mc^2$ ... the derivation of which follows:

Starting with (Eq. 5): $\qquad\qquad t_1 /t_2 = (1 - v_e^2/c^2)^{-.5}$

The Lorentz-Fitzgerald transform factor for (Eq. 5) becomes $(1 - v_e^2/c^2)^{.5}$ (to bring $t_2 = t_1$) giving t_1 /t_2 an observed value of (1).

Assuming Lorentz and Fitzgerald's supposition to be correct, one should look at *mass-in-motion as the observer on the mass sees it versus mass-in-motion as the universal observer sees it, ...*

let m_1 = mass as it appears to riding observer
let v_1 = velocity as detected by rider
let m_2 = mass as universal observer sees it
let v_2 = velocity as universal observer sees it
then.0, it follows (from Lorentz and Fitzgerald) that:

(Eq. 9) $\qquad\qquad\qquad\qquad m_1v_1$ not $= m_2v_2$ (to either observer)

So, to equate the two products, Lorentz and Fitzgerald devised their transform factor $(1 - v_e^2/c^2)^{.5}$ which would bring $m_1v_1 = m_2v_2$ to either observer,... yielding the following extension:

since,... $v_1 = s_1/t_1$ and $v_2 = s_2/t_1$ (assuming time is reference)

(Eq. 10) m_1s_1/t_1 not $= m_2s_2/t_1$ or, ...

(Eq. 10) m_1s_1 not $= m_2s_2$

Then, by substitution of the transform factor $s_2 = s_1(1 - v_e^2/c^2)^{.5}$ (assuming time is reference) into (Eq. 10) one obtains: $m_1s_1 = m_2s_1(1 - v_e^2/c^2)^{.5}$ which reduces to:

(Eq. 11) $m_1 = m_2(1 - v_e^2/c^2)^{.5}$

To re-evaluate this *relative* change in mass, one should investigate the expanded form of the transform factor $(1 - v_e^2/c^2)^{-.5}$ (which transforms $t_1 = t_2$). It is of the general binomial type:

(Eq. 12) $(1-b)^{-a}$

Hence, it can be expressed as the sum of an infinite series:

(Eq. 13) (where: b^2 is < 1): $1 + ab + a(a+1)b^2/2! + a(a+1)(a+2)b^3/3!$ + ...etc

So, by setting... $a = .5$ and $b = v_e^2/c^2$ one obtains:

(Eq. 14) $1 + (.5v_e^2/c^2) + (.375v_e^4/c^4) + (.3125v_e^6/c^6)$ + ...etc

For low velocities in the order of .25c and less the evaluation of $(1 - v_e^2/c^2)^{-.5}$ is closely **approximated** by, the first two elements of (Eq. 14):

(Eq.15) $(1 - v_e^2/c^2)^{-.5} = 1 + .5v_e^2/c^2$

so, (Eq. 11) can be re-stated as:

(Eq. 16) $m_2 = m_1(1 + .5v_e^2/c^2)$ ...(where v_e < .25c)

or $m_2 = m_1 + .5m_1v_e^2/c^2$ ...(where v_e < .25c)

and,

(Eq. 17) $m_2 - m_1 = .5m_1v_e^2/c^2$

As energy (E) is represented by:

(Eq. 18) $E = .5mv^2$ ...(where v_e < .25c)

Substituting (Eq. 18) into (Eq. 17) gives:

(Eq. 19) $m_2 - m_1 = E/c^2$ ...(assuming $v_e = v$)

Letting change in mass $(m_2 - m_1) = M$ gives:

(Eq. 20) $M = E/c^2$

or, in the more familiar form:

(Eq. 21) $E = Mc^2$

(Note: Eq. (14) should be used for best accuracy - especially where v_e > .25c)

Looking at the assumption in (Eq. 19)... (v_e) was the term used in the beginning to represent the *ether wind* velocity... This means **Einstein used fluid space** as a basis for Special Relativity. His failing was in declaring the velocity of light an observable limit to the velocity of any mass when it should only have been the limit to any observable electromagnetic wave velocity in the *ether.The velocity of light is only a limit velocity in the fluid of space where it is being observed.* If the energy-density of space is greater or less in another part of space, then the relativistic velocity of light will pass up and down through the *reference light wave velocity limit* - if such exists.

Do not fall into the trap of assuming that this *fluid space* cannot have varying *energy-density*. Perhaps, the reader is this very moment saying, *an incompressible fluid space does not allow concentrations of energy* - but he is wrong - dead wrong!

When a fixed-density fluid is set in harmonic motion about a point or centre, the **number of masses passing a fixed reference point per unit time** can be observed as increased mass (or concentrated energy). Although the density (mass per volume) is constant, the mass-velocity product yields the illusion of more mass per volume per time. Space is an incompressible fluid of varying energy **density... in this author's opinion.**

The apparent absurdity of *infinitely-increasing-mass* and *infinitely-decreasing-length* as a mass approaches the light-wave velocity is rationalized by realizing that space has inertia and as such offers inertial resistance to the moving mass. The energy of the moving mass is transmitted in front of it into the medium of space. The resulting *curl* of inertial resistance increases as negative momentum to the extent the mass is converted to radiant energy as it meets its own reflected mass in resistance.

However, to the *Star Trek* fans, take heart... just as man broke the sound-velocity limit (sound barrier) he can also break the light-velocity limit (light barrier). By projecting a high-density, polarized field of resonating *electrons* to spoil or *warp* the pressure wave of the inertial curl, the hyperlight-craft can slip through the warp opening before it closes -emitting the characteristic shock wave. Such a *spoiler* would be formed by using the electrodynamic, high-energy-density electron waves which would normally proceed before the hyper-light craft, as a primary function of propulsion.

When a similar function is executed by hyper-sonic aircraft, a *sonic boom is* formed as the inertial curl collapses on itself. In space, the light-velocity equivalent to this *sonic boom* would be in the form of *Cherenkov radiation* which is emitted as a mass crosses the light-velocity threshold sending tangential light to the direction of travel.

Ether Existence Verified

In 1913, the *rotational* version of the *linear* M-M experiment was successfully performed by G. Sagnac (see p. 65 - 67 of *The Physical Foundations of General Relativity* by D.W. Sciama, Heinemann Educational Books Ltd., 48 Charles St., London WIX8AH). In 1925, Michelson and Gale used the spinning Earth as their rotational analog to the linear M-M experiment. It also showed successfully that the velocity of light sent *in the* direction of spin around the perimeter of a spinning disc (or of the surface of earth) varied from the velocity of the light sent *against* the spin. *(refer Diagram 3 below).*

DIAGRAM 3

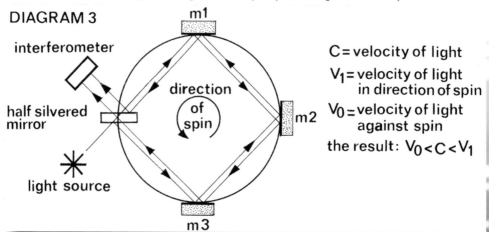

interferometer

m1

half silvered mirror

direction of spin

m2

light source

m3

C = velocity of light

V_1 = velocity of light in direction of spin

V_0 = velocity of light against spin

the result: $V_0 < C < V_1$

The Cosmic Conspiracy Final Edition 2010 © Stan Deyo 2010

Analogy of Dilemma

The *error* of the M-M experiment is *the test results are also valid for the case where* **there is an ether and it, too, is moving along with the same relative velocity and orbit as Earth maintains around the Sun.** The *tea cup* analogy can be used to explain the error. If one stirs a cup of tea (preferably white) which has some small tea leaves floating on its surface, one notices some of these tea leaves orbiting the vortex in the centre of the cup. The leaves closer to the centre travel faster than those farther from the centre (both in linear and angular velocity).

Now, one must imagine himself greatly reduced in size and sitting upon one of these orbiting leaves. If one were to put his hands over the edge of his tea leaf on any side, would he feel any tea moving past?... No. The reason is that the motion of the tea is the force that has caused the velocity of the leaf. One could not detect any motion if both himself and the tea were travelling in the same direction and at the same velocity. However, if one had arms long enough to stick a hand in the tea closer to either the centre or the rim of the cup where the velocities were different to his own, then he would feel tea moving faster or slower than himself (respectively).

Also, if one were to spin his tea leaf at the same time as it orbits about the centre, placing his hands into the tea immediately surrounding his leaf would show inertial resistance against the spin moment of his leaf.

Solar 'Tea Cup'

In the preceding analogy, the centre of the spinning tea (or vortex centre) represented the Sun, the leaf: the Earth; the tea: the ether; and the rider's hands: the light *beams* of the M-M test. In essence, what Michelson, Morley, Einstein, and many other scientists have said is that the M-M test showed the velocity of light was not *affected* by the Earth's orbital motion. *"Therefore",* they have said, *"we have one of two conclusions to draw":*

1)   The Earth **is orbiting the Sun** and there is no ether, or,

2)   The Earth is not orbiting the Sun and there is an ether but since the earth is not *moving through the ether, the ether "wind" cannot be detected.* Obviously, this conclusion is negated by Earth's observed heliocentric orbit. However, their reasoning should also have incorporated a third option:

3) **The Earth is orbiting the Sun and so is the ether;** therefore, no ether wind could be detected in the orbital vector immediately in the vicinity of Earth. In other words, the test results cannot prove or disprove the existence of an ether... only whether or not the Earth is moving relative to such an ether.

'C' Not Constant

Remember, in 1913, G. Sagnac performed his version of the M-M experiment and corrected the inconclusive results which Michelson and Morley's test had obtained. In Sagnac's *rotational analog* of the M-M test *the velocity of light was shown to vary.* Also, in 1925 Michelson and Gale verified Sagnac's results with their own rotational analog. Even more recently, similar verification has been made using a ring-laser system to detect the rotational velocity of the Earth relative to the ether.

Relativists Discard Evidence

By the time the **ether wind was proven to exist,** Einstein's theories were already winning strong support on the merits of celestial observations which closely agreed with Einstein's predicted values. As a result, the *scientific* community decided to explain the *ether wind* phenomenon as a result of Earth's spinning in its own *ether blanket* which Earth was apparently dragging through space. No explanation was ever agreed upon as to the origin or extent of this *ether blanket.* It was simply a way to sweep a discrepancy under the carpet.

Einstein Admits Error...

In a biography written just before his death, Professor Einstein is quoted as admitting he had a **fundamental error in Relativity.** It was, he said, one which (when corrected) will explain how light - an obvious wave form - can be propagated across an apparently non-inertial space. Einstein also stated that the discovery of the solution to this error would probably be the result of some *serendipitous* discovery in the sixties.

However, before he died, Einstein did manage to partially correct his error. With the help of the well-known Dr. Erwin Schroedinger, Dr. Einstein was able to construct a 'total theory' for existence. It was called the "Unified Field Theory". Although Dr. Einstein was able to lay the basic framework before his death, it is reasonably certain that a more readily-usable version of the "Unified Field Theory" was only completed by other physicists after Einstein had died.

One of the more promising contributions toward a usable unified field theory was offered by Dr. Stanley Deser and Dr. Richard Arnowitt (see p.39 of *The Gravitics Situation* in Appendix (3) of this book).

They took the General Theory of Relativity which Einstein had devised and constructed a "bridge" or "creation tensor" to link the energy of nuclear fields with that of gravitational fields by covariant matrices. The basic relationship of General Relativity which they used as a basis for their system is:

$$R_{uv} - .5g_{uv}R = 8\pi k T_{uv}$$

R_{uv} = Ricci's ten-component sub-Riemannian space, curvature tensor
g_{uv} = the metric tensor
R = the selected Ricci scalar components
k = a universal constant: proportional to Newton's gravitational constant
π = the usual constant: 3.14 etc.,
T_{uv} = the components (potentials) of the energy-stress tensor

Although Deser and Arnowitt's proposed equations were quite difficult to work with, it is rumored that subsequent linear variations have been developed - allowing major leaps in science and technology to develop.

When the correctly formulated Unified Field Theory is finally released to the public, it will be recognized quite easily; for it will have explained why the proton is exactly 1836 times the gravitational mass of an electron,... why there is no neutral mu-meson of mass 200,... why (h) is a constant... and why hc/e^2 is always equal to (137)...

The true "Unified Field Theory" will no longer be called a "theory"; it will become the "Unified Field Law". One inescapable conclusion will suddenly spring into the collective consciousness of those who grasp its meaning: *"In the beginning was the WORD (a complex wave form) ... and the WORD was with GOD, and the WORD was GOD. The same was in the beginning with GOD..."* (John 1:1).

APPENDIX No. 6: Electrodynamic Propulsion

The translation of an inertial mass from one position to another is a process usually accomplished by one of the following:

1) Pulling the mass from point (a) to point (b)

2) Pushing the mass from point (a) to point (b), or,

3) A combination of pushing and pulling the mass from point (a) to point (b).

Rockets, automobiles, and other brute force motion devices employ process (2) above.

Ramjets, turbines, helicopters, and other push-pull motion devices utilize process (3) above.

As yet, the pure attraction-only motion systems (1) find very limited use. These usually employ magnetic, electrostatic, or gravitational acceleration as a motion source.

Electrodynamic propulsion (EDP) falls into category (3). It can be accomplished by optimizing the ramjet process over the entire leading surface of the mass to be moved -if there is a medium through which to move. In the traditional ramjet, air is sucked into the front of the craft; and, with added fuel, is ignited inside the craft and expelled out the back of the craft.

The major problem in this system is the same as with push-only propulsion systems... namely, that all the leading surfaces of the rest of the craft encounter direct inertial resistance from the air that is not passing *through* the craft - but *around* it.

The philosophical concept of making little ramjet breathing openings all over the leading surface is approaching higher efficiencies to a point; however, as the ramjet needs a confining space to combust the fuel and air, all those little breathing openings would require dead (or closed) space in between them to form the confining chamber.

The optimum lead surface efficiency in a category (3) system is one where *the entire leading surface is the ramjet opening.* Such a shape is difficult to imagine;... think about it... A straight tube would *almost* give a frictionless move along the length axis; but where would the fuel and crew be placed?... what about the guidance surfaces?... If the front-end of the tube is opened out enough to shield the rest of the craft from frictional exposure, then the inside of the tube itself will offer massive frictional resistance to the incoming air.

Inertial resistance cannot be removed when one mass passes through another; however, the distribution of the resistance can be so designed as to use the air, itself, as a frictional dissipater. Thus, the optimum may be approached and *attained* by incorporating the air (or fluid medium) into the defined field of the craft.

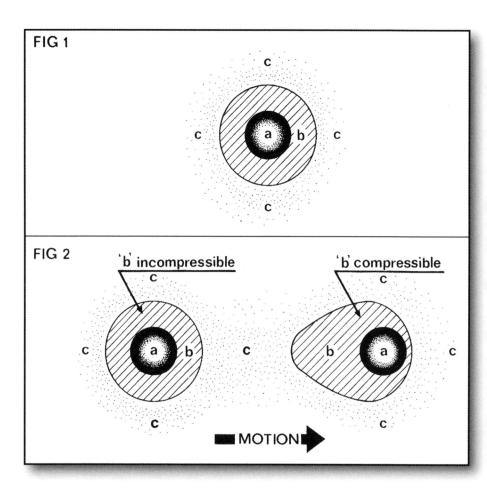

FIG 1

FIG 2

'b' incompressible

'b' compressible

MOTION

The Cosmic Conspiracy Final Edition 2010 © Stan Deyo 2010

The most obvious question, now, is how does one construct such a craft?... To answer that query, let us build such a craft one stage at a time:

1) The craft will be designed to move in fluids (i.e. it will be a hydrodynamic craft)

2) The craft and its field definition (fig. 1) will **be visualized as a regular** sphere within a sphere. The main craft is (a); the incorporated field is (b); and the ambient medium is (c).

3) The craft and its field (see fig. 2) will now be visualized as moving from left to right on the page within the ambient medium (c). *If no compression of (b) is assumed,* then the passage of the (a) + (b) field through (c) will produce frictional losses on the interface of (b) to (c)... (i.e. heat will be generated as well as other by-product radiations depending on relative velocity). Eventually the heat or radiated energy of such an exchange would be passed on to the craft (a).

To minimize such an exchange, a method of dissipating the unwanted heat must be added. Even *if (b) is assumed as compressible,* then at certain velocities the distance between (a) and (b) in the direction of motion would be so small as to negate the effect of the shielding that (b) was designed to give.

4) Therefore, let us assume that (b) is a sacrificial shield... One which is being replaced as a function of motion... (see fig. 3). In this manner the heat or radiated energy of the field (b)'s encounter with (c) is left behind the craft (a)... and is dissipated into the old sector of (c).

5) From assumption (3) motion was assumed from left to right. From assumption (4) the field (b) was assumed to be sacrificial. Let us now probe the mechanisms to produce these two assumptions.

Referring to figure (4) the craft (a) is now fitted with a point (d) from which is emitted a dense, high-voltage, direct-electric current which makes its circuit through the field (b)... (actually forming the limits of (b). In conventional terms, the point (d) is negative with respect to point (e). The shell of the craft (a) is non-conductive so that the electric moment travels from (d) to (e) via the ambient medium (c) - which by virtue of the passing electric moment is captured as (b). Two factors will now produce motion to the right.

The first is that the thrust from the accelerated fluid particles from (d) to (e) will produce a resultant to the right; and the second is that due to the Bernoulli Effect, the fluid pressure at right angles to the fluid flow from (d) to (e) is reduced;... in other words, a partial pressure reduction is formed to the right of line xy at right angles to the curved path of (de). Also, as a function of the fluid flow toward (e), from (d), there is a partial pressure increase at (e). This is caused by the collision of all the fluid particles from vectors in (dxe) with all the particles from vectors in (dye).

In figure (5) we see another side effect of this method of acceleration... At time (t = n), the craft and its field have pulsed a vector as shown by the dotted arrows. At time (t = n + 1), the craft has moved to the right of its position at time (t = n), and the region (f) is rapidly normalizing to a stagnant zone due to the vectors (see dotted arrows) colliding to generate heat and a little turbulence. In essence the craft has displaced an amount of fluid in front of itself and has moved into the space left by the displaced fluid, and has then replaced the same fluid in its original space after the craft, itself, has pulsed into the next zone. This phenomenon can be observed by watching a pneumatic tube in the older office buildings that still use them for shooting inter-office correspondence back and forth. They are sucked and pushed at the same time.

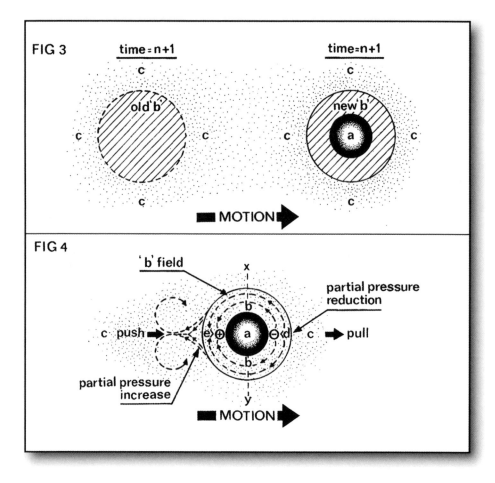

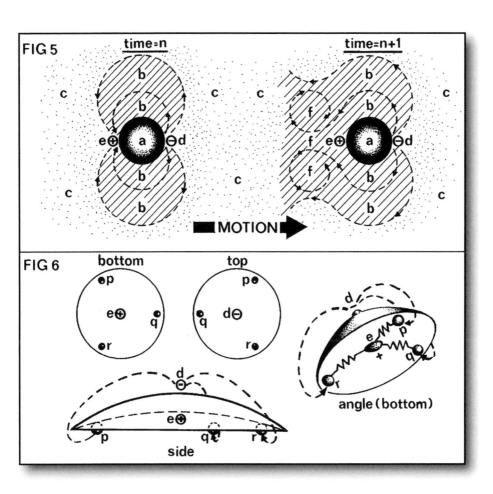

A craft operating on such a principle would leave little (if any) turbulence; it would not be hampered by high-temperatures due to friction... (there would be none between it and (c); and it would not produce high - density shock waves as it passed through the wave, velocity threshold of the medium... (the latter effect is caused because there are no forward vector components in the motion transfer to the right of line xy which eliminates the return inertial wave front that is normally encountered in the brute-force, push-only methods of propulsion).

6) If the craft is to travel in any other direction than to the right, then a method of navigational control has to be included in the mechanism. To determine a plane of orientation, a minimum of three points is required. For the ease of illustration, this discussion will use only three points... (bear in mind, however, that the more points on the navigational compass... the more accurate can be the navigational manoeuvre).

Figure (6) shows four views of the craft with the three 'nav-points' attached to the left (or bottom) of the craft. Note that the left side of the craft has now been made concave. This has been done to optimize the partial pressure increase from the colliding fluid particles. Points (p), (q), (r) are variable resistors which are electrically connected to point (e). As the three points are all closer to point (d) in terms of the electric circuit, the electric circuit from (d) to (e) can be varied so that any one or all of the three points on the bottom can be made to draw more or less current than the other two points. Since the electric current transferring (and hence the fluid transferring) from (d) to (p), (q), (r) determines the partial pressure reduction on the leading surface, then the pressure gradient across the entire leading surface can be varied at will by simply varying the resistance at (p) and/or (q), and/or (r).

For example, if the flow rate from (d) to (p) is 1000 fps; the flow rate from (d) to (q) is 1000 fps; and the flow rate from (d) to (r) is 1000 fps, then the partial pressure gradients over each of the portions of the surface controlled by one of the points are equal to one another. Therefore, if, say, the resistance of circuit (d) to (p) is made higher than the resistance in the other two circuits, then the fluid flow rate from (d) to (p) would be less than the fluid flow rates of (d) to (q) or (r); and hence, the partial pressure *reduction* over (d) to (p) would be less than over the other two sectors. This would cause the craft to turn about a moment within the (d) - (p) sector (fig. 7).

The effective lift or suction over the other two sectors would be greater than over (d) - (p), so the (d) - (p) sector would in effect look like a control surface (in vectorial function)... somewhat like a sail brake... when compared to the others. Until the flow rates are equalized again, the craft will continue to rotate as described. As soon as the desired attitude to the reference horizon is attained, then all flow rates are equalized and the craft whisks away... *top first.*

7) Notice that figure (8) shows the leading surface (the top) as a somewhat parabolic curve as opposed to the original hemispherical curve. The reason for the change is to direct more of the acceleration on the fluid at such an angle with the intended direction of motion as to obtain maximum lift for power consumed. If the shape of the surface were to be elongated more in the direction of intended motion, then the lift vectors on opposite sides of the leading surface would become more and more in opposition to each other giving less and less motion in any direction... (figure 9).

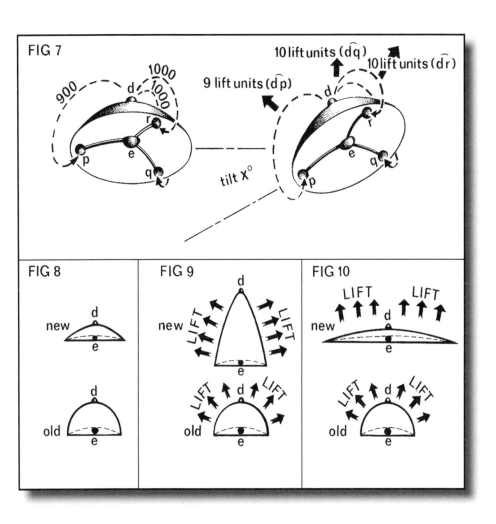

FIG 7

900   1000   1000

d

r

p   e   q

tilt X°

9 lift units (d̂p)

10 lift units (d̂q)   10 lift units (d̂r)

d

r

e   q

p

FIG 8

new   d   e

old   d   e

FIG 9

new   d   LIFT   LIFT   e

old   LIFT   d   LIFT   e

FIG 10

new   LIFT   LIFT   d   e

old   LIFT   d   LIFT   e

If, on the other hand, the shape of the surface were to be flattened in the intended direction of travel, then the angle for the travel of the electric moment becomes so acute that the charges bleed off into the medium... thus reducing the amount of partial pressure increase at (e) and also increasing power requirements drastically (fig. 10). An optimum curve has to be chosen depending upon a variety of intended or desired performance factors.

8) With such a unique method of motion comes another problem. Since the velocities attainable under such a relatively frictionless transfer process are excessive by modern concepts of safe structural velocities, a method of turning corners at speeds in excess of 20,000 mph has to be added to the mechanism of the craft. The same method must be added to the crew of such a craft to prevent structural fatigue.

The method is almost too simple. The same electric field that traverses the surface of the craft can be used to polarize all masses within the limit of the field effect.

In a conductor as the voltage and current frequency are raised over a certain value the current is observed to travel mostly in the surface of the conductor. This is commonly referred to as the 'skin effect.' Now in the craft the voltage levels will be in excess of 15,000,000 volts at frequencies up to 150KHz... (more than ample to generate the 'skin effect.' If the shell is a high-voltage semi-conductor then the current will travel along the outside of the surface and even in the fluid medium in proximity to the surface.

Once a current at such a high voltage is started in a particular direction the current tends to be very reluctant to turn sharp corners... because it is starting to have high inertial values. Since the crew and the entire craft are part of the circuit, whenever a direction change is made every molecule of the entire *polarized (unified) field is* accelerated at such a high rate of change into the new vector that the change appears uniform, thus by-passing the problem of structural fatigue due to non-uniform inertial shifts. This means the crew could be having a tea break and the pilot could turn a corner at 25,000 mph without spilling a drop of tea.

9) Partially due to hot spots in the shell circuit and 'laminar fluid lock' at the boundary layer on the surface-to-fluid interface, a pulse rate has to be induced into the transfer circuit.

For example, if the fluid flow rate were 1000 fps and the radius of the craft were such that the arc of the radius was 20 feet, then a pulse of 50hz would give a circuit power wavelength of 20 feet or the exact arc length of the point emitter (d) to the rim. By peaking the power wave at the three points on the underskirting's periphery, the turn (or curl) in the wave can be readily conducted to area (e) by the three variable resistors. As the fluid flow rate increases, the field pulse frequency will increase to maintain the same wavelength.

10) An effect which is the electrical equivalent of the "correolis effect" that makes water swirl one way going down a drain will cause the electrical field transfers of the craft to form a vortex as it moves from top to rim to area (e). Also, due to ionization potentials of the particular fluid in which the craft is travelling, there may be visible evidence of the swirling vortex. It will make the craft spin unless contra-torque is applied to hold the craft stable... This contra-torque is supplied by the returning ions on the underside of the craft. (There is, however, a great deal of contra-torque available in the secondary, energy storage mechanism of the air turbine in the practical craft).

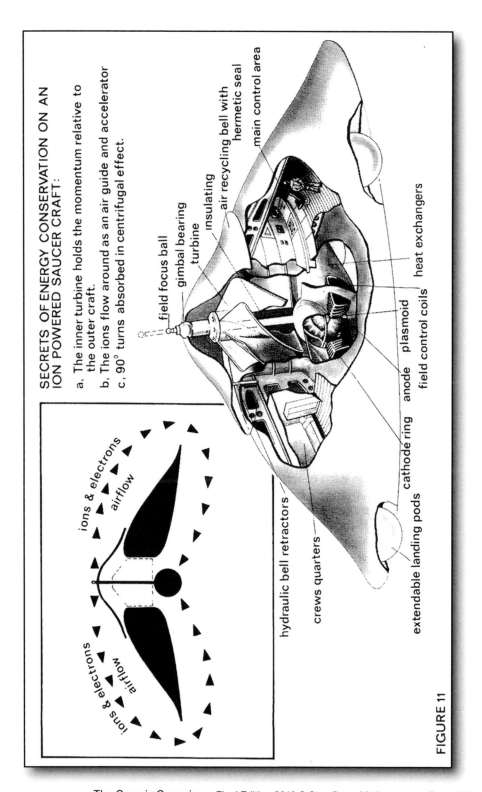

SECRETS OF ENERGY CONSERVATION ON AN ION POWERED SAUCER CRAFT:

a. The inner turbine holds the momentum relative to the outer craft.

b. The ions flow around as an air guide and accelerator

c. 90° turns absorbed in centrifugal effect.

field focus ball
gimbal bearing
turbine
insulating
air recycling bell with hermetic seal
main control area

heat exchangers

plasmoid
anode
cathode ring
field control coils

extendable landing pods

crews quarters

hydraulic bell retractors

ions & electrons
airflow

ions & electrons
airflow

FIGURE 11

Practical Ion Craft

Let us now look at the practical craft. In figure (11) is a cut-away of the craft showing: the airflow, electric ion flow, field focus ball and rod, turbine fan, plasmoid ion source, field coils, cathode ring, directional anode, navigational ion collectors landing rods, and crew quarters, etc.

There have been many attempts to fathom the workings of so-called "flying saucers". Some have actually discussed air flow mechanisms using ion accelerators. There has never been, however, a public report which showed the 'forbidden' (or unknown) secret... of how to maintain a high energy state device without a continuously equal high energy output per time.

Recall the case of the hot air balloon. It takes a certain amount of energy to heat the air inside the balloon. With 'proper' insulation, the balloonist can stay aloft for several hours on one heating. Does that sound like it takes a continuous high-density energy expenditure? Of course not.

What about the hydrogen-filled balloon. If released at ground level, it will rise to its 'specific gravitational' level. If taken from that level and drawn farther into space by a few miles and then let it go, what happens to it? It 'falls' back to its 'specific gravitational' level.

Even Leonardo da Vinci knew that. He once stated *"Gravity comes into being when an element is placed above another more rarefied element. Gravity is caused by one element having been drawn into another element... A light thing is always above a heavy thing when both are at liberty. The heavier part of bodies is the guide of the lighter part."*

If the craft and its field effect are viewed as a unit, then one will see that it has a more dense lower portion where the ions collide than it has in its upper portion... where ions are moving away from each other.

As long as the ion transfers follow that path, there will be a state of imbalance. To retain as much of the energy as possible when hovering, it is necessary to keep part of the motion of the air with the vehicle.

This is done by using a centrifugal turbine fan which recycles part of the ambient medium... (a manner of 'insulating' or containing the high-density energy source like the insulation in the hot air balloon.)

This fan also cools the containing area for the high-density energy plasmoid - which some people have seen as a dull red glow in the centre underside of their "UFOs"... and others have reported "... caused scorch marks upon landing." The three balls are landing pads in this version as it is necessary to have a certain distance between the ground and the underside of the craft to allow lift-off without taking great hunks of soil with the craft. The outer hull of the craft has positive and negative curve to compensate for the laminar turbulence (drag curls) that occur at higher velocities. The upper dome is movable vertically and even on a tilt to allow manual control of the recycled air. If the craft is to be sealed, the dome is simply dropped to form a sealed pocket. This of course means the energy expenditure from the field would have to increase to maintain altitude as the recycled air had been stopped.

The ball is vertically adjustable to change the effective voltage (charge distribution ratio) over the upper craft airspace and hence to change the field shape parameters.

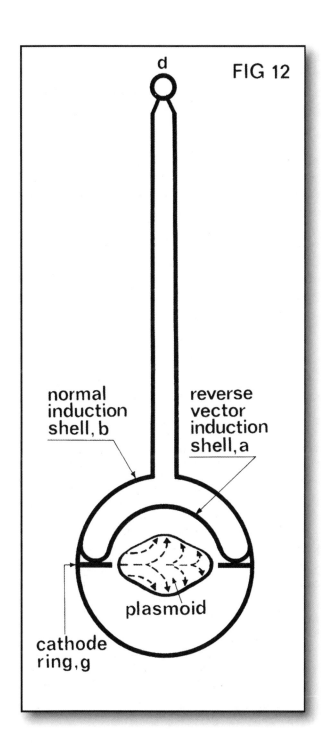

d

FIG 12

normal
induction
shell, b

reverse
vector
induction
shell, a

plasmoid

cathode
ring, g

The ball support shaft is heavily insulated to ensure proper charge distribution on the upper dome and also to effect the parameters of the field definition. There are heat exchange vanes attached to the plasmoid's containing inductor to dissipate the excessive by-product heat.

Underneath the plasmoid are shown three coils aimed at the centre of the spinning plasmoid. There would be more of these in practice but for the simplicity of the theory, three are used to show how flight direction may be controlled by balancing the current between these three coils. If the plasmoid is tilted while spinning, the field throughout the entire craft and crew tilts with it. Also, the accelerated air ions assume a correspondingly new orientation.

This gives an electrodynamic control function to the pilot which is infinitely better than those old-fashioned hydraulic systems. These field coils change relative strengths so fast that ultra high speed manoeuvres are exceedingly easy.

Just another short word on the charge path... The electric moment exits the craft at the ball; and since the voltage is so high the 'electrons' are reluctant to 'turn around and head back to the bottom of the craft.' There are some that enter through the upper portions of the craft's mass, but there are more that traverse the air to the lower third of the craft's mass before re-entry. This of course generates the air ion flow to the underside which in turn cools the plasmoid and recycles through the fan to exit from the upper dome to enhance the lift and thrust factors simultaneously.

If the craft is allowed to spin relative to the air and also to the fan, then high speed turns of thirty to forty "g's" can be 'amortized' over longer periods of time in the form of centrifugal spin... which counteracts some of the effect of being 'pushed' inwards toward the plasmoid in the momentary increases of field strength.

Communication to external sources can be effected in a number of ways; however, the most impressive is that one which modulates the field strength of the plasmoid with voice patterns. The broadcast covers a great number of radio frequencies simultaneously. In such cases 'receivers' can be the human mind all by itself. In figure (12) is a two-dimensional view of a three dimensional process. The spinning plasmoid induces an anti-vector current in the metallic (heat shielded) shell (a) which is a part of the metallic shell (b) (concentric). Because the current vector in (a) is against that of the plasmoid's, it does not 'short-out' into the plasmoid.

The reverse vector current flows up and into (d) where it is passed to the air, then to its collectors (e) and to the normal induction shell (b)... which is also connected to the (a-b) shell as shown.

Remember, the entire craft and crew are a part of the circuit. These voltages and relatively high frequencies develop a 'skin effect' all over the outside of the craft. Even though the crew are in the field, the current flow is in the outer surface because the internal charge crowding acts as an 'insulator.' No dielectric is necessary in the craft, as the current vectors act like 'phase-locking' loops.

Try to visualize the following:

1)   In figure (13-a) the top view of the plasmoid shows the old current vector (dotted arrow) and the new current vector in shell (a) (solid arrow)

2)   In each successive step the current vector is shown in its new stage versus its last.

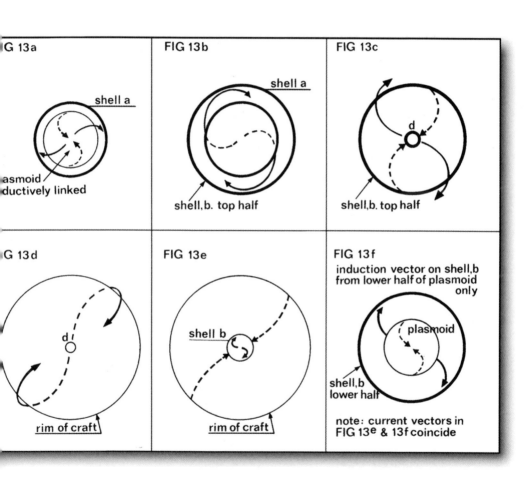

FIG 13a

shell a

asmoid
ductively linked

FIG 13b

shell a

shell, b. top half

FIG 13c

d

shell, b. top half

FIG 13d

d

rim of craft

FIG 13e

shell b

rim of craft

FIG 13f

induction vector on shell, b
from lower half of plasmoid
only

plasmoid

shell, b
lower half

note: current vectors in
FIG 13e & 13f coincide

3)   At stage (13-c) the current is anti-vectored exactly to the original current in the upper plasmoidal hemisphere (fig. 13-a, Shell a),

4)   In figure (13-d) the current has curled under the wing surface,

5)   In figure (13-e) the current has joined the induced anti-vector current from the lower half of the plasmoid (see fig. 13-f) to complete the circuit.

Throughout this exercise in building a hydrodynamic craft the fluid has not been called 'air' for the simple reason that this craft can sail in air, water, or even the fluid of space (often referred to as 'the ether or the fine structure or the quanta sea'). By varying the pulse frequency, power, and voltage levels on an electrodynamic craft, so-called 'anti-gravity', invisibility, and light-speed translation from point to point are now conceivable. Gravity has frequency... but that is another discussion all unto itself. Another discussion will detail the process for generating and storing extremely high voltage power in the form of plasmoids... (or self-containing plasmas), voltage transforming capacitors, and the rudiments or wireless broadcast of electricity to users around the entire planet... through the use of overlapping VLF standing wave power broadcast network.

Time permitting, this author will later release his papers on the order and origin of electron 'shells' and planetary orbits as functions of convergent and divergent vortexial wave forms in 'fluid space'. As a clue to those who would be interested in such a discussion, the reason that electron shell orbital radii do not apparently follow a progressively greater dimension outward from the nucleus is that they are the sum of two opposed progressions; one toward the nucleus (as a space-reflected, inertial wave form) and one away from the nucleus (as an energy-centre reflected inertial wave form). These papers will discuss the application of resonating magnetic fields to use the magnetic fields of the Earth and any other rotating magnetic body as not only sources of energy, but also new means of propulsion.

The Cosmic Conspiracy Final Edition 2010 © Stan Deyo 2010

APPENDIX No. 7: The Club of Rome's 10 Global Regions

REGIONALIZED AND ADAPTIVE MODEL OF THE GLOBAL WORLD SYSTEM

Report on the Progress in the

STRATEGY FOR SURVIVAL PROJECT

of the

Club of Rome

Mihajlo Mesarovic and Eduard Pestel, Directors

CONFIDENTIAL

September 17, 1973

1. Motivation and Objectives

The world *problematique* formulated by the **Club of Rome** is not only global in nature, involving factors traditionally considered as unrelated, but also points to the crisis situations which are developing in spite of the noblest of intentions and, indeed, as their corollary. To point out the problematique and the spectrum of critical and traumatic situations it entails is not enough; the acceptance of the reality of the problematique must be followed by changes if the concern is not to remain purely academic.

It is necessary, therefore, to present the issues within the problematique in specific and relevant terms which requires regional interpretation of the global issues. Furthermore, a basis should be provided for the resolution of conflicts (inevitably accompanying the problematique-type situations) through cooperation rather than confrontation. These factors have provided the motivation for initiation of the *Strategy for Survival project* which calls for the construction of a regionalized and adaptive model of the total world system with the following specific objectives:

(i)   To enable the implementation of scenarios for the future **development of the World System** which represent visions of the world future stemming from different cultures and value systems and reflecting hopes and fears in different regions of the world.

(ii) To develop a planning and options-assessment tool for long-range issues, and thereby to **provide a basis for conflict resolution** by cooperation rather than confrontation.

2. Basic Structure of the Model

The basic characteristics of the model are:

(i)   **The World System is represented** in terms of interacting regions with provisions made to investigate any individual country or subregion in the context of regional and global development. Presently the world system is represented by **ten** regions: **North America, Western Europe, Eastern Europe, Japan, rest of developed world, Latin America, Middle East, rest of Africa, South and Southeast Asia and China.**

(ii) In order to be able to deal with the complex of factors involved in *problematique* in a way which is sound, credible and systematic, a hierarchical structure has been adopted for the model in which each level in the hierarchy represents the evolution of the world system within a context defined by a given set of laws and principles. Specifically, the levels involved are: **Geophysical, Ecological, Technological (man-made energy and mass transfers), Economic, Institutional, Socio-political, Value-cultural and Human-biological.** Such an approach enables an optimal use of confirmed scientific knowledge and available data.

(iii)   An adequate view of the conditions in which the *problematique is* emerging and under which the solutions must be found require the recognition of the purposive aspects of the human community and adaptiveness of human beings. The model of the world system will have, therefore, two parts:

(a)   the so-called causal part, representing dynamical processes which follow historical patterns of development and

(b)   the so-called goal-seeking part which represents purposive changes under new conditions. The goal - seeking part in turn includes two levels: the decision-making or actions level and the norms level; the former represents the purposive response of the system while the latter represents the values and norms which constrain and condition such a response.

3. Progress in the Model Construction

The construction of the model as described in Section-2 and with the objectives as specified in Section-1 is certainly a rather complex task and the research is organized to proceed in parallel in several directions. The overall assessment of the model status is the following:

The model has been developed up to the stage where it can be used for policy analysis related to a number of critical issues, such as: energy resources utilization and technology assessment; food demand and production; population growth and the affect of timing of birth control programs; reduction of inequities in regional economic developments; depletion dynamics of certain resources, particularly oil reserves; phosphorus use as fertilizer; regional unemployment; constraints on growth due to labor, energy or export limitation, etc.

Specific developments which enable use of the model as described above include the following:

(i)   A **computer model of the world system** has been developed and validated by an extensive set of data. The model has two levels - macro and micro. On the macro level the model of each region includes the gross regional product, total imports and exports, capital and labor productivity and various components of final demand such as public consumption, government expenditure, and total investment.

On the micro level, eight production sectors are recognized: Agriculture, Manufacturing, Food processing, Energy, Mining services, Banking and Trade, and Residential construction. The input-output framework is used for the intermediate demands. A full scale micro trade matrix also has been developed.

(ii)   A **World population model** has been constructed in terms of the same regions as the economic model. The model has been validated by the data available. In each region the population structure is represented in terms of four age groups with appropriate delays which make possible assessment of population momentum and assessment of the effectiveness of implementation of various **population control measures**.

(iii) An **energy model** has been constructed which gives for each region the consumption and production of energy and inter-regional exchange of energy resources as a function of economic factors. Energy is treated both in composite terms and in reference to individual energy sources, namely solid fuel, liquid fuel, nuclear, gas and hydro.

(iv) A **food production and arable land use model** has been constructed which allows the assessment of a number of food related issues including: the need and availability of phosphorus required for intensive agriculture, and the **consequences of** timing and magnitude of **natural disasters** such as drought, crop failure due to disease, etc.

(v) A **major concern** in the application **of the computer model** is its proper utilization so as to avoid dependence on the deterministic aspects of model operation. In order to avoid this an interactive method of computer simulation analysis has been developed. The method represents a symbiosis of man and computer in which the computer provides the logical and numerical capability while man provides the values, intuition and experience. The method utilizes an option specification and selection program which enables the policy analyst or decision-maker to evaluate alternative options on various levels of the decision process, i.e., with respect to goals, strategies, tactical and implementational factors. **Special attention is paid to the norm-changing processes**.

4. Progress in Application

The **model has been used** both for the assessment of alternative scenarios for future regional and global developments (under different regional conditions) as well as in the interactive mode selection of policy options (specifically for the energy crises issues in developed regions).

Our efforts in the immediate future will be concentrated on further use of the already developed model. The plans include emphasis in the following three directions:

(i) Assessment in the changes over time of the span of options available to solve some major crisis problems.

(ii) **Implementation** of the regional models in different parts of the world and their connection via a satellite communication network for the purpose of joint assessment of the long term global future by teams from the various regions.

(iii) Implementation of the vision for the future outlined by leaders from an underdeveloped region in order **to assess with the model existing obstacles and the means whereby the vision might become reality**.

'Kingdoms': The Club of Rome's

10 Global Groups

GROUP 1: North America

Canada

United States of America

GROUP 2: Western Europe

Andorra	Luxembourg
Austria	Malta
Belgium	Monaco
Denmark	Netherlands
Federal Republic of Germany	Norway
Finland	Portugal
France	San Marino
Great Britain	Spain
Greece	Sweden
Iceland	Switzerland
Ireland	Turkey
Italy	Yugoslavia
Liechtenstein	

GROUP 3: Japan

GROUP 4: Rest of the Developed Market Economies

Australia	Oceania
Israel	South Africa (in Group 8 also)
New Zealand	Tasmania

GROUP 5: Eastern Europe

Albania	Hungary
Bulgaria	Poland
Czechoslovakia	Rumania
German Democratic Republic	Soviet Union

GROUP 6: Latin America

Argentina	Guatemala
Barbados	Guyana
Barbados	Haiti
Bolivia	Honduras
Brazil	Jamaica
British Honduras	Mexico
Chile	Nicaragua
Colombia	Panama
Costa Rica	Paraguay
Cuba	Peru
Dominican Republic	Surinam
Ecuador	Trinidad And Tobago
El Salvador	Uruguay
French Guiana	Venezuela

GROUP 7: North Africa and the Middle East

Abu Dhabi	Aden
Algeria	Bahrain
Cyprus	Dubai
Egypt	Iran
Iraq	Jordan
Kuwait	Lebanon
Libya	Masqat-Oman
Morocco	Qatar
Saudi-Arabia	Syria
Trucial Oman	Tunisia
Yemen	

GROUP 8: Main Africa

Angola	Burundi
Cabinda	Cameroon
Central African Republic	Chad
Dahomey	Ethiopia
French Somali Coast	Gabon
Gambia	Ghana

Guinea	Ivory Coast
Kenya	Liberia
Malagasy Republic	Malawi
Mali	Mauritania
Mauritius	Mozambique Niger
Nigeria	Portuguese Guinea
Republic of Congo	Reunion
Rhodesia	Rwanda
Senegal	Sierra Leone
Somalia	South Africa (South Africa in Group 4 also)
South West Africa	Spanish Guinea
Spanish Sahara	Sudan
Tanzania	Togo
Uganda	Upper Volta
Zaire	Zambia

GROUP 9: South and Southeast Asia

Afghanistan	Bangladesh
Burma	Cambodia
Ceylon	India
Indonesia	Laos
Malaysia	Nepal
Pakistan	Philippines
South korea	South vietnam
Taiwan	Thailand

GROUP 10: Centrally Planned Asia

Mongolia	North Korea
North Vietnam	People's Republic Of China

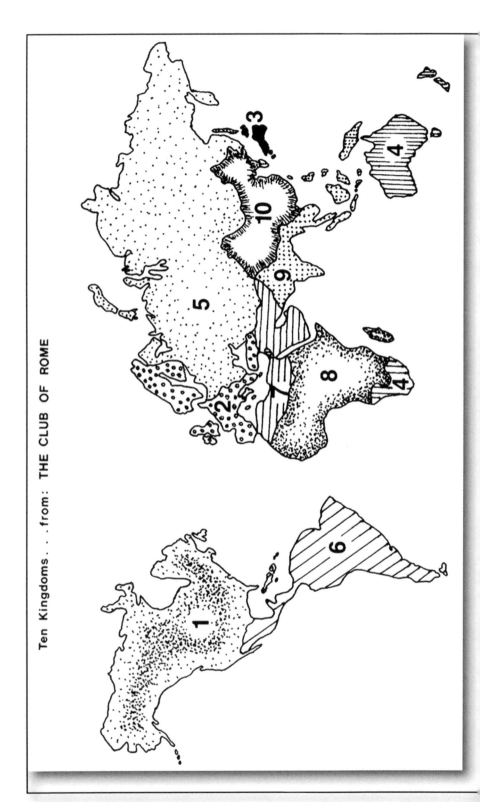

Ten Kingdoms. . .from: THE CLUB OF ROME

The Cosmic Conspiracy Final Edition 2010 © Stan Deyo 2010

THE 1992 ADDITION

It has been fifteen years since I first wrote this book. Until this edition - with the exception of a few minor editorial changes, it has remained the same as it first went into print in 1978 - a year after I finished it. Over the years, I have received communications from enthusiastic and supportive readers in over 22 countries. Some have also correctly noted errors in of my "PEG scenarios" (Partially Educated Guess scenarios); which concerned future world events (post 1978). They have noted my projected dates for certain highly-probable events had not happened on time - according to my original projections.

In the additions to this you will find I have re-calibrated my timings and enhanced my "PEG scenarios" based upon interim events since 1978. I feel that in an educated discussion this is allowed; for, when all is said and done, your letters, phone calls and tapes constitute a discussion with me on my original set of "PEG scenarios." However, the majority of my correspondents, by far, have mentioned the incredible similarities between current events and the "PEG scenarios" I first projected back in 1978.

Read well the following pages; and try to ponder their implications wisely; before it is no longer permitted - in these last remaining hours,.... in these last remaining minutes,.... in these last remaining seconds......   -Stan Deyo, April 1992

Update on the Valentich Affair

I first wrote about the apparent UFO abduction of the young Australian pilot named Frederick Valentich in 1978. Since that time, there have been numerous additions of testimony to the ongoing mystery of this young man's disappearance.

One such testimony was of a desk clerk in a hotel on the coast not too far from Cape Otway where Valentich disappeared off radar. Some two weeks after Valentich disappeared, the clerk relates that a young girl came into his hotel and asked him if Frederick Valentich was checked in there. When the clerk told her he was not and had never been booked into his hotel, she burst into tears and ran away.

Questions flooded my mind. Had she been part of a chess move planned by Valentich to force a public inquiry into the UFO film library kept at the RAAF base at Sale in 1978? Had he intended to ditch his craft and secretly meet her at this hotel on the date she had queried the desk clerk? Or, did she think he had ditched his plane; and she, herself, was not a party to any mention he might have made regarding a 'chess move.' Did the clerk tell the truth or was it a publicity stunt? Did Valentich's plane just crash into the stormy seas and sink? My answers are still inconclusive.

A couple of years ago, I had a telephone call from a man representing a housewife who claimed she had been abducted by a UFO in the Mid-western USA. He had been instructed by her to call me to see what clothing young Valentich had been wearing when he disappeared. Before he asked me the $64 question, he related her story to me.

She had been taken aboard a typical UFO of saucer shape by little grey guys with big eyes. In just a few moments they had travelled a few hundred miles to an underground base used by the alien occupants of the UFO. During her short tour of the hospital-like environment of the facility she was shown several frightening things. They had one area where human body parts were seen floating in a tank of greenish fluid. In yet another corridor, she was shown several transparent, upright cylinders which contained human bodies floating in the same greenish fluid. The humans or at least their bodies in these baths were being kept 'alive' by artificial means. One of the aliens made it known to her that one of these humans was the young Australian pilot, Valentich.

I asked the man at this point what clothing she said Valentich had been wearing inside the tube. He told me the color and material of the pants and the shirt and a few other details. He told me the woman did not want any publicity about her ordeal because of personal hardships it might cause. When he had explained the nature of these, I agreed; and I thanked him for his query.

As soon as I hung up, I rang Guido Valentich (Frederick's father). It was difficult to ask him after all those years to listen to the preceding tale to see if the clothing description matched that of his son's on the day he had disappeared. When I finished telling him the details, he seemed calmly relieved as he told me the description of the clothing was incorrect.

Since he has never revealed to the press what clothing Frederick was wearing, it was and still is a good test for an ID - unless, of course, Frederick changed clothing during his captivity prior to being put in that suspended state of animation. The search goes on.

Update on the Roswell Incidents

I am finally convinced that the story told by former US Air Force Major Jesse Marcel before his death concerning his part in the investigation of the Roswell Incident(s) which occurred on or around the 6th of July in 1947 are true. His story told of his own involvement in the investigation of one of two UFOs involved in what Dr Stanton Friedman has determined to be a mid-air crash over Roswell, New Mexico. According to Marcel's testimony and subsequently to that of at least five more credible witnesses and/or participants in the affair, two, saucer-shaped UFOs collided in the air over Roswell, New Mexico. Both suffered structural fatigue and crashed within close proximity of each other.

Supporting testimony comes from an engineer named Barry Barnett, who attended the other saucer crash site almost immediately after it landed. He was a child of five years age at the time; and his father and his uncle were with him in the desert. When they arrived at the site, they found four alien humanoids on the ground near the craft. They were about four feet tall and had the now cliché, "bug eyes." Two of the alien beings from the craft were alive; and two others were dead. His father and his uncle laid the bodies out to the side of the craft and tried to help the other two as one of them was obviously dying.

While they were at the site, a group of school children on an archeological foray into the desert with their teacher chanced upon the crash site as well. Barnett recalls that the alien who survived seemed to be projecting thoughts and visions of extreme loneliness into his young mind while they sat opposite each other near the craft. In fact, he found the sensation somewhat discomforting - like riding a rollercoaster.

The skin of the craft was found to be cold to the touch even though the temperature in the desert was typically high for the month of July. One of the other witnesses kept a piece of the hull which he later showed to his own son; whom I, myself, heard the tale from years ago while I was still a young man living in Dallas, Texas. According to the young man, the metal of the craft had been comprised of several thin layers of different metals which had been pressed together rather than being welded or alloyed. A small knife blade could peel the layers of the metal off one by one.

Other pieces of debris including what appeared to be a nine-pointed gear wheel which referenced a portion of the outer hull to the inner compartments were also souvenired by his father. I cannot explain either the gear wheel or the laminated metal scrap; but I can understand the coolness of the craft in the heat of the day. There have been numerous reports over the last fifty years of UFOs either hovering over one spot or landing for prolonged periods of time.

Some of these craft have left the surrounding environment very cold for up to two hours after leaving. Now, because of my own research into a method of converting ambient thermal energy into electricity, I am aware of a cooling effect on an environment around a thermionic converter.

Those of you who are familiar with Dr T. Henry Moray's power system developed during 1909-1943 will recall his boxes were cool to the touch while they were delivering some 5000 to 10,000 watts of power to a bank of filament globes. He thought he was getting energy from the "sea of energy" which was and is everywhere in the Universe. His device used "cold" cathode valves to convert the energy of this sea into electricity. Several very credible witnesses were allowed to inspect the inner workings of his converter; and some of them even commented on the amazing coolness inside the converter as it supplied the equivalent of five to ten one-bar electric heaters.

Also, you may recall the case of a Mr Harry E. Perrigo who invented a converter to extract electricity out of the air during the period 1916-1927. He found his device produced more electricity when a breeze was allowed to circulate through the room or when a warm body stood close to the antenna of his device.

In my humble opinion, the craft that young Bennett found to be cool had a thermionic converter attached to its hull. By converting the ambient thermal and nuclear mechanical energies into electrical and/or gravitational power, the alien craft cooled its environment while powering its systems.

Furthermore, if a generic UFO craft had purely "anti-gravity" as its propulsion, it would still encounter horrific inertial resistance (friction) as it moved through various media; thus producing massive turbulent wakes and sonic booms.

On the other hand, if the craft were to use "asymmetric" toroidal entrainment of the medium around its hull, then it could extract the heat out of the ambient medium while it literally sucked its way forward - leaving no turbulent wake. This type of propulsion is already known and covered by several US patents in regard to torpedo propulsion systems for the US Navy.

In summary, the Roswell incidents appear to indicate the craft utilise an electromagnetic field-effect in the propulsion technique; and, also, that they produce a "smoke ring" type of curl to the ambient media around their craft to remove the inertial feedback from their own forward motion. Finally, the cooling effect is indicative of a thermionic exchange which could be the main power source for the craft.

Captured UFO Craft

Where I could originally only supply a few names (like Dr. Edward Teller and Dr. James Maxfield) of those people who had been and still were responsible for much of the UFO cover-up in the USA, other contemporary authors such as William Cooper and John Lear have supplied a much longer list of those names under the code names of "MJ 12," "MAJESTIC 12" or "MAJORITY 12." These same authors and Bob Lazar have also pinpointed an advanced technologies research area in the Arizona/Nevada region code-named, "Area 51" or "Dreamland" which supposedly houses flying saucers and living aliens of various descriptions deep in underground caverns.

According to the young physicist, Lazar, he has worked on the propulsion systems of four, fully-operational, alien, saucer-shaped craft which had been captured by the U.S. Government in an effort to 'back-engineer' them into terms of our current industrial technology. His discussion of the matter/antimatter propulsion technique used by the craft is very close to something I mentioned regarding funding at NASA's JPL in 1975 for, "matter/antimatter mutual annihilations as a propulsive force."

Alien Infiltration of Man-made Saucercraft Projects?

In my latest book, *The Vindicator Scrolls*, I have included an interview with an informant from within British security who has confirmed my hypothesis on the man-made 'flying saucers.' This same testimony coupled with the inexplicably sudden cessation of all formal publications and nearly all media coverage of over 100, known, electrogravitic research projects in progress in May of 1958 makes one wonder if the 'elder source' did not infiltrate and take control of mankind's fledgling flying saucer research and development programs circa May of 1958... Yes, one wonders (see page 22 original version of this book).

In the *"The Vindicator Scrolls"* I have included a great deal more on the possibility that "Atlantis" did exist; and that it was the place where cloning between off-worlders (aliens) and man might have occurred *(see page 110 of the original Cosmic Conspiracy preceding)*

Aliens To Reveal Themselves By 1993?

I only have to wait a year or two to see if Dr Robert Jastrow (director of NASA's Goddard Institute for Space Studies, 1978) was correct when he stated he believed this planet would be contacted by aliens by the end of 1993.

Certainly, these emerging pieces of the UFO puzzle either support my *"conspiracy to invent aliens hypothesis"* or my *"conspiracy to withhold advanced technology hypothesis."*

Either way, I have been vindicated to a degree; because someone has gone to a lot of trouble to throw up a smoke screen over the whole UFO scene. What we discuss must be a threat to something very important to attract so much new information laced with so much disinformation (or 90% Truth + 10% "Bull Dust").....

The "Death Ray"

Much has been mentioned in the media concerning "Star Wars" type weapons. The American "death ray" which I disclosed has since been admitted by the U.S. Department of Defence; but only after the release of the movie, *The Falcon and the Snowman*, based on information from Christopher Boyce. The weapon partially ionizes a channel of air between the gun and the target using a laser beam tuned to a frequency which excites nitrogen molecules into conductivity. A series of short-period pulses of curling plasma toroids (doughnut-shaped rings) are shot toward the target by travelling along the surface of the ionized channel of air which passes through the hole in the middle of the toroids. I dubbed this system, "PTL" for Pulsed Toroidal Laser;" since I am not certain of its project sign.

The PTL method was found to be superior to the straightforward, high-energy Carbon Dioxide Gas Laser (CGL or "Prometheus' Fire") techniques which had severe functional difficulties. The CO_2 laser beams were very powerful; but when they struck any metallic targets, they melted the metal so rapidly that the initial cloud of metal ions caused most of the power of the beam to reflect away from the target. This cloud of ions was called the "plume effect." Furthermore, when the beam was targeted over several kilometers, its accuracy dropped due to interaction of the magnetic components of the beam path with those of the Earth's natural magnetic field.

The PTL device was designed to overcome both the "plume effect" and magnetic interference. The plume effect was overcome by using the natural curling motion of the toroidal plasma to scoop out the ionised metal at the target. This scooping effect proved far more destructive to the structure of the target.

The high-speed spin in the toroid also helped to stabilise the inertia of the internal magnetic components of the host laser beam. This toroidal technique was utilized in many more spectacular technologies far removed from weapons systems.

Unified Field Theory

And, while I am on the subject of toroidal fields, on page 32, in the discussion entitled, "Time and Space," I included my opinion on fundamental changes necessary to the classical view of nuclear structures.

Since then, Ephraim Fischback of Purdue University[1] has found a major flaw in gravitational theory resulting from a re-interpretation of the *"Eötvos"* data. Furthermore,

---

1   NEWSWEEK, Jan 21, 1986, p. 111

physicists John Schwarz of Caltech and Michael Green of Queen Mary College in London have recently published, *"The Theory of Everything."*[2]   Their paper is at the forefront of some 100 papers a month which are being presented in a new approach to formulating a "Unified Field Theory." Their theory uses one-dimensional "superstrings" which wind, curl and coil at infinitesimal sizes—producing toroidal fields quite similar to my "smoke ring" analogy.

The 'Face' on Mars

Research supplied by a well-known science newsman in both his 1988 address to several hundred space scientists and engineers at NASA's Goddard Space Flight Center and his 1990 address to the several thousand scientists and engineers at NASA's Lewis Research Center in Cleveland, Ohio strongly suggest that intelligent life left not only a solution to our unified field problem encoded in the Cydonia monuments of Mars; but also, they left information for us about a new source of energy in the form of the dynamic fields inherent in all spinning stellar or planetary bodies. This "new" source of energy is a form of converting apparently chaotic or random motions of mass into harmonic energy exchanges. his persistent analysis of many NASA photos taken of the Martian surface have led him and his team to suggest pyramids (some five-sided) and a huge face carving of a humanoid on the Martian surface were left as primer textbooks to us.

From these images and their mutual placement to each other, he has been able to devise a preliminary mathematical concept of spinning masses which has enabled him to accurately predict the correct latitude of a great spot on Neptune as a function of inertial curl. His hypothesis is closely akin to my own observations of spinning inertial systems and their relationship to mass and energy.

Geophysics and Weather

Mount St Helens has erupted. The San Andreas Fault has given California a hard shaking in the San Francisco shake; and Californian experts are predicting a uniquely severe one in the next ten years. The world's weather has gone absolutely berserk. Seasons are inconsistent, rainfall has been dropping off severely in North Africa and in portions of the old Soviet Union. It now seems some American scientists are saying the ozone layer hole(s) might not be due solely to aerosol propellants; but, as well, to changes in solar radiation frequencies and the solar wind.

Remember, the majority of the aerosol users live in the northern hemisphere - yet the hole in the ozone layer first appeared in the southern hemisphere. In 1978 on page 46 of this book I warned of yet another possibility in relation to the ozone layer. I warned of the possible alteration of the ozone layer by chemical and physical means to allow excessive UV penetration over targeted areas on Earth. The source of my data for this was Dr Gordon McDonald of Dartmouth College. How much of the current UV problem is intentional for, say, population control or other man-made initiatives?

Dr Adam Trombly of Aspen, Colorado, recently gave us another possible factor in the ozone layer depletion scenario. According to him, the solar wind mass has varied up to ten times its level of 1988 indicating a severe change in solar output levels; and it has caused external heating of our upper atmosphere to the extent that a number of near-Earth satellites' orbits are prematurely decaying from the increased drag of the atmospheric blanket.

Also, recent findings by Dr Trombly's avant garde group of physicists at The Institute for Advanced Studies (Aspen) have caused them to formulate radically new mathematical models for the structure and operation of the Sun; and those models are moving

---

2   TIME, January 23, 1986, p. 26-27

closer and closer to the one I suggested in this book some 14 years ago;.... our Sun may be about to throw off a shell and then collapse into a brighter - but smaller - star. Let us not forget, as well, that skin cancer is on the increase as a function of the increased UV exposure through the hole(s) in the Ozone Layer.

Update on the Planetary Alignment

A number of my readers have written to me requesting an update to the planetary syzygy plot I included on page 54 of this book. To that end, I have generated thirty years of additional plots using the same program. These plots are labelled, "Grand Planetary Syzygy 1986-1995," "Grand Planetary Syzygy 1996-2005" and "Grand Planetary Syzygy 2006-2015;" and are located on the following pages.

I also plotted the major earthquakes which occurred in the ten-year period between 1980 and 1989 (see Major Earthquakes Chart); and, discovering there was a partial correlation between the earthquake events and the planetary alignment events (albeit an inverse one), I produced a combined graph (see 1980-1990 Earthquakes and Alignments Chart) to allow you to perform your own analyses. It appears that major earthquakes occur just after a peak alignment of planets. A valley always follows in the alignment plot just after a peak; so this might indicate some correlation between inertial "whipping" vectors as alignments form; then re-form into less tight alignments.

The sudden change in the tides of the Sun and, hence, in some inertial mechanism linked to the spin of Earth might trigger crustal stress zones on the surface of our planet. Much more research is needed in this area; but, like all essential things these days, funding for such has to be prioritized.

Grand Planetary Syzygy 1986-1995

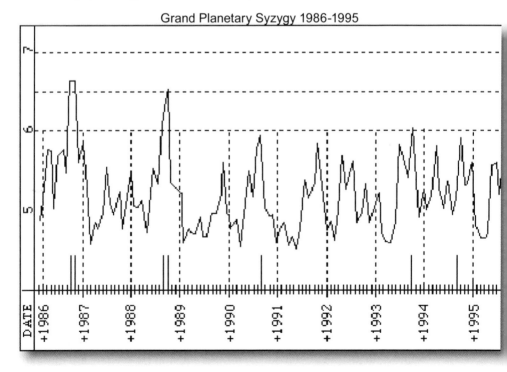

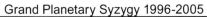
Grand Planetary Syzygy 1996-2005

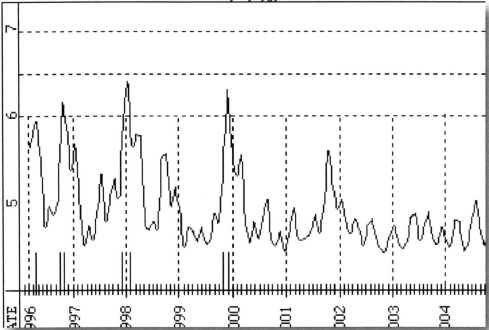

Grand Planetary Syzygy 2006-2015

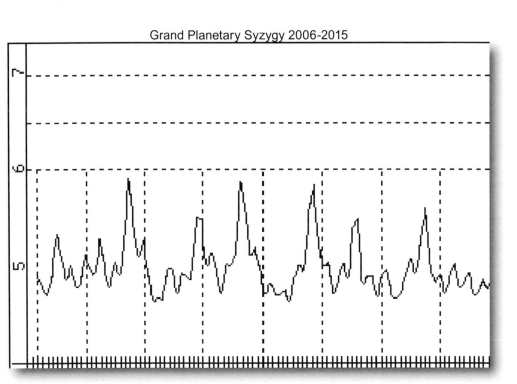

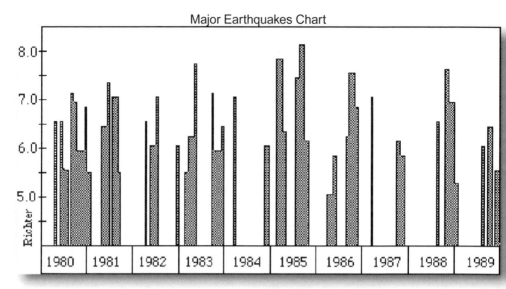

Major Earthquakes Chart

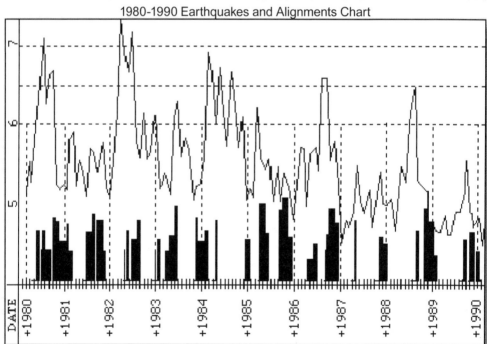

1980-1990 Earthquakes and Alignments Chart

UPDATE ON THE GREAT SEAL OF THE UNITED STATES OF AMERICA

On page 69 of this book, I gave a description of how the Great Seal of the United States of America had been designed and delivered to Thomas Jefferson at 'Monticello' which was his home near Charlottesville in Virginia. This description was drawn from a Rosicrucian publication on the subject; which stated the date of the Congressional approval for the design of the Great Seal was June 10th, 1782; and, furthermore, that the Seal was cut and delivered to Jefferson on June 17th 1782. The official versions of this event differ; as the following illustrates.

The official version ('OV') of the design and delivery of the Great Seal was published by the US Department of State in April of 1957. In this version, Franklin, Adams and Jefferson were appointed as the original design committee. The 'OV' went on to say that it took six years and three committees to negotiate the final design for the Great Seal. Apparently there were many divergent ideas as to what meaning the final design would convey amongst the committees.

The 'OV' went on to state that most of their ideas, "tended to transcend the miniature confines of a seal and the strict conventions of heraldry. Franklin, for instance, proposed a device showing Moses dividing the Red Sea for the children of Israel and the waters closing behind them to overwhelm Pharaoh in his chariot. The motto read, 'Rebellion to tyrants is obedience to God.' Jefferson favored showing the children of Israel in the wilderness under the guidance of a cloud by day, a pillar of fire by night."

According to the Encyclopaedia Britannica version[3] or ('EBV'), the first committee consulted with an artist by the name of Pierre Eugéne du Simitiére from Philidelphia to design the obverse (or front) of the Great Seal; while they allowed the reverse to be designed by Benjamin Franklin; who stated the eye in the triangle represented "the Eye of Providence in a radiant triangle."

The 'EBV' stated the second committee comprised of James Lovell, John Morin Scott and William Churchill Houston consulted with the Treasurer of Loans under the Continental Congress, Francis Hopkinson, for his artistic input. The 'EBV' continued, saying the third committee comprised of Arthur Middleton, John Rutledge and Elias Boudinot called upon the services of one William Barton to complete the seal design because of his expertise in both heraldic manner and artistic drawing. Barton promptly submitted two designs - the second of which was accepted by the committee on May 9, 1782.

On June 13th, according to the 'OV', the drawings of all three committees and all their motto suggestions were jointly submitted to the Secretary of the Congress, Mr. Charles Thomson; who then made a few changes to the artwork and added the words, "E Pluribus Unum" to the obverse and "Annuit Coeptis" and "Novus Ordo Seclorum" to the reverse. Thomson also re-introduced the Latin date of "MDCCLXXVI" (1776) to the reverse side. The final design of the Great Seal of the United States of America was approved by its Congress on June 20, 1782. A brass model was cut soon after; and the first public use of the Great Seal was recorded on September 16, 1782.

Which of these three versions is entirely correct? Well, obviously, the Rosicrucian dating differs with the Congressional Record of the time. As far as the mysticism involved, Charles Thomson is the one who seems to have added so many 'interesting' graphics and phrases. No matter who made the changes to the Great Seal, the symbolism is most pagan in my opinion....

Great Seal Phrasing

As stated in my original Cosmic Conspiracy, the Great Seal has two unusual phrases on the reverse side; as well as an unusual set of graphics. Since I first wrote on this subject, numerous letters and pieces of information on the subject have come my way. The Thomson phrases are, "Annuit Coeptis" and "Novus Ordo Seclorum." The 'OV' claims these phrases are to be interpreted, "He (God) has favored our undertakings" and "A new order of the ages." This may be the way they intended these phrase to be interpreted; but it is not what the Latin words say..... (far from it!).

The first phrase, "Annuit Coeptis" literally means, "Year of Beginning." The second phrase, "Novus Ordo Seclorum," can be translated several ways using the following definitions (where the words in bold are the preferred usages): "Novus" = new, anomalous,

---

3   Encyclopaedia Britannica (1965 Edition), Vol. "S", under "SEALS", pp. 235-238

young, fresh, strange or inexperienced; "Ordo" = order, class, arrangement, line or series; "Seclorum" = (now, this word is a difficult one for the every Latin scholar I have approached on the subject; because it appears to them to be some sort of a slang spelling of the Latin word, "saeculum" meaning, "of the centuries (ages)." However, I feel it is a slang word made from two words.

The first word, "sec" could be any of the following: 1) an abbreviated form of "secundum" meaning, "in favor of, according to, after, next to, behind;" 2) an abbreviated form of "secundus" meaning, "following, next, second, inferior, favorable or fortunate"; or 3) "sec" meaning, "to divide, cut or set apart from." The second word, "lorum" means, "lash, whip, strap or leather charm."

Now this means the phrase might be interpreted into two, diametrically opposed meanings. The proponent version would read, "Year of beginning: new order (class) set apart from lash (whip);" while the opponent version would read, "Year of beginning: new order (class) in favor of lash (whip)." Which version correctly describes the United States of America today? Since it is for the moment the strongest military force on the planet and since it is apparently the defender of freedom on the planet at the same time, you will have to be the judge... Remember what I stated on page 73 about the Sanskrit name, "Mahamudra" which means "Great Seal" and mystically stands for "the union of all apparent duality."

However, if the USA is part of that 'Babylon' mentioned in the prophecies of The Revelation of John, then, from the same book, you are strongly admonished to remove yourself from it to avoid the severe judgements which are soon to befall and utterly destroy it for its ungodly behavior....

As an ex-patriot Texan (that's semi-American) I would not like to envisage the destruction of my own blood relatives in the USA particularly parents, fellow siblings, children and grandchildren; but, I believe they are living right in the middle of 'ground zero' if my understanding of the prophecy is correct. It is a hard scenario to espouse.... very hard.

New World Order

Widespread fears of a global economic shut-down necessitating some new global order are rampant in the wake of the US-Iraqi War. Additionally, the dissolution of the USSR and the open call for the establishment of a "New World Order" (the same one I correctly warned about) by President Bush and other world leaders is further proof of my original PEG scenario on this new world order.

The world stands precariously balanced on what would appear to be a modern version of the Great Depression of the 1920s while the energy barons of the Middle East wheel and deal their way into the driver's seat of world affairs in the "New World Order."

Let us not forget that Dr Henry Kissinger's peace plan has been adopted as the 'bible' of the current Israeli-Arab negotiations which are the primed detonator of 'Homo Destructus" preparing to annihilate himself...

Furthermore, as I originally discussed in 1978, there would arise a great leader from among the population of Earth - one whom even the most wise, most knowledgeable and most understanding of the Hebrew elect (such as those of the Lubavitch Movement) would be fooled into calling, "HaMesshiah" (The Messiah). He will arise during a time of extreme crisis on the planet; and would appear to have all the "right stuff" to effect "peace and safety" for the people of Earth.

Even now, I are seeing the stage set for that earthman to be crowned by not only the Hebrew elect; but also by most of the population of this entire planet. There are only a few of us who still steadfastly espouse caution in accepting such a charismatic leader; but I may be completely vindicated on this issue in just a very short time.

Remember,... remember... when your time comes....

Russia appears to be moving toward the Western style of rule; as it declares communism dead. In fact, since the dissolution of the Soviet Union, the northern sectors of the former USSR have petitioned NATO for admission. The two Germanies have re-united; and the infamous Berlin Wall now sits in thousands of homes as curios of an era suddenly passed.

Anti-Semitism (as opposed to purely anti-Judaism) is returning to the re-united Germany and is spreading around the world in the wake of the Arab-Israeli disputes. Yugoslavia is moving towards a "Western" stance as we discovered in the Club of Rome Model; and we still do not know why Tasmania was listed as a separate country (page 89) unless it had something to do with world heritage legislation in regard to the forests and lakes.

As I write these words, the Australian government is announcing the inflation figures are the lowest in at least 30 years - under 4.5%. This is a sign of limited - if not - zero economic growth in the Australian economy. True to the words I originally reported,

"economic growth as well as population growth must be stopped cold sometime between 1975 and 1990 by holding world investment in new plant and machinery equal to the rate at which physical capital wears out...... It is hard to see how growth could be halted or even substantially slowed without a world dictatorship...... Many of the intellectuals now are so hungry for order that they would be willing to see the end of democracy and some new kind of Napoleonic order coming in."

Have I not warned in my original *"PEG scenarios"* of a 'New World Order' which will try to issue you a world citizen's ID number or mark; - an ID which you will need; because it will be your only source of earning or spending money in the "New Age."

The next time your politicians or your religious leaders try to force you to accept a personal identification number (ID Card) or a special ID mark which must be presented for virtually all monetary transactions involving income or disposal of income, remember not only that I warned you in 1978; but also that my warning was based upon the writings of a Hebrew prophet named, "Yohanan" (John) who in 95 AD scribed, "The Revelation of John" as it was revealed to him by the highest authority, THE YAHWEH ELOHIM: OUR CREATOR......

Did I not state that it would appear the United States of America was either a pawn of or the seat of a modern "Babylon" (a term synonymous with 'Chaos' or 'Confusion')? Today, other American scholarly students of John's writings are saying the same thing.... America is a (if not, the) major part of a modern Babylonian System.

Are you ready for the "New World Order?" Remember one of those 'PEG scenarios' I wrote , "...frozen in an energy crisis, saddened by the Watergates of the world, dying from environmental pollution, starving from food shortages, frightened of a global nuclear war, sick of moral decay, afraid of the daily news, bankrupted by global money fluctuations, unemployed from economic depressions, crowded by the ever present birth rate, frightened by the suspicion that a global weather catastrophe was about to happen... mankind, that great sluggish mass of which you and I- in reality - are a part, would have been ready for 'Alternative Three.'" I am yet to be vindicated on this one; but from reading the newspapers, I do not think I will have to wait long.

Energy, Energy Who's Got The Energy?

What Is Energy?

'Energy' may be defined either as the ability to do work or as the ability to produce motion of a mass in relation to an inertial coordinate system. 'Work' may be defined as the exchange of energy from one mass to another over a finite amount of time. 'Power' may be defined as the amount of energy exchanged from one mass to another per unit length of time.

To put this into more recognizable terms, when you use a two-element electric heater in your home, you are consuming two kilowatts (2,000 watts) of power; and, since a kilowatt is 1,000 joules of energy per second, it is the same as saying your heater is exchanging 2,000 joules of energy each second. In one hour's operation, 3,600 seconds would have elapsed; and the amount of energy exchanged in that entire period (the work) would have been 3,600 seconds x 2,000 joules per second or 7,200,000 joules.

What is the Value of Energy?

What is the value of energy? The entire known Universe is a sea of energy; and without that sea of energy we would have no galaxies, no stars, no planets and no sub-atomic masses. Energy is the potential or kinetic relative motion between all things known to exist. Without it,..... without it, we could not exist.......

It is second in priority only to our Creator. Nations are measured by the energy they command. Leaders hold control over their populations by controlling the flow of energy. Imagine what would happen to your country if there were no more petrol (gasoline), no more electricity, no more natural gas and only limited amounts of coal.

Airlines would stop; transport would be severely relegated to local distances with archaic means of motivation. Cars would not function; economies would collapse overnight; and food supplies would be greatly reduced because of transportation limitations. There would be no computerised systems in operation. Military systems would be seriously impaired and returned to a very primitive form of operation. There would be no lighting except by candle or crude wind or solar generators; and the radio and television industry would not exist. There are literally thousands of things you would miss severely.

Is this not putting a very high comfort value on your energy sources? Would this not persuade you to follow your leaders at virtually all costs; as long as you could have your creature comforts? You tell me, "What is the value of energy?"

The Electricity Bill

In the previous example with the two kilowatt electric heater, if you had left the heater on for just one hour, your electric bill would have reflected you had consumed two kilowatt-hours of work. which is a shorthand way of saying you would have consumed two kilowatts of power continuously for one hour. Based upon present electricity prices that work might be have been worth anywhere from US$.06 to US$.40 depending upon how cheaply your power company had obtained its electricity. Thus, if you were to operate your small electric heater for ten hours per day over thirty days, you would have to pay from US$18 to US$120 for it!

This would not include your use of any air conditioners, bore water pumps, clocks, clothes driers, coffee pots, compact disk players, computers, dishwashers, electric blankets, electric cars, facsimile machines, fans, hair curlers, hair driers, hand driers, hot water heaters, irons, lights, microwaves, ovens, pool filtration pumps, radios, refrigerators, security systems, stereos, stoves, tea kettles, telephones, televisions, typewriters, vacuum cleaners, video games, video recorders, washing machines, and the myriad of other appliances which serve us daily.

With the addition of these, the average home could easily spend between US$70 and US$460 per month for electric power. What is the electricity bill? It is a reflection of the value you and, hence, your society place on energy. If it is a great value; it is because it is a treasure to you both.

Community treasures should be protected from mismanagement, entropy and burglary. These three rules imply equal responsibility to the consumer as well as to the supplier of the highly treasured energy. Consuming nations (or "Looters" as Ayn Rand of Atlas Shrugged fame named them) should stop their waste of energy; while producing nations should ration their sales until a better energy source can be found. However, we all know the diverse cultural priorities which masquerade as laws derived from the various indigenous religions[4] prevent us from achieving a peaceful energy management program in time to avert a global conflict of major proportions.

Middle East Oil

It may be the oil producers are practising some sort of rationing already by raising their prices; but it is of little use if the process is unilateral; and the consuming nations do not curtail their energy excesses. Energy in the form of oil and gas certainly costs us more and more as the amount of viable reserves diminishes. In many countries the gasoline (petrol) costs are four to six times what they were just fifteen years ago.

Is burglary the only way the consuming nations can see to get their oil quota? The Gulf War over the Iraqi invasion of Kuwait was fought over the oil which will be needed to power the western nations in the near future. Unless the United States can secure a strong presence in the region, its dwindling oil reserves will be exhausted before the end of this decade. Even with a high rate of both primary and secondary recovery from its strategic oil reserves, the high capital cost required will make the cost of energy too high for both domestic and commercial applications.

It would appear the United States has not invested enough capital in the production of oil and gas from its vast reserves of coal. Had it done so; there might be no energy crisis today. However, had it done so, the pollution by-product in the manufacture of the synthetic hydrocarbon fuels may have defeated the object of the exercise. Perhaps the search for an efficient, non-polluting and renewable energy source should have taken a higher priority than it did. As a result the consuming Western Nations are tied to the whims of the Middle Eastern oil suppliers. One child has been overly self-indulgent; the other, over generous. It's just about time for their Dad to settle the argument.

Consider the following statistics carefully; and a frightening probability will begin to emerge. The accompanying chart [5] entitled, "Global Energy Consumption and Distribution" highlights the fact that 79.5% of the world's energy consumption is attributed to the Western Nations whose primary religion is either Roman Catholic Protestant Christian; furthermore; that 16.8% of the world's energy consumption is attributed to

---

4   "Religion" here refers to those complex, man-made institutions which have been erected from very basic understandings of the origin and maintenance of the Universe at the Hand of The Creator, Ha Yahweh Elohim. These institutions have evolved and are still evolving into a confusion of great proportion; and at each evolutionary progression they get farther and farther from the simple truth which was given to mankind in many ways through many prophets, many rabbis and only one Messiah: There is only One Creator; and our choice is either to work with The Creator's Plan or against The Creator's Plan. To work with The Creator's Plan, one must learn and follow the instructions for proper behavior within that Plan; to work against The Creator's Plan takes no effort at all.

5   1990 Britannica Book of the Year, published by Encyclopaedia Britannica, Inc. Chicago, USA. Data extracted and computed from 'Britannica World Data: Energy,' pp. 822-826 and 'Britannica World Data: Religion,' pp. 778-789.

the Eastern Asian Nations who have either no major religion or have various mixtures of Buddhism, Shintoism and/or Hinduism; and that 3.7% of the world's energy consumption is attributed to the major Islamic Nations. Israel is set apart from all the nations because of its emerging independence from all the nations.

97% of the Western Nations' global energy consumption is represented by USA 26.06%, USSR 15.58%, Canada 4.32%, United Germany 5.14%, France 3.12%, UK 2.98% Italy 2.12%, Brazil 2.09%, Poland 1.41% , Sweden 1.36%, Australia 1.26%, Spain 1.26%, South Africa 1.17%, Norway .99%, Mexico .98%, Czechoslovakia .85%, Yugoslavia .78%, Romania .73%, The Netherlands .69%, Belgium .58%, Finland .56%, Venezuela .52%, Argentina .50%, Switzerland .45%, Austria .43%, Hungary .39%, Colombia .34%, Denmark .30%, Greece .29% and New Zealand .26%. These Western Nations represent only 28% of Earth's population.

92% of the Eastern Asian Nations' global energy consumption is represented by Japan 6.7%, China 5%, India 2%, North and South Korea 1.2%, Indonesia .3%, and The Philippines .2%; and these Eastern Asian Nations represent about 42% of Earth's population.

96% of the Islamic Nations' global energy consumption is represented by Turkey .43%, Iran .36%, Saudi Arabia .35%, Indonesia .33%, Pakistan .32%, Egypt .31%, Iraq .22%, Kuwait .18%, Malaysia .17%, Libya .14%, United Arab Emirates .13%, Algeria .13%, Nigeria .09%, Syria .07%, Morocco .07%, Bangladesh .06%, Oman .04%, Qatar .04%, Tunisia .04% and Lebanon .04%; These Nations represent about 16% of Earth's population.

Israel's global, almost insignificant, energy consumption is .17%; with .10% of Earth's population. *(Note: Israel's share of the world oil store is stated reserves; however, reliable sources have informed me of two major oil fields which have been added to Israel's reserves. One is offshore underneath the Mediterranean Sea and the other is very deep beneath the Jezreel Plain in northern Israel. The latter is said to be equal to all the known reserves in the Middle East Arab fields.)*

Global Energy Consumption and Distribution

GROUPED NATIONS	% WORLD ENERGY USAGE	% WORLD OIL STORE	% WORLD NUCLEAR STORE	% WORLD COAL STORE	% WORLD GAS STORE	% WORLD HYDRO STORE
Western	79.50%	28.00%	84.00%	84.00%	58.00%	95.00%
Eastern Asian	16.80%	4.50%	5.95%	13.00%	2.00%	1.20%
Islamic	3.70%	67.50%	.05%	3.00%	40.00%	3.80%
Israel	.17%	.01%	.00%	.00%	.00%	.00%

Global Energy Distribution

Now if we look for both the most widely used and most readily convertible form of energy, we will find that oil is the leader in both instances. If the Western Nations had shown the foresight to develop more economic secondary oil recovery techniques; and if they had invested more capital in generating portable energy sources from coal and gas; then they would not be in the unenviable posture in which they find themselves today.

The accompanying chart[6] entitled, "Global Energy Consumption and Distribution" also highlights the fact that only 28% of the world's energy distribution is in the hands of the Western Nations; furthermore; that only 4.5% of the world's energy distribution is in the hands of the Eastern Asian Nations; and that 67.5% of the world's energy distribution is in the hands of the major Islamic Nations.

---

6   1990 Britannica Book of the Year, published by Encyclopaedia Britannica, Inc. Chicago, USA. Data extracted and computed from 'Britannica World Data: Energy,' pp. 822-826 and 'Britannica World Data: Religion,' pp. 778-789.

99% of the Western Nations' global energy distribution is represented by: USA 3.2%, USSR 7.1%, Canada .8%, UK .6%, Brazil .3%, Australia .2%, Norway 1.3%, Mexico 6.6%,Venezuela 7%, Peru .3%, Trinidad & Tobago .4%, Panama .6%, Aruba .3%.

84% of the Eastern Asian Nations' global energy distribution is represented by: China 2.7%, India .5% and Indonesia 1%.

99% of the Islamic Nations' global energy distribution is represented by: Saudi Arabia 20.6%, Iraq 12.1%, Kuwait 11.1%, Iran 7.6%, United Arab Emirates 6.8%, Libya 2.7%, Nigeria 1.9%, Algeria 1%, Indonesia 1%, Egypt .6%, Oman .5% Qatar .4% and Malaysia .4%.

Israel's global energy distribution is officially almost non-existent; however, this may prove to be a gross understatement.

Global Energy Strategies

The Western Nations use 79.5% of the energy resources of the planet. They control 28% of the oil, 84% of the nuclear energy, 58% of the natural gas and they generate 95% of the world's hydroelectric power. They will run out of crude oil within 15 years. Most of their military and industrial bodies are heavily committed to crude petroleum extracts. Their nuclear energy is not too portable. They have not manufactured enough new plant and equipment to utilise natural gas in sufficient quantities to replace the crude petroleum extracts in a hurry. The power from their hydroelectric schemes is not terribly portable either unless high-density electrical energy storage can be achieved rapidly.

Strategy: Prevent the Islamic Nations from joining the nuclear power producing nations to keep them from being able to produce and use nuclear arms against the Western Nations. Switch to gasoline production using coal extraction processes as soon as possible. Switch to natural gas as quick as possible. Reduce energy waste and penalise energy abuses. Establish a strong, permanent military presence in the Middle East to guard the oil fields against any damage or unfavorable change of ownership. Try to maintain tight control of all Islamic negotiables stored in Western banking systems. Limit Islamic ownership of major Western Nations' key corporations. In essence, try to control the Islamic Nations like a cowboy riding a wild steer at the rodeo.

The Eastern Asian Nations use 16.8% of the global energy resources of the planet. They control 4.5% of the oil, 5.95% of the nuclear energy, 2% of the natural gas and produce only 1.2% of their power by hydroelectric schemes. They will also run out of oil within 15 years; as their populations are now fast becoming consumer-oriented societies; and as such demanding more and more energy.

Strategy: Make friends with the Western Nations to get their industrial and military technologies where possible. Establish some trade relations with the Islamic Nations. Develop rapid mass transit railways and waterways within the region; and link to the Middle East by rail using the Old Silk Route through the Hindu Kush region to make shipping of Middle East oil and large amounts of Potassium Chloride[7] (fertiliser from the Dead Sea) to Eastern Asia a more cost efficient operation.

Develop nuclear weapons capabilities. Train large field armies of several hundred million men to organise and control the populations; and equip them with weapons and transport that do not require excessive gas or petroleum energy to produce or operate. Keep a watchful - but patient - eye on the Western Nations (particularly the fragmenting USSR) and their struggle with the Islamic Nations. Be prepared to grab the surviving oil fields in the Middle East should a war weaken the Islamic and Western Nations.

---

7   The Dead Sea contains over 5,600,000,000 tons of Potassium Chloride (Potash) which represents about 88 times the entire world's annual usage of this fertiliser! It would be of some value to the Asian agricultural community. Encyclopaedia Britannica, Macropaedia, Volume 16, p193; Vol 4, p133.

Build water distribution networks to channel the freshwater supplies from mountain streams to the fields. **Develop rainwater storage systems for the Indonesian island group and assist them in taking Australia from the Western Nations when the time is right.**

Develop Australia as an agricultural base. Utilise its mineral resources such as uranium, iron, vanadium, aluminium, nickel, gold and diamonds; and develop its oil and gas fields.

The Islamic Nations use 3.7% of the global energy resources of the planet. They control 67.5% of the oil, 40% of the natural gas, almost no nuclear power (excepting Turkey which Saddam has been thinking about for some time) and generate only 3.8% of their energy by hydroelectric schemes. They have such limited supplies of water, that they must depend on foreign industries to manufacture most of their needs at present.

Strategy: Carefully ration oil sales. Buy control of as many key Western Nation corporations as possible using proxy owners to avoid detection. Acquire nuclear weapons technology from the mainly Islamic southern regions of the former USSR (like Azerbajdzan, Armenia and Georgia), from Russia itself in exchange for oil, from Germany, from Turkey and from the Eastern Asian Nations through trade if possible. Use Palestinian terrorists to assemble tactical ("suitcase") nuclear bombs inside key US cities in preparation for a strike at the heart of the Western Nations. Organise all Muslims into a Jihad (holy war) against the "Infidels of the West." Adhere strictly to the fourth instruction of the Koran in relation to a Jihad against Jewish and Christian infidels who submit themselves to Islamic rule without accepting Islam; and, thusly, charge them the appropriate polls and land taxes.

Israel uses .17% of the global energy resources of the planet. She has limited energy resources according to all published sources; however, her strategic reserves of oil, gas and oil shale may surprise a lot of folks. Freshwater is growing very scarce as more Jewish folk return to Israel to live. The Islamic Nations want the State of Israel to be shut down. The Western Nations want to use Israel as a foothold in the Middle East to stay near the oil supplies. The Eastern Asian Nations want the fertiliser from the Dead Sea; and they want to control not only the Middle East oil fields; but also the Jezreel Plain: the gateway to the Western Nations and the bread-basket of Israel's agriculture.

Strategy: Firstly, to put her faith in Elohaynoo: Melech HaOlam and His Torah. Next, to develop oil refineries and prove the oil fields within her borders. To seek out and develop freshwater sources within her borders; such as artesian basins, and economic desalinization processes. To develop nuclear power and defence capabilities somewhat quietly; but to let the Islamic nations know how severe the retribution would be if they were to try to destroy Israel; to continue the reafforestation schemes; and to begin re-claiming land throughout the country by using the immigrant labor force which grows daily.

To develop more efficient ambient temperature thermionic conversion processes to utilise solar radiation and other low temperature heat and Brownian motion energy sources. To develop lightweight cements and concretes using polystyrene bead "aggregates" to give in-built thermal and acoustic insulation properties to the buildings. To develop trade relations with China in fertiliser for agricultural produce agreements. To view with caution any attempts to put the Palestinian issue "on hold' for a while by making a temporary Peace Treaty with the Palestinians which would give them a semi-autonomous rule of a select portion of Israel during the trial period. To avoid her people being "numbered" by anyone on their bodies with a consumer ID number.

To keep a watchful eye on the portion of the USSR that was once ruled by the Kossacks. To prepare now for that coming Northern Army invasion by establishing hidden "bolt holes." Lastly, to put her faith in Elohaynoo: Melech HaOlam and His Torah. Amayn.

Energy Control Would Give Someone Total Power

The Roman Union

Let us consider another 'PEG' scenario based upon the preceding chart. The Catholic/Christian block (let's call it the Roman Union) has control of most of the world's banking system, its large-scale food production, its advanced commercial technology, its massive industrial complexes and its very advanced weapons systems. However, if we ignore the strategic oil reserves of the Roman Union, it does not have enough ready, economic energy to last it more than 15 years - even if what it does have is equitably shared amongst its member nations.

Under this assumption, the situation appears to be approaching a crucial decision point. However, if the strategic reserves of the Roman Union were to be considered, another 15 to 35 years might be granted; but since that information is closely held by the member nations, it is an unknown factor.

The Persian Union

The Middle Eastern block (let's call it the Persian Union) has control of a vast reservoir of petroleum and natural gas. Assuming an annual population increase of 1% and a pro-rata, linear, petroleum consumption, they have enough oil to last them about 666 ± 34 years. They do not have enough food production; nor do they have advanced commercial technologies. Furthermore, they have to buy or capture much of their advanced weapons systems. They are for the moment at the mercy of the Roman Union in the world banking scene; however, when they formally declare the Persian Union they will be able to break that control decisively.

The Five Star Union

China is the Five Star Union. Its flag depicts five yellow stars on a field of solid red. It is no longer dormant. It is stretching its arms; it is wiping the slumber from its large eyes; it is hungry. For the time being, it has neither vast energy stores nor significant influence on the world banking system. It has few advanced weapons systems; and its commercial technology is still in the human assembly line stage. Its strength lies in its military might - not in its weapons; but in its manpower and in its mobility using the thousands of steam locomotives it has been building to use along the Old Silk Route to the Middle East where there will be plenty of petroleum, plenty of gold and plenty of bargaining power for the needs of its ravenous population.

The Northern Union

This is a union of Russia and a few nearby allies including Germany. Russia has little or no sway with the world banking system; but Germany has considerable sway with it. This Northern Union has advanced weapons systems and a growing mutual jealousy of the Israelis. It wants the immense strategic oil and gas fields which Israel has not declared; it wants the fertile fields of the Jezreel Valley and the fertilizers of the Dead Sea. It wants to use Israel as a watch-tower against not only the Persian Union with its emerging 'king"; but also against the Five Star Union with its growing hunger for energy.

Who Will Get The Energy?

Well, I only write conditional scenarios; I am not a prophet; but I do anticipate the possibility of the following moves on the energy stores in the Middle East:

1) The Roman Union will eventually be overthrown both economically and militarily by the Persian Union. The Persian Union will have gained considerable numbers of advanced nuclear weapons systems from the dissolution of the USSR. They will use these against the Roman Union - particularly against the USA; as it has already become the 'praetorian guard' for the emerging Roman Union.

2)   Since Iran (possibly the seat of the Persian Union) will have gained some of those advanced weapons systems from the southern countries of the former USSR, it will probably seek to control its immediate neighbors to ensure that the energy stores of the Persian Union will stay in Islamic hands.

I suspect the predominantly Sunni Muslim countries of Iraq, Kuwait and Saudi Arabia will be overthrown by the predominantly Shi'i Muslims of Iran to create a fertile environment for the sprouting Shi'i philosophy. This move would also give direct control of over 50% of the world's oil stores directly to Iran which would control the Persian Union.

3)   Instabilities within the Persian Union and the threat of the it controlling the Middle Eastern oil and gas fields will prompt the Northern Union to invade Israel after the Roman Union has been defeated by the Persian Union. This Northern Union occupation of Israel will probably end quite rapidly; as I suspect the Northern Union will suffer from extreme energy shortages and, subsequently, other catastrophic setbacks in the region.

4)   The Five Star Union will sit back and wait until the Roman, Persian and Northern Unions have sufficiently weakened each other to make it relatively certain that the Five Star Union can steam into the Middle East and take the lot away from the smaller, weakened unions. I think this will probably set the stage for the largest, most destructive war in the history of mankind: WWIII or perhaps, 'Armageddon.'

5)   All things may happen just because we have not set aside adequate stores of fossil fuels and have not developed viable, alternate sources of energy. Only an act of the GOD will help us avoid such a bleak scenario; but because of mankind's headlong race for energy, we have set a pattern of self-destruction. What would have happened had we developed that cheap, simple and viable alternative energy source....

Each of the nations in the following table is shown with the estimated depletion date ("Expiry") of its petroleum stores; however, these figures are excluding both strategic and secondary recovery reserves:

Global Petroleum Clock

NATION	EXPIRY	NATION	EXPIRY	NATION	EXPIRY
Algeria	2024	Indonesia	2009	Philippines	1994
Argentina	2003	Iran	2067	Poland	1999
Australia	1997	Iraq	2091	Qatar	2014
Austria	2002	Italy	2011	Romania	2007
Brazil	2002	Japan	2001	Saudi Arabia	2088
Canada	2001	Kuwait	2189	Spain	1993
China	2011	Libya	2055	Syria	2006
Colombia	2005	Malaysia	2006	Tunisia	2036
Czechoslovakia	2008	Mexico	2048	Turkey	2010
Denmark	2004	Morocco	2015	UK	1995
Ecuador	2001	Netherlands	1994	U. Arab Emirates	2087
Egypt	2004	New Zealand	2005	USA	1998
France	1997	Nigeria	2021	USSR (1989)	2002
Germany	2004	Norway	2017	Venezuela	2085
Greece	1992	Oman	2008	Yugoslavia	1997
Hungary	2009	Pakistan	1998	Zaire	1998
India	2008	Peru	1998		

What Sources of Energy Could We Convert?
Solar Energy: Heat, Wind and Waves

An enormous amount of solar energy reaches our outer atmosphere every year. In fact, about 70,000 times more solar energy hits our atmosphere than all the generated electricity mankind currently consumes per year. The amount of energy arriving from the Sun each year is 5000 x 10^{18} BTU. To make this figure easier to handle, we can just call that amount of energy 5000Q.[8]

Only 1500Q to 2500Q of this energy actually reaches the surface of the planet. The rest is either reflected away or absorbed by the upper atmosphere in the production of various chemical changes in the gases or in creating winds. 900Q is absorbed by the land and the vegetation on land. The sea absorbs the other 1600Q of energy using some of it to produce swells and a lot of it to feed its indigenous marine life forms. About 90Q drives the winds of the planet; and about .17Q is found in flowing water streams.

If we were to convert only .25Q (.01%) of the energy which actually reaches the surface of Earth into electricity with only 25% efficiency, we would find we had enough electricity to give every human being on this planet as much power as the average American takes for granted every day right now; and without environmental pollution!

To convert and supply the electricity for the entire Earth we would only need to dedicate an area some 350 kilometers by 350 kilometers (122,500 square kilometers[9]) in size for the solar collection arrays. This means that to supply the power per million people we would only have to reserve an area 4.5 kilometers by 4.5 kilometers (20.25 square kilometers) for the collecting arrays.

Mankind only utilizes some 2% or 1.8Q of the wind energy in various energy exchange devices and processes; and another 2% of its electrical energy is derived from hydroelectric systems; however taking too much from these ecologically balanced systems could severely effect the agricultural systems and the general ecology of the planet. Lunar gravitational tides could be utilized by various anchored pumping devices to generate electricity. Ocean wave actions could provide another source of energy using similar types of anchored pumps.

Fossil Fuels: Coal, Oil and Gas

The quantity of coal available at workable depths is 10^{12} tons. Annually, we consume about 2*10^9 tons[10] of it; and reliable estimates of the total recoverable coal[11] energy = 32Q to 150Q. At our present use of coal, we would have about 500 to 1000 years of coal left - as long as our demand for coal remains constant and as long as nothing happens to remove the major coal reserves from easy access; as would be the case should nuclear fall-out pollute the coal sites in the emerging Roman Union.

We consume about .04Q energy of crude oil per year. This means that if we could convert all the coal to gasoline and if we could recover 25Q energy worth of the oil and gas from wells, shale, tar and coal, we could project the fossil fuels energy shut-down date to be some 120 years from now based on a 5% annual consumption increase.

---

8   Encyclopaedia Britannica, Macropaedia, Volume 6, p. 857
9   This means a desert area about twice the size of the Navajo Indian Reservation in the mid-western USA could collect enough power to supply the entire planet with over seven times its current energy requirement.
10   Encyclopaedia Britannica, Macropaedia, Volume 4, p. 790
11   Encyclopaedia Britannica, Macropaedia, Volume 6, p. 858

Nuclear Fuels: Uranium, Plutonium, Deuterium and Tritium

Uranium is the most common nuclear fuel. Plutonium can be produced as a by-product of certain types of nuclear reactors called "breeder reactors;" so it is really a subset of the uranium reacted. Over 44,000 tons of uranium are produced annually on Earth. The energy demand for uranium-based power supplies will be about 165,000 tons annually by the year 2000.[12] The known and estimated reserves of uranium are over 2,000,000 tons (excluding the sea-water sources) which would give us only a 10 to 40 year supply; whereas, the sea-water sources would give us 1,000,000,000 additional tons of uranium in solution; which would easily yield a 1000 to 6000-year supply. The hazardous side of such reactors is the by-product radiation.

There will soon be a way to produce nuclear energy using deuterium (heavy water) and tritium (heavier water). This process requires an incredibly high temperature to cause a fusion reaction to occur in the fuel pellets made from deuterium or tritium. Although this process is theoretically possible, no one seems to have solved the plasma containment problem to allow a continuous energy output from such a device. A major attraction of this type of power is that a failure in the containment field would only momentarily let a relatively small amount of heat escape before the reaction would cool down and stop - leaving no residual radiation. The source of deuterium fuel is virtually infinite on Earth with an abundant supply (10^{15} tons) of it in the sea-water on the planet.

Who Controls the Major Sources of Energy?

As we have seen from the preceding charts and tables, the Islamic factions certainly have a large portion of the oil reserves in their possession. Even if they were to share that oil equally with all the consuming Western Nations, it would only last 45 to 50 years. Israel has over 11,000,000,000 tons of oil shale; and they have already built efficient power plants to prove the feasibility of the oil shale as a primary power source.

The nuclear energy production technologies, conversely, are held in the hands of the Western Nations. It is known that one pound of uranium yields as much energy as 3,000,000 pounds of coal. If we were to shift energy production to uranium, we would only need about 1100 tons of uranium per year to meet the energy demands of the entire planet. The major sources of economic concentrations of Uranium in various ores are: Zaire, Australia (3,100,000 tons), South Africa (273,000 tons), France, Czechoslovakia, Canada, Western USA, USSR Brazil, India, Madagascar and China. The Queensland uranium deposits discovered in 1970 in Australia are of massive proportions; and would appear to be a major asset to that country - although its environmental and labor groups have repeatedly thwarted any serious attempts at mining the ore (which may not be such a bad thing). In less economic concentrations; but still in considerable quantity, sea-water contains over 1,000,000,000 tons of uranium in solution.

The Common Factor in Most Energy Conversion

We get energy from coal by using steam turbines, smelting furnaces, domestic heating stoves, ovens and thermoelectric converters. Oil and Gas conversions are accomplished with petrol piston-driven engines, petrol turbines, gas turbines, gas piston-driven engines and thermoelectric converters.

Geothermal energy is utilized with steam turbine generators - although this type of energy is not an efficient source as Earth's surface is not good heat conductor and thermoelectric converters. Hydroelectric sources only supply 2% of current demand; and using all the waterways would stop flowing water.

---

12   Encyclopaedia Britannica, Macropaedia, Volume 18, p. 1037

Solar energy is converted naturally and by man in many different ways. Photosynthesis of solar energy-fed plant life which formed an integral part of oil and gas production. Mirrored furnaces produce high-grade steam for turbine generators; while low-grade steam turbine assemblies using black body absorbers convert the energy more efficiently albeit more slowly.

Thermoelectric converters in the form of silicon wafers (solar voltaic cells) at low efficiencies of around 16% are also used to convert solar radiation to electricity. Nuclear systems either use steam turbines or thermoelectric or thermionic converters to convert heat to electricity. All of the foregoing fuel conversion processes use thermoelectric conversion of the energy in varying degrees of efficiency.

Synergetic Energy Exchange Devices

Synergetic Energy Exchange Devices (**SEEDs**) are (in theory) devices or processes which through resonant or harmonic processes form a seed (or catalytic) matrix within a random field of energy; which, by its presence within the field, causes vectoral polarization of its chaotic energy exchanges into a coherent oscillation; thus transforming the random field into an ordered matrix after the order of the seed matrix.

In still simpler terms, SEEDs are ways of generating order out of chaos using less energy than is contained in the ordered chaos. A "black hole," in astronomical terms, would constitute such a SEED. The "black hole" is described as an object in space whose mass has become so concentrated that its gravitational field will allow neither its own electromagnetic radiations (of which visible light is a small portion) nor mass to escape from it; and furthermore, as one whose field is able to suck other entire star systems and other electromagnetic radiations into itself; thus, creating the illusion of a black hole in the space surrounding it because nothing appears to escape it. An interesting thing about these black holes is that, recently, avant garde astronomers have theorized they may re-emit energy in various extremely high electromagnetic frequencies such as gamma radiation. This would mean they suck-in incoherent energy sources and re-emit them as coherent energy beams (through their polar regions).

A laser does something similar in that it receives incoherent light from its strobing source; and then gradually over a few microseconds orders the flow of the photons into a lengthwise oscillatory traverse of the crystal or the gas in which the lasing action is taking place. In this example, the seed matrix is the atomic lattice of the lasing crystal. Both ends of the crystal are silvered; but one end is only half-silvered; so that it appears semi-transparent. When the reflected photons reach the threshold or maximum reflective surface area on the half-silvered end of the crystal, they make one more traverse of the crystal and then most of the photons slip between the silvered spaces on the semi-transparent end.

In the case of the laser, however, the coherent energy in the laser beam itself is only a few percent of the energy which went into creating it. The losses are large; and, hence, the efficiency is woeful. The important point to remember is that the random photons were ordered until they reached a breakthrough level at which time most of them punched through the containing surface at one end. This is a diodic (or one-way) effect which acts as an energy doorway ("Maxwell's Demon" to the physicist).

Now suppose you could make a device that would convert random atomic and molecular vibrations into a coherent energy source within a tuned electrical circuit. This device would be able to convert random motions (often called Brownian motions) or random photons of electromagnetic energy or phonons of heat energy into electricity at the ambient temperature of the media around it. For instance, it would be able to cool a glass of boiling water; and in the process re-charge an electric battery with the heat energy it had removed from the water.

"Is this possible?" you ask. Yes, of course it is. As just noted, the form of energy conversion common to all the major sources of energy is either a thermoelectric or thermionic device. Nuclear reactors onboard orbital and deep space probe craft use an impractical, high-temperature differential of about 1000°C between two semiconductive plates to convert the heat of a nuclear reaction directly in electric current. The efficiency is only 12%; so there are no rave reviews about the cost-effectiveness of such devices. In fact, most classical thermionic or thermoelectric converters have an efficiency around 12% ± 5%.

I am talking about a more efficient form of a technology that already works; and, furthermore, about a form which will work over a wide range of temperatures and frequencies. The sad thing about this technology is until we either engineer such a device ourselves or the "powers that be" allow it, we will not have the pleasure of using one.

Our Search For That "Free Energy" Device

Many inventors have claimed the successful design and construction of a "perpetual motion" or "free energy" machine over the last two hundred years; but none of these devices seems to have survived either the test of a proper scientific evaluation or the lure of a high-priced sale to some private or corporate entity which would keep the technology from the public. There may well be some members of the modern college of "free energy" enthusiasts who have actually succeeded in extracting energy from sources not presently defined in applied physics; but until I able to see demonstrable proof that they can take the power out of such a device itself and feed it back into the input side of the device as the only source of power, I will not be convinced.

Most of the modern "free energy" generators fall into the generic category of "over unity" devices; which means they generate more than 100% of the power it takes to start them up and to run them. Most of them use exotic magnetic arrangements where coils of wire are wrapped around magnets. The magnetic fields are then varied mechanically in various manners to produce current flow in the coils. The inventors of these devices usually make claims that some "zero-point energy" source in the etheric fine structure of space is responsible for supplying the over-unity power in their devices; and, although this may be the case, I still have not seen sufficient evidence from any of the cases I have investigated that they have been able to disconnect their devices from the initial power source once they have started to produce "free energy."

I have a close friend and business associate named Ed Scott. Ed is an electronics engineer. We have spent the last six years researching this subject together and prior to that, collectively, for some thirty years, although independently.

During that time, we principally examined the works and philosophies of the late Dr Thomas Henry Moray[13] and Dr Nikola Tesla[14] down to the most minute detail.

We also correlated their findings with the works of José Yglesias[15], Harry Perrigo, Roy J. Meyers, Hans Coler, Stanley Meyer[16] and other less well-documented cases. A common factor linked most of these systems; as they all required a tuned circuit of some description to extract this elusive "free energy" from the environment. Each had a slightly different approach; but each appeared to have accessed some extra energy using a resonant oscillation which we refer to as, "the seed matrix.".

---

13   See "Sea of Energy" written by Dr T. Henry Moray, COSRAY Research Institute, 25 South 4th East, Salt Lake City Utah, USA 84115.

14   Refer Tesla Book Company, PO Box 121873 Chula Vista, California, USA 91912, Telephone 1-619-426-8213.

15   For more info on José Yglesias, Harry Perrigo, Roy J. Meyers and Hans Coler see "Space Energy Receivers" © 1984 by Lindsay Publications, Bradley, Illinois USA.

16   See article Electronics World + Wireless World, January 1991, page10-11.

With several of the inventor's work we found positive evidence of a cooling in the immediate environment around their antennae structure. In some of Dr Moray's later notes *(part of which we inherited from a deceased estate in the US)* we found discussions about Brownian motion and its relationship to various interface phenomena involved in his energy conversion devices. For instance, it was reported by several reliable witnesses in sworn statements that Dr Moray's devices were cool to the touch - even internally - while they were delivering some 5,000 to 10,000 watts of power to an array of filament light globes.

Dr Tesla built and patented[17] a simple device using a highly-polished flat plate as a "collector" which he used to collect various frequencies of radiation including infrared (or heat) from the Sun and from artificial sources. This plate was connected by a wire to one terminal of a capacitor which had an electrostatic motor shunting the terminals. Even though he tuned the plate and the capacitor to allow free resonance the capacitor still took a two days to accumulate sufficient charge to move the motor. Dr Moray used Tesla's work on this device as the basis for his own patents later.

In another report on the work of Harry Perrigo, we read that his device worked better when fresh air was allowed to circulate through the room and his antenna. Furthermore, he stated that if a warm body like a person stood close to his device, it produced more power. Conversely, if he closed the room off and removed himself and other heat sources from proximity to the device, it diminished its power production in significant amounts.

Stanley Meyer's work in separating water molecules into its component gases using a pulsed resonant circuit of high voltage gives a similar clue. He uses concentric, parallel plates of stainless steel which form a capacitive element in a high-frequency, parallel resonant circuit tuned by an external inductance. He includes a diode and a current limiter which function somewhat like the half-silvered mirror on the laser crystal we have just discussed. The voltage between his plates rises until the water breaks down and carries a momentary high current which causes the water to separate into hydrogen and oxygen gas. The system apparently requires far less energy to split the water molecules than a normal electrolytic cell. Again, the astonishing fact is that the water never heats up during this process!

Based upon the work of all these scientists, Ed and I have recently designed our own primitive valve in the process of creating a "free energy" device that will demonstrably work and will efficiently convert "heat" energy in the forms of electromagnetic radiation and molecular motion directly into electricity.

Our design will use a cylindrical metallic pipe which has a wire running down its centre and parallel to its length. The whole assembly will reside in a sealed chamber at slightly reduced atmospheric pressure; and as a whole will be somewhat like a cold anode valve.

A high-voltage, high frequency pulse will be imparted to the wire to cause a natural resonant frequency to ring between the wire and the surrounding metal tube. The object will be to establish a seed matrix in the form of electrons oscillating back and forth between the wire and the tube wall. The heat and vibration of the tube molecules will ensure a certain number of random molecular vibrations in the wall of the tube will be coherently vectored with electrons heading back toward the wire in the centre.

The distribution of random vectors follows a standard "bell curve" which means at least a small portion of the time there will be an concentration of vectors toward the centre wire. These will add speed to the returning electrons by virtue of a mechanical

---

17   See US Patent number 685,958 "A Method fro Utilizing Radiant Energy" dated 21 March 1901 and US Patent number 577,671 dated 23 February 1897 - both granted to Nikola Tesla.

collision. This increased speed will cause an electric field increase at the centre which will initiate a sudden discharge of current across the diode in series to the load circuit. As the diode will prevent current flow back into the system, it will mean the energy imparted to the electron by the vibration of the molecules in the tube wall will have been converted into electricity and siphoned off.

This will cool the wall of the tube; and will also create a vector hole in the molecular vibration of the wall everywhere the energy was siphoned off by an electron. This will cause more of the molecules of the wall to fall into the hole left by the converted or lost energy; so increasingly more energy is converted per second as the "bell distribution curve" begins to split into two peaks - one on either side of the norm. This would explain why Dr Moray stated repeatedly, "the more energy you draw off, the more energy my system can supply."[18]

What Would Happen If We All Had Such A Device?

What would happen if the average person could get his own cheap energy directly from his environment using an efficient thermionic converter? To start with, he would find his electric bill would become almost nothing; because he would not be consuming a resource that costs anything. he would be able to make his own air conditioner which produced the electricity to cook, light and power his household in the process. He would be able to make his own, simple electric car which would never run out of energy and which would be easy to service.

Freshwater could be obtained from sea-water (as could deuterium to produce more nuclear heat) by using virtually "free" if not very cheap energy to distil the water. Another method of obtaining freshwater might be to make an electric dew condenser along the lines the Australian Aborigines have used for centuries; but using electricity to increase the amount condensed from the air even during the daytime. This freshwater could be re-cycled; and could also be used to water plants growing in electric greenhouses.

If he were to utilise the old electric power grid currently used in his community, the average person could provide back-up power to other users by dumping his excess electricity onto the grid. Also, by using the new Vanadium[19] fluid power banks, he could store large amounts of off-line energy for use by customers who want to use power without cooling their immediate environment.

The downside of such a miracle device would be the economic effects it could generate in the community. As more people began to make their own electricity, more people would become unemployed in the fossil fuel-dependent industries. Also, more accidental deaths and injuries would occur as the home-inventors

---

18   If you wish to read more about my thoughts and activities along this line, you might read my last book entitled, The Vindicator Scrolls; where a discussion on our atmospheric energy pump is presented.

19   Reference to: "Vanadium fluid-powered battery" invented by Dr Maria Skyllas-Kazacos and Dr Miron Rychcik of the Department of Applied Science at the University of New South Wales in New South Wales, Australia in 1987. Project was funded by the Australian National Energy Development and Demonstration Council. The world license for this process has been taken up by the Agnew Clough Corporation in Perth, Western Australia. Coincidentally, Agnew Clough own Australia's only vanadium mines and plant. Process suspends crystalline vanadium in two containers of fluid. One contains positively-charged vanadium while the other contains negatively-charged vanadium. When the two fluids are allowed to pass through special battery cells, electricity is produced. The fluids can then be re-charged and re-used over and over. About 70 litres of these fluids can supply enough charge to run a car for 150 kilometers before a fluid exchange is required.

began to proliferate their own functional - but perhaps unsafe conversion devices. The energy companies would have to shut most of their refineries except for those needed to produce plastics, lubricants and rubber; which would cause a major loss of revenue and of political power.

The government would lose the ability to monitor individual energy consumption; and it would suffer a loss of tax revenues; however, it would also have the opportunity to set up power sharing grids whereby producer/users could sell their excess energy to the government for its own uses or as a form of personal taxation. If energy were to become freely available to all, the government would not be able to establish the energy quantum as the basis for a new economic order; because anyone could get energy; hence, its value would not be a negotiable item.

Certainly these are only small problems when one considers the benefits to the community. Manufacturers would be able to produce cheaper products because their energy would be basically "free." They would also be able to provide better working conditions to their staff. The anti-pollution lobbies would get off their backs. Molecular physics would be affordable; so that precious elements such as gold, silver and deuterium could be extracted from sea-water using "free" electricity. Atomic transmutation of elements on a larger scale could be effected using the cheap electricity; thus could antimatter be created and stored cheaply for use in subsequent storage of energy.

The community would see the creation of jobs for those who will manufacture new converters. Solutions to reclaiming brackish water would be self-evident with "free" (very cheap) energy. Vegetation could be grown faster to replenish our oxygen supply. The reduction of hydrocarbons and chlorine, bromine, fluorine compounds in the air would result as there would be a growing decline in the use of fossil fuels; and energy exchangers would maintain thermal equilibrium or at least consistency in the environment.

What Does It All Mean?

The warmth of the Sun holds back the chill of the times for a moment; as I sit here underneath an old White Gum tree. The shadows of the tree leaves and branches sway gently to and fro in the calm of the afternoon. The odd, small twig falls to the tin roof of my writing room with a small, clanky thump. My stomach is full; and my eyelids feel heavy; but, I tell myself, this is allowed now; because my work is finished. The 1992 update to The Cosmic Conspiracy is complete.

Here and there in the garden around me, cheeping sounds of the smaller birds add to the reverie of the moment; and, occasionally, a rising crescendo invades the calm as a Kookaburra[20] goes hysterical over some "in-joke." When startled by a sudden breeze or a dog straying too close to its property, the neighbor's Peacock shouts an unmistakable, "OH NO!"

Rustling leaves spray my ears with almost hypnotic white noise; and intermittently, I perceive the sounds of our children laughing and playing in their cubby house in the front yard. In the distance I hear the doppler-shifted pulses of a motorcycle passing through the neighbourhood; and high, overhead I hear the lazy drone of someone's single-engined aircraft playing in the clouds. My mind wanders higher and higher....

I know that somewhere out there, whizzing around the Earth at a snail's pace are other humans engaged in their occupations as astronauts. Perhaps there are other alien life forms around or near to them, watching, waiting,... even baiting them. A door slams in the wind with a crash that brings me back to Earth!

---

20   An Australian bird known as the "laughing jackass" because of its fiendish laugh.

Questions flood to my mind in an adrenalin rush. I ask on a personal basis: Why do I have an increasing stack of unpaid debts in my "bills payable" folder on the desk. Why can't we afford the time or the money to effect badly needed home and car repairs? Why is it our children look wishfully at toys and clothing they know we will not be able to afford for a long time? Why is it we find it harder and harder to buy adequate food for the family even with the multitude of "bargains" at the stores? When will the Bank be forced to foreclose on our overdue home loan?

And on a global basis: Is there a threat of yet another war in the Middle East? Will there be an Indonesian invasion of Australia? Are there any man-made governments capable of solving the current world problems of famine, drought, flooding, energy shortages, nuclear weapons stockpiling, war, the Arab-Israeli (Palestinian) conflict, ozone layer depletion, forest destruction, water pollution, weather anomalies, earthquakes, economic chaos, dolphin genocide, whale genocide, gorilla genocide and human genocide?

Finally, at the end of the rush of all these mind-numbing queries, I ask myself, where is the safest place to take my family in anticipation of the future? In human terms I realize there is none. No place on Earth could be spared the trauma of the coming man-made and GOD-ordained catastrophes on Planet Earth. We would not be able to escape the coming problems with sudden variations in the solar radiation levels. We would not be able to escape the ravages of war. We would not be able to escape the financial and military collapse of the Western Nations (the Roman Union). We would not be able to cope with an alien invasion. We would seek death as an escape; and, yet, sadly, it would elude us. And yet, as soon as these thoughts race through my mind, I realize the answer is not to be found in human terms but in the Terms of our Creator.

As I walk in the garden, a gentle breeze caresses my skin with the cool of the day. I sense, for the moment, we are caught in that pregnant pause, that deceptive calm which is the "Eye of the Storm." I ponder this; but, at the same time, I realize I have done the right thing by warning all of you about these things; and, yet, there is no joy at being a harbinger of this doom. Sadness wraps itself about me; as I realize so many of you either have not heard or have not accepted the alternative to the horror that awaits the world in the next decade or so. I also rest assured that my household and I have done our very best to battle the great odds against us; so that we might share even these few ideas with you.

The watchmen cry out to their people, "Get out of Babylon!," "Join GOD's team!" Sadly, oh so sadly, not many can hear their warnings; because they sit on comfortable pews within "churchianity" which somehow exempts them from any serious involvement; or they piously attend their synagogues (not to be confused with their Temple) in ritualistic mindlessness; or they do not study the writings of the prophets; or they do not accept any personal liability for our global mess; or their televisions are too loud; or their music drowns out the warnings; or, worst of all, their pride exceeds their common sense.

I can only speak for me and my household. As for us, we have chosen to serve our King, The Creator, The Elohim ("GOD"). In obedience to His Word and to the best of our abilities, we are involved in trying to make the world a better place in which to live until He returns for us; which means we also realize the need for our Creator's "operating instructions" to the Planet Earth. Until all sentient beings on or around Earth accept this need and His Word, there will be no lasting peace, order, harmony or joy.

Will You Do Something About It, Now?

Sign Page 127?

THE 1998 MILLENNIUM ADDITION

Here we are, twenty-one years since this book was first written and we find ourselves pacing to and fro like expectant fathers waiting for the prophesied, "end times" to begin. We are definitely at the doorway to the 21st Century and, possibly, to the beginning of the Great Tribulation spoken of by the Hebrew prophets of ancient times. Please don't misunderstand me, I don't believe we are to expect great changes in the Earth **just** because we are at the turn of the century; however, I do believe we can expect greater changes in the Earth and its ecosystems because pre-cursive events are already starting to occur - regardless of the year.

Many of these events point to an imminent start to what most North American Indian cultures call, "The Purification Time" and to what sincere students of Biblical "end times" prophecies believe to be an immediate start of the Great Tribulation Period in which the earth will be "purified" or "judged, purged and made new." They realize that even if we could assemble the resources needed to stockpile food, medicine and water for our families to 'weather' the coming catastrophic changes, we could not guarantee we would be able to avoid such changes as: asteroid impact, increased solar radiation, nuclear war, severe earthquake, nuclear winter, magnetic pole shift, epidemic disease, tidal wave or even, 'alien' invasion.

Sure, we do realize there would be some wiser places than others to make our domiciles or to establish our businesses in anticipation of the coming changes; but no man can cast guarantees of peace and safety in concrete. Having taken this gloomy position, I must add that I do believe there is one sure way to prepare for what is coming. You may disagree with me; but it is the Way my wife and I and many others have chosen. I make no apologies for my boldness nor my single-mindedness in what I am about to say. Holly and I are not members of any organization, cult or 'weirdo' group. We are just two plain folks sharing what we have gleaned from the Bible and other Judeo-Christian students of the Bible.

We have made an unequivocal acceptance of Jesus of Nazereth - The Messiah - as our Champion, Protector and real Friend. However, we do realize, even followers of our Messiah are not guaranteed exemption from physical death prior to and during the coming tribulation of these times. His followers are all guaranteed they will inherit the Kingdom of Heaven and will gain everlasting life in a new kind of body which will be constantly renewed and thus will never die. Only a small percentage will actually never die from their physical bodies in the normal way. To me this makes it imperative that every soul make it their highest priority to prepare for the life hereafter; as that is - unquestionably - the ultimate preparation.

It does appear that catastrophic death is highly probable for most of the world's population over the next seven to eight years. There will be NO SAFE PLACES on Earth. However, I DO BELIEVE you can be taken to a place where my Messiah ("Moshiani" in Hebrew) will shield you from the final judgments. All you have to do is to make a conscious commitment to accept my Messiah's gift of being saved from the coming judgment of GOD on an evil and unrepentant world (pp 120-127 of this book).

This does not, however, mean that you should not put away some biological necessities like food and drinkable water in case of civil emergencies and loss of community resources.

This makes good sense at the best of times; but especially so, if one happens to survive any of the great judgments which I believe are coming. If you don't survive, there will probably be someone after you who will use what you left behind; so the effort will not be wasted - especially if a written record of the Gospel and the promises of GOD are left with the provisions.

In the event of a serious disaster, the average western world's city has only about eight days of food on hand. Food reserves would disappear even before this; as hoarding and looting began to spread. If you think this is an exaggeration, consider that just two days **before** a big hurricane strikes the Florida coast, the shelves in the food stores are emptied by the locals in panic buying.

Folks how to prepare for diseases, famines, droughts, earthquakes, volcanoes, war, asteroids or alien invasions. We tell them to prepare as they would for any loss of power, gasoline, heating oil, water, food and cash for up to sixty days. We have become so dependent upon our communal resources for these commodities that we are totally unprepared for any real emergencies. This puts undue stress on our disaster relief budgets. It is just sound community spirit to store emergency provisions at your home to help keep order after any disaster which might hit your area..

Regardless of whether you believe any of these catastrophic events are about to happen, you should at least prepare your households for the normal run of emergencies such as floods, quakes, volcanic eruptions, power failures, ice storms and droughts. Many have done so already in Canada and have easily endured the weeks without electricity and, hence, the loss of electric heating, gasoline pumps, TV, refrigeration, electric lights, hot water and so forth. To a lesser degree, folks near Auckland in New Zealand have also been put to the test as the power lines to their central business district have been cut-off for at least 8 weeks and maybe more by a fluke accident. Folks whose jobs were in the city would find it very comforting to have their emergency rations at hand while they have no jobs to go to during the power outage.

If you wish to get FREE advice on what supplies to get and how to use them in an emergency, please visit our web sites on the Internet at one of these URL addresses:

http://millennium-ark.net/News_Files/Hollys.html

http://standeyo.com/News_Files/Hollys.html

http://standeyo.com/

If we have to move our web and the above URLs stop working, just use any of the standard web search engines to look for "Stan" or "Holly Deyo" or "Noah's Ark". If you do not have access to the Internet, we are compiling a survival preparation manual as quickly as we; but as of August 1998 it is some five months away from completion.

Remember, we do not advocate people should move to the safest places; because we don't know where any are. However, we feel it would be judicious exercise of common sense to move away from the more obvious hazard regions. If folks choose to try to move to less dangerous places, that would be fine - in our opinion; but we maintain the only absolute preparation for the great disasters which are coming is a strong and unwavering faith in GOD through Jesus of Nazareth,... **Moshiahnoo**.

-Stan and Holly Deyo, 4 July 1998
email: standeyo@standeyo.com
email: holly@standeyo.com

Groundwork For The Great Deception

After you've read this next section, many of you who have accepted a lot of what I have said so far may think I've lost my marbles. This may well be true; but I reiterate, the things I have written in all editions of this book are MY OPINIONS based on "facts" which I, (like anyone else with an opinion) have CHOSEN to accept as axiomatic. My conclusions could be proven invalid by anyone wishing to disallow certain premises from which I have predicated my position on these issues. Having made this clear, the first issue I wish to discuss is that of the huge UFO/alien deception that is being and has been foisted upon the people of Earth.

If I did not think this was very important, I would not take up this precious space to expand on a premise which I first stated from the first edition of this book. One of the greatest disasters which the Christians ("Moshiani") of this age have to face is that of being deceived into following the wrong 'deliverer' or 'messiah' during the coming crises. The real Messiah, said it this way in Matthew 24:3-13: *"And as he sat upon the mount of Olives, the disciples came unto him privately, saying, Tell us, when shall these things be? and what shall be the sign of thy coming, and of the end of the world? And Jesus answered and said unto them, Take heed that no man deceive you.*

For many shall come in my name, saying, I am Christ; and shall deceive many. And ye shall hear of wars and rumors of wars: see that ye be not troubled: for all these things must come to pass, but the end is not yet. For nation shall rise against nation, and kingdom against kingdom: and there shall be famines, and pestilences, and earthquakes, in divers places. All these are the beginning of sorrows.

Then shall they deliver you up to be afflicted, and shall kill you: and ye shall be hated of all nations for my name's sake. And then shall many be offended, and shall betray one another, and shall hate one another. And many false prophets shall rise, and shall deceive many. And because iniquity shall abound, the love of many shall wax cold. But he that shall endure unto the end, the same shall be saved".

For those of you who do believe the Biblical prophecies are about to become realities in our time, I draw your attention to a clue in the preceding verses. It would appear that the wars and rumors of wars, the famines, pestilences (plagues, epidemics) and earthquakes in diverse places are already increasing as I write these words. The warning above is aimed at Judeo-Christian believers who will still be alive and on Earth even after these catastrophes occur.

In the last three decades, a lot of interest has been generated in an event called the "Rapture" of the Christian believers. In the Bible there are convincing verses for the case of GOD removing a small group of believers from the Earth prior to a time when it would be almost impossible to live. This removal is to be a secret and swift 'catching away' of the believers inside of cloud-like mists to a secret meeting with the Messiah in the air whether they be at work, play or asleep. The general wisdom on this event is that it will happen around the time of the Great Tribulation which is prophesied to happen on Earth during the end of this age.

However, there is disagreement over whether the believers will have to endure any of the Tribulation period before they are spirited away to safety in the "Rapture". The preceding passage from Matthew could imply that the "Rapture" of the Christians may occur later in the Tribulation period. Therefore, we feel the wisest position is to assume the latter and to welcome the "Rapture" if it comes earlier.

As I look around the world today. Many Arabic Muslim nations are preparing for war in the Middle East at this very moment. I wonder if China will continue to threaten Taiwan, hence, the United States indirectly?

It is now certain that Iraq and Iran will join forces with each other and possibly with Syria, Lebanon and Russia to attack Israel and, hence, the US defense forces in the region someday soon. The Bosnian Crisis is just on hold. And, Ireland's 'peace' is shaky at best. These are only four of several **rumors of war or actual wars** in this decade.

We have seen the rise of AIDS, antibiotic-resistant e-coli, golden staphylococcus, mutated tuberculosis, bubonic plague, leprosy, smallpox, physteria and ebola. And, now, the US, Canada, Australia and Israel are threatened with biological warfare using Iraqi-produced anthrax and bubonic plague derivatives. These are but part of the **pestilences** about to be released on mankind. The recent flu season in the United States has hit 122 major cities so badly that they were declared as epidemics.

The Asian financial structure has continued to crumble as the rest of the world's markets hold their breath. Japan has just had its elections and their economy is still on a downward slide. It's economic woes are a major factor in the growing economic crisis which is developing into the most serious of all the modern, man-made disasters.

When it does collapse, the ensuing global economic collapse could be the justification for offering the people of Earth a new, computerized, economic order. This new system would offer digital "money" to all those who have the, allegedly, "foolproof" bio-mark ID on their body. The Bible says those who would receive this personal ID number on their body would also develop a mysterious rash of sores on their skin (possibly, in reaction the ID implants). These, too, can be considered a part of the **pestilence** that are prophesied to prevail in those times.

More **earthquakes** are occurring in **populated** areas than ever before. I have analyzed seismic data collected from both the U.S. Geological Survey's National Earthquake Information Center and from the International Data Center which monitors global seismic disturbances for nuclear test ban treaty violations. From these data I have been able to see disturbing patterns in the magnitude and number of seismic events globally. The concentrations of the quakes and the emergence of new quake zones do impress us - especially when they are associated near or underneath massive volcanic sites which have long been dormant.

Significantly, the Sun has officially increased its total energy output over the last 11 years by .036%. However, even more disturbing news has just hit the media in February of 1998. The rate of energy output is fluctuating beyond the median increase as you will see in the following article:

> Small and subtle changes in the Sun's radiation have potentially strong effects on the Earth's climate and could be boosting effects of global warming, scientists said Friday. Changes in ultraviolet radiation and electric charges in "solar wind" affect the atmosphere on Earth, they told the American Association for the Advancement of Science. Joanna Haigh said she had a computer model showing shifts in ultraviolet radiation had corresponding effects on the protective ozone layer. She said she had seen changes of 2% to 3% in the amount of ultraviolet radiation coming from the Sun, and a corresponding shift of 2% to 3% in ozone levels.

Now we are seeing a tremendous increase in crop failures and in the spread in the number of **famine**-stricken regions throughout the world. As a result, in the United States and Australia we have seen evidence of significant food prices increases. The US Government has already cut back severely on its internal food stamp programs and the subsidized food banks. The numbers of homeless and starving are growing in the over-crowded cities of the world - even in the 'fat' cities of the advanced, western nations. **Famine** is a commodity these days.

If current events are, indeed, heralding the arrival of the prophesied 'tribulations', this next decade in our history will indeed be the "**beginning of sorrows**" mentioned above. Thus, the warning, "take heed that no man deceive you" also has tremendous importance to all of us who believe and follow the Word of GOD.

When times get bad, people seek fortune tellers and soothsayers to ease their fears and brighten their hopes. Sadly, history is full of stories about such people who have used these questionable abilities to gain materially from those around them in such times. In such a terrible, terrible time as is defined in the Bible, the soothsayers will be more numerous and more dangerous in their **deceptions**.

However, since we are in an age of technological miracles, one could be forgiven for thinking it would be a lot harder to deceive people with technological "magic". Deceiving those of us with college degrees or doctorates or long titles after our names would just be unthinkable... Yet, these soothsayers, astrologers and channelers are making a fortune out of well-educated people seeking solace from their prognostications! Heck, even two of the last three U.S. presidents were advised by personal astrologers! **Un**real!

There have also been a number of men in recent times who have already **deceived** many faithful Christians, Muslims and Jews alike. That "Ti Do" character who led 40 folks to suicide by convincing them he was the messiah was one such deceiver. Jim Jones was another - as is "Reverend" Moon. David Koresh was a charismatic leader who also caused many faithful to falter and stumble in their walk of faith. He managed to convince them he was a special messenger from GOD - but, then, he, too, began to teach and practice doctrines which were not in the holy scriptures...

There have been many others of less notoriety who have **deceived** folks into lifestyles in one separatist cult or another which variously awaited "elder brothers" (the Urantia adherents) or "our space brothers" (the Adamskiites) or "the Pleiadeans" (the Billy Mier believers) or "The Galactic Federation" (Nidle's ground crew).The list is long and confusing... and growing.

Some of the newer candidates claim to be faithful to GOD and to be adherents of Jesus' teachings - but then they say and do things which are contrary to those of one who truly follows the Word of GOD as it is found in the older Bibles. This is an important point. The Bible advises believers and true seekers of GOD's not to listen to a man's words of faith - but to observe his actions to see his true beliefs.

We have received many inquiries regarding our opinions on the philosophies and teachings of various modern seers or prophets of one description or another. We answered these folks by saying the only way we could test any of the folks they suggested was to wait and see what their predictive accuracy rate was and to watch their actions for signs of an agenda contrary to that which they preach.

In the end, GOD is the judge of all souls. Holly and I have listened to and tested many folks' predictions. We considered people like: Edgar Cayce, Dannion Brinkley, Elizabeth Clare Prophet, Ed Dames, Richard Hoagland, Robert Ghostwolf and Robert Morningsky among many others. Of the ones who have actively promoted themselves as seers, none seem to have had a faultless hit rate so far.

Some of these people have "sold" philosophies on alien invasions, alien assists to mankind, secret UFO bases and technologies, conspiracies of all kinds, safe places to endure the coming earth changes, methods of raising one's spiritual consciousness and vibrations, remote viewing (using mental techniques for prophetic viewing of future events and investigative viewing of contemporary events), talking with spirits of dead prophets, witchcraft, shamanism, astrology, channeling of beings from other dimensions or from great distances in this Universe and a host of other equally questionable activities.

Those of you who know us realize that Holly and I speak honestly in answer to such queries. We are concerned when we hear anyone who receives "channeled" messages from some disembodied spirit or from some alleged "alien" or "star" person on the way here to help us evolve properly. The Bible tells us to AVOID dealing with disembodied spirits (necromancy) and familiar spirits. It warns us to avoid them not because they aren't there or that they don't have some knowledge that the average human doesn't... BUT because they are **deceivers** and do not always tell the complete truth. We have found they frequently lie with a purpose; a purpose which is to turn the listener from the real reason humanity was created by GOD.

Whenever people try to bend me to some of these more exotic deceptions I ask them to examine themselves to see if they are willing to bet their life (eternal or otherwise) on what some seer, channeler or clairvoyant tells them. The looks I get are quite interesting; but, they do realize in the end that I am serious in my query. The name of this game is definitely, "You Bet Your Life, Buddy".

We are concerned when we hear anyone claim that Jesus was a "great prophet" **but** was not the Son of GOD or that He wasn't the Messiah or that he was never crucified to death and never arose from the dead or that He is not the rightful King of Creation (as the Bible clearly states). We are concerned when people acknowledge the existence of Jesus **but** deny His Godliness. We are told that deceivers directed by Satan would come in the last days proclaiming these things. We are told they would come in their own names instead of Jesus' name; and there would be some of them who would claim to be the messiah and would gather people to them in secret places, in deserts or elsewhere where not all could follow.

We are concerned when we hear people claiming to be of the "light" and thus implying they are of the one, true GOD; for the Bible clearly warns that Satan, himself, will descend to Earth in the last days of this age and will do so posing as a "messenger of light". Remember, his original name before the rebellion in the heavens was Lucifer or "light bearer". He was apparently a splendid being emitting the colors of the rainbow from his presence. We love rainbows; but we are concerned at the mystical attributes given to the colors of the rainbow by the "new age" movements and philosophies.

So, loud alarm bells go off when people send us email or write to us or sometimes speak to us with phrases like: "in love and light" or "blessings of spirit" or "he is an ascended master" or "my spirit guide told me.". or "my star sign is..". or "the christ consciousness.". or "we are god..". or "their vibrations were good" or "elder brothers from space" or any other phrase which elevate man's rank above where it should be and demote God and Jesus the Messiah to human standards.

I would be remiss here if I did not also mention that we have been greatly disappointed in the mainline, "Christian" or "Jewish" religions. We have discussed Biblical scriptures with Anglican priests, Catholic priests, Jewish rabbis and many other leaders of Protestant or catholic derivation. Many of these religious leaders seem to be teaching what is socially acceptable to their congregations as opposed to what people should be warned concerning the persecution of these times. It appears far too many have become judgmental and have directed their attentions to the small details of their faith while missing the larger and more important issues written for today. This is my opinion and not a judgment. We are saddened by what we see and wish we could help these leaders of the faith to overcome the deficit in their works; but it is too late now.

It is a sad thing; but since their income depends on the tithes and offerings of their congregations, many of these leaders tend to preach safe positions. This omits the scriptural teachings which deal with the "end times" or the "end of this age". This omits the warnings against deceptive and ineffective teachings to their congregations by a plethora of "new age" philosophies which have grown-up suddenly like weeds in the garden. This omits the warnings about avoiding the numbering of every living human on Earth in a new, global monetary system. This omits the warnings concerning the arrival of non-human entities (alleged aliens) on Earth in these last days. This omits getting their congregations prepared to survive for a short time outside their established, socioeconomic systems. These acts are **deception** by omission in my opinion.

In my opinion these modern deceivers are ALL directly or indirectly a product of "New Age" disciplines or philosophies. Without fail they are herding the "sheep" of our culture (that's most folks) into many glittering, prismatic gateways which all lead to the same spot: the soul abattoirs of the one individual who is behind all the chaos and misery we have in the world today, Lucifer (Satan).

Do not think for a minute that Satan is a myth. He is real and on his way to Planet Earth according to the 12th Chapter of Revelation. The ancient prophetic writings of the Hebrews are extremely detailed in their discussion of the events which will soon grip our world because of Lucifer's rebellion; and I marvel at the number of people who still do not have a clue they are being herded into a one-way path to death. The book of Matthew did say this great deception would be so convincing that only the very elect of this planet would see through it. I trust you will be among those who see through it.

For a moment, as you read, try to accept the following argument as real... just for a moment, try to remove your objections and fears of peer group mockery over your believing anything so wild. Try. Your real life may depend on it.

Fertilizer For The Great Deception

Let us consider our world. There are so many apparent disasters on the tomorrow's horizon that a person feels utterly inadequate to solve them. Here are some of the typical concerns we hear these days:

1.   How can we stop industrial damage to our ecological system when so many economies are totally dependent upon these same processes which are destroying the ecology? And, furthermore, when so much irreversible damage has already been done?

2.   How can we tame the earthquakes which release so much energy that we could not even hope to quieten or defuse them? Even if we did have the power needed to defuse them, how could we avoid quakes when we cannot accurately predict them far enough ahead of time to even move populations to lower-risk areas?

3.   How can we avoid losses due to volcanic eruptions when we have at least 50 around the world which are active and potentially destructive? Should we start moving settlements now from such volcanoes as: Montserrat, Popcateptyl, Mammoth, Rainier, St Helens, Ruapehu, the Kamchatka volcanoes, Vesuvius, Etna and so forth? If Mammoth does blow, how can folks in the downwind shadow east of its resurgent dome hope to avoid the fallout of the coming eruption when the last time it erupted (760,000 years ago by conventional wisdom) it deposited heavy ash fields over a five-state area and formed the Grand Canyon in the process?

4.   How can we avoid the disaster of an Earth-impacting asteroid without adequate deep space tracking? How can we hope to target these asteroids without accurate mathematics to solve the head-on intersect courses of our missiles? How can we deflect such high-momentum objects away from an Earth collision without fracturing thus causing a rain of "shotgun" impacts?

[In spite of what you have been told publicly, I must tell you the chances in favor of Earth getting hit by an asteroid in excess of 1km diameter in the next ten years are getting higher by leaps and bounds. More asteroid packs are being discovered each month; and, already, some 5,000 new ones have been catalogued in first six months of renewed scanning the entire heavens from our position in the Universe. Of these it is certain that at least seven will cross our orbit on a collision course with Earth. We are told not to worry about these; however, because the time of their impact seems to be nearly a hundred years from now. Gosh, don't you feel relieved now? They haven't finished the scan for all of these near Earth objects (N.E.O.) yet; but we are told we can relax because those they have JUST now discovered won't impact us right now]

In fact, one US Air Force Space Command officer whom we know personally is so concerned about the possibility of such an occurrence that he has just retired his commission from the Air Force to establish a civilian organization called CPANEO or "Citizens Protection Against Near Earth Objects" to prepare the various civil bodies and the populace at large for just such a disaster! He knows the risks as they really are!... And, believe me, he is not going to be quiet about it. Look for more from our brother in the faith, Major Wynn Greene (recently retired).

5. How can we hope to avoid the disasters associated with our failing ozone layer? How can we prevent the increasing numbers of skin cancers from these penetrating UV wavelengths which get through the "holes" in the ozone layer? How can we treat the multitude of new diseases and revived, older ones which easily attack humans when the UV exposure has dropped our auto-immune systems into increased susceptibility?

6. How can we avoid the disastrous "super storms" and "super droughts" which are forming more frequently due to increasing temperature disparities in the air, the ocean and the outer layers of the planet which are certainly linked to increasing solar radiations?

7. How can we avoid the solar energy output variations? Since 1991, observers at the Kitt Peak Solar Observatory and other research groups have been aware of at least two new spectral bandwidths of radiations coming from our Sun?

[One of these observers has sent me photos and a paper on his concerns over the absorption of these new frequencies by marine life forms; and I have verified these solar variations with the USAF at its Australian solar observatory.]

8. How can we avoid the disastrous famines which are already moving into "advanced" countries as I write these words? How can we feed so many people when so many religions and cultures refuse to control their birth rates - causing an unavoidable shortage of resources of all kinds?

9. Oh, and worst of all, how can we hope to even defend ourselves against the preceding scenarios of doom when we are hell-bent on destroying ourselves and our environment with wars and more wars... always escalating towards that "ultimate war"?

Look at the Middle East "Powder Keg" over Palestine. Look at China; it is like an awakening Ghengis Khan as it plays Israel against the emerging Arab Consortium and, hence, America against Russia because of mutual defense treaties. In true Mongol tradition, China prepares itself for war on a massive scale while encouraging its target nations to fight between themselves. Then, when the targets have beaten themselves to a weakened state, China plans to step in to claim victory over all the nations of the world.

Shall we just leave it up to our politicians? Shall we put our faith and trust in people who know little more than we ourselves about obtaining peace with each other and our planet? Shall we trust these people who have become nothing more in our eyes than puppets dancing to the manipulations of invisible, nameless "puppet masters?"

Are you ready to jump off this "Late Great Planet: Earth" yet? ha! ha!... Folks, if you aren't, you haven't been paying attention! Good Lord! Have we got problems here! We are so ripe for the picking that we are ready to fall off the branches into the hands of the first itinerant harvester to come along with a promise and a basket.

The Great Gray Deception

With all the foregoing problems confronting mankind at the moment it is not surprising that we see so many groups and writers calling for harmony (peace) and, hence, "security" in the world both between each other and between mankind and the environment. They and those who agree with them would welcome just about anyone or any group they thought capable of giving them these desires.

There have been power groups through the ages of mankind's history. In the beginning, there were relatively few humans on the planet; so biological resources were not an issue. As time passed and mankind began to increase in numbers, the competition for those resources in localized areas became more serious and deadly. Tribes formed and were bonded by intermarriage of family groups. In time, these tribal structures gave way to larger groups called nations. In the process, the wars for control of those resources became more impressive as greater numbers of combatants faced each other on the battlefields.

When, in the fullness of time, the entire livable areas of the world had been colonized by nations of peoples, the numbers of warring nations crystallized into larger and larger "allied" nations with fairly common resource management ideas. Each successive war reduced the number of non-aligned nations until today we find ourselves witnessing the emergence of ten super-groups of nation states allied by geographic, philosophic and economic factors as shown in pages 197-200 of this book.

It is easy to see that the next "unification" war could destroy the Earth. If the Earth is to ever be unified under one 'tribal' law, it would seem the only chance we have is to find a way to convince these remaining, 'tribal groups' into trying a new, global governmental structure for long enough to give it a fair chance of success. Any alternate is too terrible to consider.

I find myself asking the same old questions over and over on behalf those of you who don't know what's really been happening around us. Is mankind ready to accept **any** order or group or power which can solve its planet and "guarantee" it peace and security? What sort of power or group would **all** mankind - regardless of race, education, creed or color - accept and follow? Common sense dictates it could not be from an existing nation, culture or religion - as this would eventually breed instability in any emerging world order.

It is a matter of public record that just prior to the planned "Bay of Pigs" invasion in Cuba, a high-level group inside the CIA had considered the use of a religious hoax to make Cuban civilians cooperate with the invading American troops. The plan was to use the cover of night to drop pamphlets written in Spanish amongst the people of the region of Cuba where the troops were to land. These pamphlets were to give quick and simple announcements to the Cubans. They were to be told that the "Virgin Mary" would appear in the clouds and that she would be as a sign to them that their "deliverers" were coming to free them from the oppression of the communist government. The believers were to help the troops sent by the "Virgin". To augment this, it was planned to project a large image of the traditional catholic concept of the "Virgin Mary" on the night clouds just long enough ahead of the landing forces to be of psychological assistance to the American forces. Obviously, this never eventuated; but it **was seriously considered** as a valid means of getting the people to assist invading troops in the name of a divine being in their religion.

I have a Cusco CD playing; and its hauntingly ancient, Central-American melody transports my mind back to a more recent past; where, I can see the beginnings of the greatest conspiracy of all time. I can see it germinating; then growing; and eventually emerging from within the minds of powerful men in the more recent, American history.

I feel the sadness, the needs and the challenges which permeated the leaders of the immediate, post-WWII period in America. As I put myself back into that period, I can see those powerful men who really wanted to avoid another war of any kind on Earth.

I can see that most of them wanted to establish a global order which could give everyone equal rights and resources - regardless of race, religion or political position. But, I can also see others with personal agendas for obtaining absolute power over their fellow man. Sadly, I can see the dominating avarice of the latter group consuming those of the former group - those of the altruistic paradigm.

I condemn the actions of these power barons of the fifties who initiated the original form of the modern deception which I shall be explaining. However, I do have a healthy respect for the cleverness, bravery and fortitude of these men. I could not help but feel affinity to many of them after reading some of their correspondence files. Many of them really meant to do a good thing but could not reach consensus between their sub-groups nor totally remove their subjectivity. So, while you read the following exposé, know that I report it in respectful sadness for these men who, for the most part, no longer even draw breath on Earth.

In 1958 a well-known, American industrialist, Agnew H. Bahnson, wrote a fictional novel entitled, "**The Skies Are Too High.**"[21] In this book (the only one he ever wrote), he outlined a plan for establishing a powerful 'advisory' group which he initially called 'The Council of 8'.

In his proposal, this group was, firstly, '*to provide a new atmosphere of communication between nations and was to influence the policies and the outlook of these nations;*' and, secondly, '*to bring the influence of a broader point of view* [than was] *presently available from other sources to* [bear upon] *the President and his administrative assistants on a continuing basis.*'

Strangely, in the notes of a speech he made just after the release of this book, I noticed he had crossed-out the number '8' and had replaced it with the number '12' making the 'Council of 8' read as the 'Council of 12' (somewhat similar to Majority 12 or Majestic 12?).

He wanted this council to continue from one administration to the next to ensure the continuity of influence on each succeeding leader in every participating country. It was to advise on education, economics, ideologies and international relations (much as the G7 functions today). To start things moving on both sides of the iron curtain, Bahnson even sent a copy of his book and a letter of explanation to Nikita Kruschev in 1959. Bahnson's attempts to establish an advisory 'Council of 12' in each country for the purposes of maintaining peace and encouraging joint development between the nations of Earth appeared quite sincere. However, there is strong evidence to say that he, inadvertently, jeopardized the secrecy of a similar group which had already been formed by the U.S. and agreed to by other nations prior to the publishing of his own proposal.

In Bahnson's personal correspondence files (which his heirs allowed me to peruse) were letters to the White House (answered by Eleanor Roosevelt), to Einstein and to many leading physicists like John Wheeler and Bryce and Cécile deWitt. Bahnson

---

21   **The Skies Are Too High** by Agnew H. Bahnson Jr., (Random House, 1959; Bantam,1960).

was truly an altruistic man, a man of peace and long-term vision. He wrote only this one book and it was to popularize an idea which he had formulated in many private discussions with friends and associates of like minds. His book was a rather linear concept of a way to fool the people of Earth into peace and security using advanced, hidden technology - much like that which he and the late, **Dr Thomas Townsend Brown** researched at the Bahnson Company's high-voltage, propulsion laboratory in Salem, North Carolina in 1958-1960.

What is unknown to many is that the whole premise of Bahnson's book was a close parallel to a large-scale plan held by his friend, **Dr Gordon Gray**, then serving in the White House for Eisenhower. I am not sure whether Bahnson knew what his friend, Dr Gray was doing at the time; but Bahnson and his idea for a new global order were so similar to those of Dr Gray's special "discussion and study" group at the National Security Council, that it is possible Bahnson became a liability to a much greater version of his own proposal.

Just six years after Bahnson's book was released, he was killed in the accidental crash of his own private plane while landing at his local airfield - one which he knew like the back of his hand. I often wonder if Bahnson's plane crash was really an 'accident.' Perhaps, when you have finished reading the data I have compiled on the mysterious Dr Gordon Gray's 'project', you, too, will view our world with less naiveté and more caution.

Dr Gordon Gray

Thomas Townsend Brown

Again using the Internet, I was able to peruse the "History of the U.S. National Security Council, 1947-1997" where I found the following important details concerning the somewhat ethereal trail left by Dr Gordon Gray:

Eisenhower Administration, 1953-1961

The President's **Special Assistant for National Security Affairs**, a post held under Eisenhower by Cutler, Dillon Anderson, William H. Jackson, and **Gordon Gray**, oversaw the flow of recommendations and decisions up and down the policy hill, and functioned in Council meetings to brief the Council and summarize the sense of discussion. The Special Assistant was an essential facilitator of the decision-making system, but, unlike the National

Security Adviser created under Kennedy, had no substantive role in the process. The NSC staff managed by the Special Assistant grew during the Eisenhower years, but again had no independent role in the policy process.

In 1954 **NSC 5412 (National Security Council** memo #5412) provided for **the establishment of a panel** of designated representatives of the President and the Secretaries of State and Defense to meet regularly to review and **recommend covert operations. Gordon Gray assumed the chairmanship of the "5412 Committee"** as it was called[22], and all succeeding National Security Advisers have chaired similar successor committees, variously named "303", "40", "Special Coordinating Committee," which, in later Presidential administrations, were charged with the **review of CIA covert operations**.

When Eisenhower briefed President-elect Kennedy on the NSC system, and when **Gray briefed his successor McGeorge Bundy**, they emphasized the importance of the NSC machinery in the management of foreign policy and national security affairs. They might have been more persuasive had they pointed to the fact that the NSC system was essentially limited to policy review and was not used to manage crises or day-to-day foreign policy.

Kennedy Administration, 1961-1963

President Kennedy, who was strongly influenced by the report of the Jackson Subcommittee and its severe critique of the Eisenhower NSC system, **moved quickly** at the beginning of his administration **to deconstruct the NSC process** [Was this sufficient motive for someone to kill Kennedy?] and simplify the foreign policy-making process and make it more intimate. In a very short period after taking office, the new President moved to reduce the **NSC staff from 74 to 49, limit the substantive officers to 12**, and hold NSC meetings much less frequently while sharply curtailing the number of officers attending. The Operation Coordination Board was abolished, and the NSC was, at the President's insistence, pulled back from monitoring the implementation of policies. The coordination of foreign policy decisions was ostensibly left to the State Department (and other agencies as necessary).

Circumstantially, the preceding NSC historical report, makes it appear that Dr Gray was in the right positions at the right time to have constructed the covert operation I discovered and to have kept it very quiet in the 'interests of national security.'

We have all seen schoolyard fights between one group and another and sometimes between several groups with different agendas. They all seem to hate each other unmercifully until either a tougher group from down the block shows up to whip them all or until a school authority figure steps in to punish them all.

When either of these events happens, we see the schoolyard groups quickly unite either to try to whip the bigger, 'badder' 'foreign' or 'alien' group from down the block or to avoid the school's punishment by pretending mutual friendship until out of sight or earshot of the officials. It is like the old Yiddish adage, 'When you have two Jews locked in a room, how many opinions do you have on a given issue?... The answer is, "Three".

Why? You have the individual opinions of each person and then you have the collective opinion". It's all right for you to fight with your brother; but when a bully attacks your brother, you and your brother unite for mutual defense - overlooking your family disputes **for the moment.**

---

22   Could this have been the origin of the nickname, "the Majestic 12"? **5412 = 1954+ Majestic 12**?

And, so it would be on Earth if we were to encounter a common enemy or a common authority. Assuming they had the power to wipe us out, it would make our different cultures, tribes, religions and countries unite to defeat a common foe. This concept is not new at all.

What is new is that at least one group on Earth has actually set such a plan in motion on a global scale back in 1954. Its plan was designed to create a convincing illusion of both a common foe ('a bad alien race from space'?) and a common messiah ('a race of elder brothers from space'?) with a view to duping the peoples of Earth into a new world order of peace and security.

To have even hoped to pull off such a daring scheme, that group would have had to have access to power at the highest government levels. It would have required large sums of money... LARGE sums. It needed to have leadership that was at least partly visible in the community. This meant there would be some small clues or traces of their leadership and their game plan - if a person knew where to look.

A good mastermind for a clever deception such as an alien threat to the planet would need to know a lot about human behavior and behavioral modification. The clever **Dr Gray** would appear to have been qualified for such a task. I unearthed more information about this mysterious man.

He was trained as a **lawyer**; his political affiliations were somewhat neutral; and from 1937 to 1947 he was a **newspaper publisher**. Then in 1948 he became the Assistant Secretary to the US Army from whence he rapidly advanced to being the **Secretary of the Army** (1949-50). Next, in 1950, Dr Gray was appointed as **special assistant to President Truman on National Security Affairs**. The experience and connections from these jobs alone would have made him a leading candidate; but it didn't stop here.

Upon leaving the chief position of the Army in 1950, Dr Gray was appointed President of the University of North Carolina - the oldest state university in America. Interestingly, as such, the University has been the training ground for many a federal politician and high-level appointee since its formation in 1795. Dr Gray held the title of president of the University from 1950 to 1955 even though the last four years were in absentia; as he was away in Washington by appointment of President Truman and, later, president Eisenhower.

While Averell Harriman was President Truman's special assistant for national security affairs in 1951, Dr Gray was appointed to direct the activities of the **CIA's Psychological Strategy Board** - America's psychological warfare department during the "cold war". This organization was in charge of American propaganda, economic, and political activities for the "cold war" period. Remember, it was this same CIA that later planned that fake appearance of the "Virgin Mary" over those Cuban villages in the Bay of Pigs invasion. This has just got to have rung alarm bells in your mind.

Dr Gray later changed hats again to become **the Assistant Secretary of Defense for International Security Affairs** from 1955 to 1957. At the end of 1957, he served as the Director of the Office of Defense from 1957-1958. Doing the same job as Harriman had for Truman, Dr Gray held the posting of **special assistant for national security affairs** to President Eisenhower from 1958 to 1961; where he acted as the President's intermediary, strategist and confidant, in the Whitehouse's relations with the CIA and the U.S. and allied military forces.

Included in his work with these organizations as well as the NASA at this time were classified exchanges pertaining to satellites, the missile program and the Space Council (NASA's historical file on Dr Gray is four feet thick!). During this time, Dr Gray was a golfing buddy of George Bush's dad, Senator Prescott Bush. Dr Gray was definitely well-connected.

Yes, Dr Gray and his friends had access and control of the highest levels of the US Government; hence, access to public money. Also, they had the security authorization to conduct "space research" without detailed nor complete public accountability for the various places the money was spent. I have heard personal testimony from two high-level officials of the NASA which revealed that many space-related parts and devices were made by sub-contractors who had no idea where these parts would be used or what function of these parts would be when in the final assemblies - wherever they might be put together.

The secrecy allowed parts to be made for advanced air and spacecraft technologies right under the noses of the taxpaying workers at many aerospace companies without them knowing what they were making. This means many if not all the so-called "UFOs" from 1950 onwards could well have been made right under our noses in our own aerospace corporations!

Many secret organizations were established inside and outside of the US Government with Dr Gray's support and direction. He was not alone in these efforts; but he was a (perhaps, 'the') major player in the policy-making roles. His background was one of a man who was cross-trained in all the "right stuff" needed to pull off a super-secret covert deception on a planetary scale. One has to be impressed by his postings within the power structure (or should I say, the "real" power structure) of Washington.

While it appears he may have been the principal architect of what we might call the "mother of all deceptions," were his intentions for the betterment of mankind? Perhaps he and his associates truly meant to establish a lasting peace on Earth by a very clever deception. Who can say at this late date. Dr Gordon Gray died on the 25th of November 1982 at the age of 73. It is not certain who now controls the deception which I am about to explain.

However, I can be certain about the existence of one such clandestine, US-based, saucer development group. You see, in 1971, it was men from what I call "The Gray Project" who approached me to join their technical group in Dallas, Texas. Their group, Dr Gray's group, had discovered my work on electrogravitics and thermionic energy conversion. To keep me quiet at a critical time in their preparation for their "Gray Deception" they then spirited me away to Australia; as I discussed earlier in this book. It was, coincidentally, a man named "Robert Gray" who handled my transfer to Australia through the Australian consulate in San Francisco under the instruction of Dr James R. Maxfield in Dallas - a man who, by his own admission to me, was a member of the "Gray Project".

In my last book, "**The Vindicator Scrolls**" [23], I also addressed Dr Gray's probable participation in a covert operation to produce an artificial "alien" invasion and the establishment of an "invisible college" of advisors to the leaders of the world. If the sources I have quoted are correct, then Dr Gray was active in the great deception as early as 1947. Consider what I wrote:

"The 'Majestic-12' document which was apparently 'leaked' to the public by Stanton Freidman and William Moore (a self-confessed agent of a clandestine US intelligence agency) appears to be a photocopy of an 'eyes only,' 'Top Secret' briefing document prepared for president-elect Dwight Eisenhower on the 18th of November 1952. The document relates a few of the classic UFO sighting and crash cases before stating that the entire project must be kept under wraps to prevent a public panic over the possibility of alien invasion.

[It is my opinion that all these 'leaked' documents were a product of the Gray Project to point folks to accepting an both an alien threat and an alien salvation]

23   **The Vindicator Scrolls** by Stan Deyo, ©1989, WA Texas Trading, (based on pp.213-219)

The document states, 'OPERATION MAJESTIC-12 is a "TOP SECRET" research and development and intelligence operation responsible directly and only to the President of the United States. Operations of the project are carried out under the control of the Majestic-12 (MAJIC-12) Group which was established by special classified executive order of President Truman on 24 Sept 1947, upon recommendation by Dr Vannevar Bush and Secretary James Forrestal.

Members of the Majestic-12 Group were designated as follows: Admiral Roscoe H. Hillenkoetter, Dr Vannevar Bush, Gen. Nathan F. Twining, **Gen. Hoyt S. Vandenberg**, Dr Detlev Bronk, Dr Jerome Hunsaker, Mr Sidney W. Souers, Dr Donald Menzel, Gen. Robert M. Montague, Dr Lloyd V. Berkner, **Mr Gordon Gray** and **Secretary James V. Forrestal** who died on 22nd May 1949.

Gen. Hoyt Vandenberg   Dr Detlev Bronk   Sec James Forrestal

[Gray and Forrestal are of primary interest here. Dr Gray is (obviously) because of his link to Agnew Bahnson and former Secretary of Defense Forrestal because he apparently committed suicide when he tried to expose the MAJIC-12 in a diary which went missing at his death.]

Secretary Forrestal resigned his post as Secretary of Defense in March of 1949 suffering from what physicians called a 'depression similar to battle fatigue' only to be immediately admitted to the Bethesda Naval Medical Center, Maryland where, soon after, he (apparently) dived to his death from a second-floor window.

I have interviewed one source who actually worked inside the assembly areas for the man-made 'flying saucers'. His testimony further strengthens my case for a grand, man-made deception at work in the current UFO arena. This source was a former member of a section of British Intelligence which was responsible for maintaining security on four of the human research and development bases in 1949-1962. The testimony and evidence which I received from this source was witnessed by one of my closest associates at the time who was just as astounded as I was by the man's revelations of the 'truth' behind the whole UFO sub-culture.

We were allowed to record the session in writing; and he did allow us to draw illustrations of various objects and devices which he described to clarify certain points for us. We were allowed to ask any questions we wished except for the names of the leadership.

According to him, immediately after WWII, a power group surfaced on the international scene. The Gray Group (as I have dubbed it) had power over most of the Allied nations and their occupied territories and defeated countries. It recruited scientists, industrialists, security personnel, administrators and engineers from a host of countries which included, specifically: America, England, Canada, Russia, Germany and Yugoslavia.

The leaders of these various countries were not permitted access to any of the underground development centers without special approval by the Gray Group. Failure to comply with that condition meant immediate execution for the offending leader or his representatives. *[This could very well have been the MAJIC-12 Group.]*

These facilities were periodically given blueprints of technologies for which they were to develop the expertise to manufacture. Their instructions were not to research a new technology; but to back-engineer existing -yet foreign- technology which was periodically delivered to them by a courier - usually male, mid-twenties and of thin build and carrying a slim briefcase. When I asked my source who the Gray Group were, he facetiously quipped he didn't know; and that, for all he knew, they could have been little 'green men.'

When I queried this statement, he simply said he meant they had never been told who was in charge. He had been told about the capture of two little green men who had survived the crash of their saucer craft just off the English coast near a little village of some 1200 people during WWII. Perhaps he meant the technology could have even been coming from a connection with these beings. He said for some earth-based group to have perfected technology which was given to them without the industrial capability to even test it first had to indicate the use of a borrowed encyclopedia from a more advanced culture. Even his opinion reflected a belief in the presence of real aliens at that time. Maybe he was right; but if he was I am sure there were two 'alien' threats - one man-made, the other, real.

Bizarre as it may sound, it is also possible the drawings were 'channeled' through human from another place or dimension where beings spoken of in the Bible as 'disembodied spirits' or 'demons' or 'rebellious angels' dwell. It is possible such beings controlled the minds of men in high places to generate the 'alien' technology and the whole deception.

I did ask my source to explain how the saucer propulsion worked; and he surprised us by saying three versions of propulsion were developed over the post war period. *[This agreed with other information I was given a few years before by another agent from the Gray Project who allowed me to view part of the file his group held on me at the time.]*

The first type of propulsion had been an extension of the German-based research using air which flowed over the top of the saucer and through its porous metal to return to the intakes of a turbine spinning in the middle of the craft. This was fast but because of the drag on the air in the pores of the skin, it was found to be horribly inefficient and the 'g' forces on the crew were hard to handle.

The second (also based on captured German technology) found a way to produce ionized plasma gas flows over an entire wing surface coated with a high temperature, ceramic insulator like barium titanate. The plasma gas was then re-circulated through a bottom opening in the center of the craft where it was again spun out over the outer surface by direct current pulses in a bank of toroidal coils. This craft was usually known as the Lorentz-O design because of the name of the force which moved the plasma around the craft in the magnetic fields: The Lorentz Force.

[I must add here, many early UFO researchers were looking for a man named "Lorenzo" who supposedly held the secret of UFO propulsion. It was not a man but a device, an "O" shape utilizing the Lorentz forces to stack an electromagnetic drive field... ha ha!]

The third is most impressive because it introduced time-dilated, acceleration curves. It created a convoluted, toroidal-shaped field using pulsed, direct-current, electromagnetic fields to propel it. This method also allowed uniform acceleration to all mass within the propulsion field so that 'g' forces were handled by amortization over several seconds of 'fast time' relative to the observer and 'normal time' relative to the crew. The field even altered the passage of time inside its field in relation to the outside world. The craft went so fast that navigation was initially a great problem for its crew. They kept overshooting destinations.

I asked my source to tell me how to construct a test version of either or both of the last two types of propulsion systems. He obliged by drawing schematics of both for me and by telling me the appropriate mixtures and types of gases to use in testing the plasma version. I subsequently tested the plasma version; and it worked perfectly in a small laboratory model. The second version is a much more hazardous and trickier bit of physics for the private experimenter. *[I have since tried to make this one work in the lab and have not succeeded to the degree that I am confident to brag about it.]*

At least six underground facilities were pinpointed for my friend and myself. One was in Norway on the Russian border, another in Germany inside a large German bunker which has been closed off to the public since the War, another in Saudi Arabia about 120 km from the Jabal Tuwayq Mountain Range (coincidentally, near the center of ancient Atlantis just southeast of ancient Sumeria), one in Ecuador, one at the South Pole under the ice and yet another in Australia in the outback in an aboriginal reservation.

The Psychology of the Great Deception

By now, most of you will have seen or heard about the movies, **Independence Day: 4th of July** and **Contact**. They are the latest and most impressive of a long list of Hollywood offerings which have helped to psychologically "condition" the television-viewing and movie-going people of the world for an encounter with "alien" beings who will be (ostensibly) far superior to mankind in technological knowledge, physical strength and problem-solving ability. Each of the following movies or series have exhibited heavy conditioning (whether intentionally or otherwise) toward two objectives:

1) instilling fear in mankind of the retributional power of any arriving alien race possessing "obvious" technological superiority to us or...

2) instilling respect for that same race as potential 'messiahs' or 'deliverers' of mankind from its great host of environmental and cultural problems.

2010: The Year We make Contact	*Star Trek: First Contact*	*Alien (the series of four)*
Fire In the Sky	*Star Trek: original TV series*	*Battlestar Galactica: TV Series*
Flight of the Navigator	*Invaders: TV Series and movie*	*Close Encounters of the 3rd Kind*
Last Starfighter	*Cocoon 1 & 2*	*Men In Black*
Stargate	*Contact*	*Official Denial*
Starman	*Dark Skies: TV series*	*Predator 1 & 2*
The Arrival	*Day the Earth Stood Still*	*Species 1&2*
X-Files TV series and Movie	*E.T. (Extra Terrestrial)*	*V-the Movie and the TV series*

For nearly 50 years we have been swamped with literature, photos and films showing us various types of small 'flying saucers' or 'alien scout craft.' In the beginning, these fast little space craft were impressive because of the almost magical powers they and weapons they seemed to possess. In those days, you could mention "UFO" or "flying saucer" to your friends and they would conjure visions of a super-race with ultimate technology. Seldom, however, did the media present data on the 'mother ships' from whence any and all of these little spacecraft seemed to originate.

Which type of craft would strike more terror into a populace from just its size and appearance? Would it be the small tactical, scout craft capable of carrying super nuclear warheads (See Scout Craft Image)?... Or, would it be a large, "mother", spacecraft (see Mothership Image) hovering over their cities casting a 1-mile diameter or a 1-mile long shadow over them? Both craft could carry nuclear weapons of horrific power; but the bigger craft would be the preferred psychological tool if one wanted to scare the dickens out of a populace without demonstrating a nuclear weapon. The smaller craft would only seem threatening if they demonstrated their weapons whereas, the larger craft would seem threatening by implication of its massive size and obvious technological superiority.

The huge ships create a psychological superiority over a whole city because we "know" that we don't have anything that size. When we have to look up to someone we tend to respect them because they are bigger. Consider how many tall men and women are leaders in elected government or in responsible public positions such as policemen, firemen, rescue workers and so forth. If we have physically look up to someone, we may feel intimidated or subjugated to them or we may also feel a sense of security with them...

Scout Craft Image

The Cosmic Conspiracy Final Edition 2010 © Stan Deyo 2010

Mothership Image

So, if someone wanted to really impress and intimidate earthmen, they could make a lot of extremely large aircraft and park them over major population centers and at, say, one to four miles up. Most of the people would assume such technology is not possible on Earth today.

AND, those people would be wrong! During WWI, German technology in moving large aircraft lead the world; for it was they who developed the dirigible (blimp) to the stage of the Hindenberg. "Ah yes, but look what happened to the Hindenberg," you might exclaim! True, the Hindenberg did explode because it was filled with flammable, hydrogen gas instead of inert, helium. However, the Germans had asked the United States for Helium gas and had been refused the privilege of buying it from America; because America had correctly assessed Germany's aggressive posture in Europe. It would have been madness to equip Germany with helium dirigibles in that environment.

Still, the German blimp technology was superior to all other competition at the time. During WWI, Germany had even used its blimps as airborne aircraft carriers. They had trapeze bars hanging underneath for biplanes to hook onto. Whenever the blimps came over the cities of Europe, the people stood out and were impressed as those great lumbering hulks drifted overhead taking minutes to traverse the overhead sky.

They could not help envisioning those huge bags of space filled with gas as large, **solid** objects. It was an illusion; because only a tiny amount of the entire blimp was actually transportable cargo or crew. However, as long as the blimp was in motion relative to the surrounding air, it could support a lot more weight than it could at a dead standstill. This was due to the extra lift generated from the shaped surface of the blimp and due to the blimp's angle of inclination at any given moment.

So, German scientists have had some 80 years to experiment with blimps. If that work was continued by those of them who were recruited after WWII to work at the various classified bases like the one at the South Pole, then the blimp of modern times might truly be an impressive thing to the unsuspecting mind. Their work might well have produced a 'mother ship' which appears to have more mass than it really does simply because of the huge unseen cavities of gas within its structure. This is, of course, assuming an earth-based group wanted to deceive the people with a super craft which appeared to be a massive 'mother ship.'

I have often argued this point with my friends and university physicists who have tried to tell me that we cannot provide enough thrust from any known rocket or engine to keep an aircraft hovering in one spot for over 30 minutes. It would just require too much weight in onboard fuel; and if it were nuclear, the thrust to mass ratio would be so small that it would not impress.

I have ruined these arguments every time when I have explained that I could use a hot air balloon or a dirigible to keep an aircraft aloft for hours - perhaps, days - without using prodigious amounts of fuel to do it. They have all responded, "Oh, well that's different. You're talking about specific gravitational differences there. That doesn't count!" To this I would usually respond by asking them, "If your life depended on being able to get an aircraft to hover over an area for a day without using heaps of fuel would you then use the blimp technique - regardless of the semantics of how it was being accomplished?" Yes,... well,... the answers were usually 'hrrummps' and red faces because they could not argue the point.

There it is. We have already got the technology to build huge airships and smaller spaceships, for that matter, which are mostly space inside variously filled with lighter-than-air gases or with nothing at all in the vacuum of space. So, if an enterprising group were to secure a base where they could develop such technology away from prying eyes and if that group were able to secure the resources needed, it is feasible that a huge fleet of blimp 'motherships' in the shape of a saucer or a large flat triangular warship or a huge, fat cigar could be made and tested.

Furthermore, with clever compression and decompression devices, such ships could be submerged under water while waiting to be used. Then, when it was time, they could be raised up as their gas cavities were filled. Now, they would need to move rather rapidly and not be swayed by prevailing winds near the Earth's surface to give the appearance of being an invincible city in the air from some alien civilization.

This could be effected by using a number of smaller craft to power the vehicle from inside its architecture where they would not be seen or heard by people below. As the Gray Project scientists showed me, we have had a undersea and under ice base down at the South Pole since the late fifties. As they also showed me, they developed three main types of saucers and propulsion of our own since the mid fifties.

So, if that 'enterprising group' were to use the more exotic saucer craft we have made, they could make a large, lightweight aircraft do some very impressive things - ESPECIALLY if the large blimp-type craft were to spin air over its surface like the smaller craft used to do in the early saucer models of the fifties. In fact, since the new electrogravitic craft generate a field which extends some 400 feet away from the craft, it is feasible that a supra-field could be established by placing the smaller craft at a certain distance from each other and by linking their drive frequencies in phase.

The result would be a field which would encompass all the smaller craft and all the space between them. To the untrained eye, it would seem these craft were flying in a tight formation. To the trained observer, it would be seen as a man-made approximation of geese flying in formation. It saves energy and each goose is locked into his position by the air wash from the goose ahead of him in the formation. So, if it were possible to include a skin over these man-made formations of small craft, then the appearance of a much larger entity would manifest. It would behave like the geese - as a new body made of many smaller bodies working in concert to produce motion in what appears to be a body corporate.

What we would see is a role reversal from the days of the first aircraft carrier blimp. Instead of it carrying the smaller aircraft, **it** would be carried by the smaller aircraft. In fact, extra smaller saucer craft inside such a 'megacraft' could be used to fly to and from the 'mothership' as though it were the home base instead of the 'Trojan Horse.' Thus, if the smaller craft needed to re-fuel, some of them could buzz away to a secret depot while the others stayed inside the big ship to keep it aloft and 'flying the flag.' The illusion would be complete if loud speakers hidden inside the 'mothership' would make droning, bass throbbing in the human hearing range. Furthermore, if the smaller craft made a show of their own weapons capabilities, the people would assume by association that the big mothership' would have a 'big' weapon on board.

This, then, is how that 'enterprising group' could create the illusion of a huge and powerful technology. Having seem how arrogantly blind so many of my fellow physicists have been when confronted with the issue of specific gravitation as a viable means of levitation of large craft, it would not surprise me in the slightest to see them be the first to run out underneath those great empty blimps yelling, "It is IMPOSSIBLE! You just cannot exist! You don't fit into my physics bible!"

No alien threat or salvation would be complete without real alien critters whether humanoid or reptilian. Our 'enterprising group' would find this task a little challenging - though not impossible. It would require the development of advanced genetic engineering. It would require those same isolated bases where the products of genetic engineering might be replicated and conditioned into believing that they really are from some other world - here on a mission to mankind. If this plan were to work as well as the alien mothership and small flying saucer deceptions, then the contact period with mankind would have to be held at a minimum and eventually stopped almost entirely to prevent discovery.

Remember what Matthew 24:3 said, "And as he sat upon the mount of Olives, the disciples came unto him privately, saying, Tell us, when shall these things be? and what shall be the sign of thy coming, and of the end of the world? And Jesus answered and said unto them, **Take heed that no man deceive you**".

The verse **did not say**, *"Let no alien or alien god deceive you"*. It **did say** let no **man** deceive you; however, other verses also speak of a non-human "arrival" on Earth. Thus, having been inside the Gray Project, I can only say that earthmen have certainly had plans to deceive the people of Earth with a fake alien threat since the early 1950s-regardless of whether or not real aliens beings from other dimensions or somewhere else in this Universe have recently appeared on the scene again.

I have had a number of small clues given to me by sources inside the Gray project; and from them I have deduced it is highly probable that the plan which started as the brainchild of the well-meaning Dr Grays of the world **has been usurped by a real, non-human culture** which even the Bible in some sections says will appear on the Earth in and around this time to deceive mankind into total submission to them. My last word on this subject is to beware of everything new which comes your way in the near future. Do not allow yourself to make knee-jerk reactions to rumors and especially to offers of material salvation by the new, global power structure (whoever they may claim to be).

The Face on Mars Farce

Now while we are on the subject of aliens and deceptions, it seems only proper to have a brief look at a the claims of some folks recently that Mars had a pre-human civilization with extremely advanced technologies. One science journalist gave two speeches to NASA personnel regarding suspected "alien" artifacts on Mars. One address was in 1988 to several hundred space scientists and engineers at NASA's Goddard Space Flight Center and the other was in 1990 to several thousand scientists and engineers at NASA's Lewis Research Center in Cleveland, Ohio. Both addresses strongly suggested that intelligent Martian life left a solution to our unified field problem encoded in the Cydonia "monuments" of Mars; and, furthermore, that they left us information about a new source of energy in the form of the dynamic fields inherent in all spinning stellar or planetary bodies.

The journalist's persistence in his analysis of many NASA of the Martian surface have led him to suggest pyramids (some five-sided) and what appears to be a huge face of a simian and humanoid cross on the Martian surface were left as primer textbooks to us. From these images and their mutual placement to each other, he was able to devise a preliminary mathematical concept making a new statement on spinning masses which has enabled him to accurately predict the correct latitude of the great spot on Neptune, Jupiter and other Earth-based landmarks such as the pyramids in Egypt.

HOWEVER, I feel I must inject my own hypothesis to give a balancing view to that journalist's "face" hypothesis. In 1997-98, I retrieved all the available image data I could find regarding the so-called "face" and "pyramid city" concepts put forth by the journalist. This included digital images of Cydonia (a Martian region) from his web site and from associates of his and from NASA itself.

One of the images I gathered (**see Viking Orbiter 1975 photo no. 673b56**) showed a lot more of the Cydonia region than the images the journalist and his associates had used to argue their premise of advanced culture signs there. I could immediately see that something wasn't right about their premise (a premise, by the way, that I used to support as well). One evening as I was looking at the images, it suddenly hit me. The pyramidal shapes were not unique to the little area of the image which they called the "city at Cydonia". In fact, for many miles to the right of the "face" and "city" locations there were pyramidal structures as well. More of the so-called pyramids were scattered over a much greater area than the Cydonia region. Furthermore, they were part of a larger, flat plateau that seemed to have broken at its left side and created a mess of debris which contained more of the crystalline (or pyramidal) structures.

Suddenly, I saw a different possibility for the creation of those pyramidal shapes. The flat area could have been the remains of a formerly curved exterior surface of a small moon or planet which had exploded some distance from the Martian surface.

As I analyzed the "impact" area near to Cydonia, I realized that this was not an isolated event on the Martian surface. There were signs of a much greater calamity all over the planet. One could see a large triangular area in black on the Martian surface. It gave the distinct impression that a large object hit the surface at a very shallow angle and exploded leaving the fan-shaped deposit along its impact zone (**see Cydonia Location**).

From further study of the entire Martian surface as supplied by the NASA and JPL imagery, I have deduced that a small planet (perhaps one which used to occupy an orbit next to Mars and between Mars and Jupiter) exploded from some (as yet) unexplained series of events on an astronomical scale. The dust and the fractured igneous rocks from that planet were blown into and outside of its orbit. Thus, leaving the asteroid belt of today and numerous wandering groups of asteroids in the solar system... AND a huge deposit of red dust and sharp-edged igneous rocks **(see Martian sharp-edged debris image)** all over the planet Mars.

These rocks were and still are sharp-edged. This implied that they were deposited slow enough to allow them to impact without melting their edges on the way through the thin Martian atmosphere. From this and other clues, I deduced the debris must have been traveling fairly slowly in respect to an impact vector to Mars due to the effects of its original orbital elements when it approached Mars.

In effect, the debris from the other planet or moon must have been so massive and spread out like a canopy that it (relatively speaking) slowly descended to the surface. This would have caused a change in the spin rate of Mars. If the debris had come from a planet or moon not connected to mars by orbital components, then the addition of the extra mass would have caused Mars to slow down its spin thus lengthening the Martian "day". If the debris was from a moon already orbiting Mars, then it would have sped up the spin rate like a ballerina's spin when her arms are drawn into her body.

I believe this is why the red dust is non-uniformly deposited across the Martian surface. This may also be why large pieces of that "moon's" out layers impacted upside down on the Martian surface to create those large, raised, flat-topped mesas near Cydonia...

It could explain why these areas exhibit fracturing toward one end more than the other. They would not have hit exactly parallel to the Martian surface; and, thus, would have "snapped" the end arriving last. This would have caused the fracturing and splattering of fragments several miles along its impact trajectory. Later as the red dust landed the fractures would have become less obvious as the dust began to mound around the edges and to fill the gaps.

Furthermore, from the way the left side of the flat area had fragmented on impact, I deduced the pyramidal structures could have been rolled off the flat surface in a random pattern from the forces of the impact.

You see, when a planet or moon is young, it has a molten core spinning vortically at various rates. As the moon or planet grows older and cools, the core starts to change in much the same way as a geode does on Earth when it is flung from a molten eruption and then cools over the next few days or weeks. When one cuts a geode in half, one sees a layer of quartz crystals inside the geode (**see Planetoid Core Stretched Out**). Sometimes on top of this layer or instead of this layer there are found amethyst and beryl (emerald) crystals. The core is empty as the water of hydration has been incorporated into the organized, crystalline layers and has thus taken a lot less space.

On a planetary scale, those crystals would be miles across JUST like the "pyramids" of Mars. One can imagine how a curved chunk of a small moon with its crystallized "pyramids" would have descended to the Martian surface having its fall braked by the curved shape of the chunks old surface. It would have eventually struck the ground and would have whipped the far left end so hard that the crystals would have been ejected from the inside of the surface as it flattened out on impact. They would have tumbled farther along the impact path leaving the "cities" we now observe.

I spent the next few days developing a way to estimate the diameter of such a hypothetical moon or planet if its explosion produced a surface fragment of the dimensions I saw before me in the image. Using trigonometry and making some basic assumptions, the object's diameter appeared to have been from 150 to 225 miles. This would have been a very small moon or planetoid, even.

On another image (**see Cassini Crater and Cydonia Region**), I found Cydonia on the "western" ejecta (secondary) rim of a huge impact crater named "Cassini"... (When a meteor impacts with so much energy, it makes two "rims" - one close around the meteor impact hole and a second one much farther out as a result of the mass ejected from the initial impact hole in a splash-like rebound.)

What I saw in the Cassini Crater modified my hypothesis for the origin of the Cydonia area. The primary impact crater at Cassini could have been made by an object roughly 300 miles in diameter. The secondary expulsion of impact debris may have thrown a backwash of pieces from the impact object back over the secondary rim to the west forming Cydonia from a piece of the surface and core of the object. The speed of the ejecta from the expulsion of matter from the primary site would have been much slower than that of the initial object. This would explain the 'graceful' landing of that huge sheet (**see Mars Viking Orbiter No. 673b56 image**) which broke into pieces on its 'west' end only.

Look closely at the images I have included here to illustrate my hypotheses. I could be wrong; but I believe what I have deduced is very close to the way it must have happened. My explanations for what may have happened on Mars to have generated the Cydonia topography is not as exciting as a "Martian race" hypothesis; but I do believe it is closer to the truth than a "hyper-dimensional physics lesson". Neither of my hypotheses supports the notion that a Martian "city" ever existed at Cydonia. The Martian "astro-geology" around Cassini just cannot be ignored.

Mars Viking Orbiter 1975 Photo No. 673b56

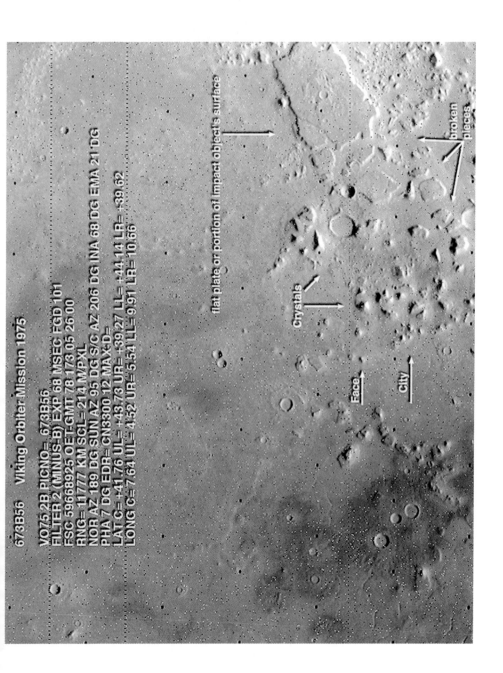

Cydonia Location

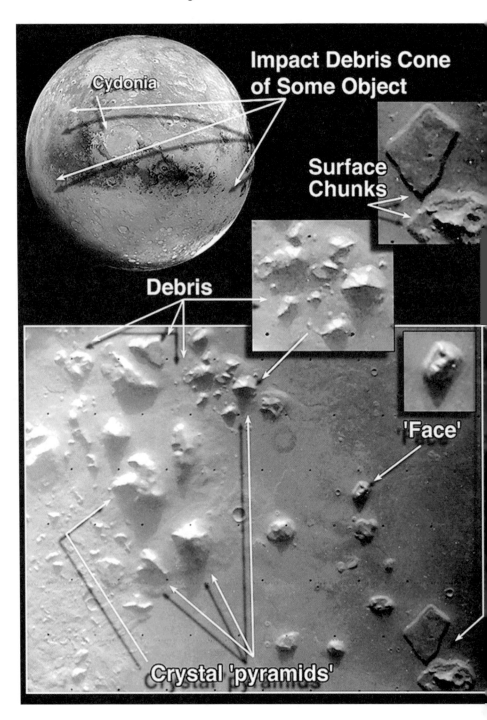

Cydonia

Impact Debris Cone of Some Object

Surface Chunks

Debris

'Face'

Crystal 'pyramids'

Martian Sharp-Edged Debris

Planetoid Core Stretched Out

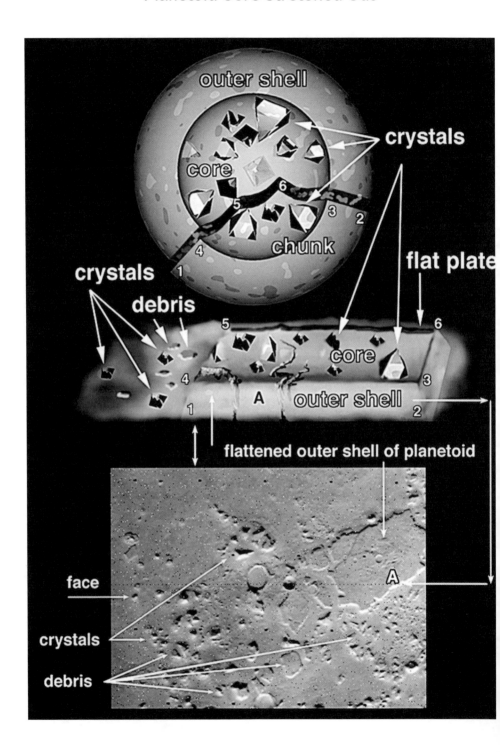

The Cosmic Conspiracy Final Edition 2010 © Stan Deyo 2010

Cassini Crater and Cydonia Region

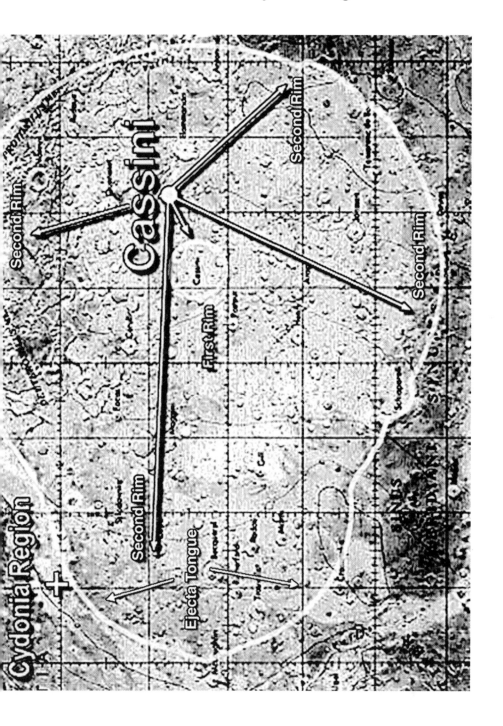

The 'Face' on Mars (High Resolution)

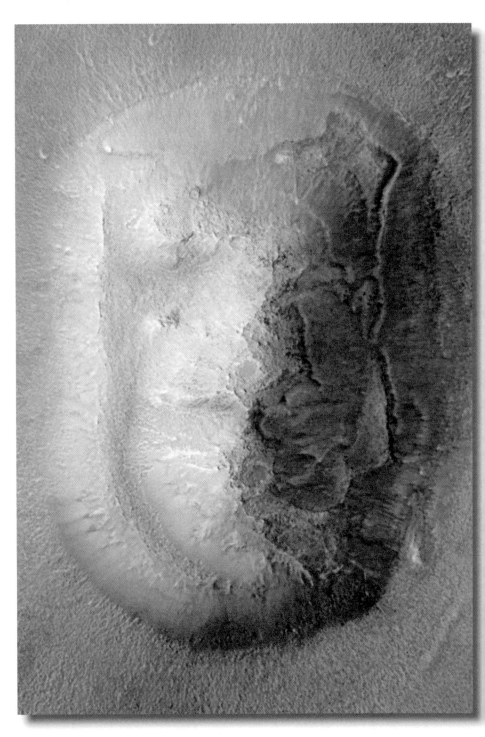

The Cosmic Conspiracy Final Edition 2010 © Stan Deyo 2010

The Great Seal of the United States of America

On page 69 of this book, I gave a description of how the Great Seal of the United States of America had been designed and delivered to Thomas Jefferson at 'Monticello' which was his home near Charlottesville in Virginia. This description was drawn from a Rosicrucian publication on the subject; which stated the date of the Congressional approval for the design of the Great Seal was June 10th, 1782; and, furthermore, that the Seal was cut and delivered to Jefferson on June 17th 1782. The official versions of this event differ; as the following illustrates.

The official version ('OV') of the design and delivery of the Great Seal was published by the US Department of State in April of 1957. In this version, Franklin, Adams and Jefferson were appointed as the original design committee. The 'OV' went on to say that it took six years and three committees to negotiate the final design for the Great Seal. Apparently there were many divergent ideas as to what meaning the final design would convey amongst the committees.

The 'OV' went on to state that most of their ideas, *"tended to transcend the miniature confines of a seal and the strict conventions of heraldry. Franklin, for instance, proposed a device showing Moses dividing the Red Sea for the children of Israel and the waters closing behind them to overwhelm Pharaoh in his chariot. The motto read, 'Rebellion to tyrants is obedience to GOD..' Jefferson favored showing the children of Israel in the wilderness under the guidance of a cloud by day, a pillar of fire by night"*.

According to the Encyclopedia Britannica[24] version ('EBV'), the first committee consulted with an artist by the name of Pierre Eugéne du Simitiére from Philidelphia to design the obverse (or front) of the Great Seal; while they allowed the reverse to be designed by Benjamin Franklin; who stated the eye in the triangle represented *"the Eye of Providence in a radiant triangle"*.

The 'EBV' stated the second committee comprised of James Lovell, John Morin Scott and William Churchill Houston consulted with the Treasurer of Loans under the Continental Congress, Francis Hopkinson, for his artistic input. The 'EBV' continued, saying the third committee comprised of Arthur Middleton, John Rutledge and Elias Boudinot called upon the services of William Barton to complete the seal design because of his expertise in both heraldic manner and artistic drawing. Barton promptly submitted two designs, and the second was accepted by the committee on May 9, 1782.

On June 13th, according to the 'OV', the drawings of all three committees and all their motto suggestions were jointly submitted to the Secretary of the Congress, Mr. Charles Thomson; who then made a few changes to the artwork and added the words, *"E Pluribus Unum"* to the obverse and *"Annuit Coeptis"* and *"Novus Ordo Seclorum"* to the reverse. Thomson also re-introduced the Latin date of *"MDCCLXXVI"* (1776) to the reverse side. The final design of the Great Seal of the United States of America was approved by its Congress on June 20, 1782. A brass model was cut soon after; and the first public use of the Great Seal was recorded on September 16, 1782.

Which of these three versions is entirely correct? Well, obviously, the Rosicrucian dating differs with the Congressional Record of the time. As far as the mysticism involved, Charles Thomson is the one who seems to have added so many 'interesting' graphics and phrases. No matter who made the changes to the Great Seal, the symbolism is most pagan in my opinion...

---

24 Encyclopaedia Britannica (1965 Edition), Vol. "S", under "SEALS", pp. 235-238

Great Seal Phrasing

As stated in my original *Cosmic Conspiracy,* the Great Seal has two unusual phrases on the reverse side; as well as an unusual set of graphics. Since I first wrote on this subject, numerous letters and pieces of information on the subject have come my way.

The Thomson phrases are, *"Annuit Coeptis"* and *"Novus Ordo Seclorum".* The 'OV' claims these phrases are to be interpreted, *"He* (GOD) *has favored our undertakings"* and *"A new order of the ages".* This may be the way they intended these phrase to be interpreted; but it is not what the Latin words say... (far from it!).

The first phrase, *"Annuit Coeptis"* means, *"Year of Beginning".* The second phrase, *"Novus Ordo Seclorum,"* can be translated several ways using the following definitions (where the words in bold are the preferred usages): *"Novus"* = **new, anomalous**, young, fresh, strange or inexperienced; *"Ordo"* = **order, class**, arrangement, line or series; *"Seclorum"* = (this word is difficult for the every Latin scholar I have approached on the subject; because it appears to them to be some sort of a slang spelling of the Latin word, "saeculum" meaning, "**of the** centuries (**ages**)". However, I feel it is a slang word made from two words.

The first word, "sec" could be any of the following: 1) an abbreviated form of "secundum" meaning, "**in favor of, according to**, after, next to, behind;" 2) an abbreviated form of "secundus" meaning, "**following**, next, second, inferior, favorable or fortunate"; or 3) "sec" meaning, "to divide, **cut** or **set apart from**". The second word, "lorum" means, "**lash, whip, strap** or leather charm".

Now this means the phrase might be interpreted into **two**, diametrically opposed meanings. The proponent version would read, "**Year of beginning: new order (class) set apart from lash (whip);**" while the opponent version would be, "**Year of beginning: new order (class) in favor of lash (whip)**". Which version correctly describes the United States of America today? Since it is for the moment the strongest military force on the planet and since it is apparently the defender of freedom on the planet at the same time, you will have to be the judge. Remember what I said on page 73 about the Sanskrit name, "Mahamudra," which means "Great Seal" and mystically means "the **union of** all apparent **duality**".

Based on the Babylonian numbering system hidden in the Great Seal (see page 70 of this book), if the USA is the 'Babylon' mentioned in the prophecies of *The Revelation of John,* then, from the same book, you are forewarned of the severe judgments which are soon to befall and utterly destroy it for its unfaithful behavior to God...

I am not the only one calling attention to the possibility that America could be the "Babylon" of Biblical prophecies. Just a few weeks ago, Richard Coombes, (also a faithful American like myself) published a full and erudite book giving another VERY plausible way one can identify the Babylon of the Revelation. He titled his work, "America, The Babylon[25]". I have just read a copy the author sent me and I am astounded at the ramifications for America! YOU MUST READ THIS BOOK. This man has laid a very strong case using the Bible, history and current circumstances to identify the nation and the city of Babylon. Even General George Washington had a vision which said America would be almost totally destroyed by a dark host of invaders after three great wars had tested America - supposedly, The Civil War, WWI and WWII.

As an ex-patriot American and always as a Texan, I would not like to envisage the destruction of my relatives in the USA - particularly my parents, siblings, children and grandchildren; but, I do believe they are living right in the middle of 'ground zero' if my understanding of the prophecy is correct. It is a hard scenario to espouse.... very hard.

---

25   This book sells for US$19.95 in the States. ISBN # 1-890622-33-8 from TLCS, 3336 South 7th Street, Terre Haute, IN, USA 47802. Main Phone: (812) 234- 3905, Fax Number: (812) 238- 0108

Hey, "Frog," Is the Water Boiling Yet?

And one sweating frog turned lazily to the other and asked through the steaming haze over their pot of almost-boiling water, "Does it seem warm in here to you?"

In the Republic of Ireland a local war rages (ostensibly between Catholic and Protestant "Christians"). In Bosnia, another seething war festers between "Christians" and "Muslims." In parts of India "Christians" and "Hindi" are close at the brink of war. In the Middle East, Syria, Iran and a host of other Islamic nations are aligning themselves with the "anti-Israel" PLO to destroy the state of Israel once and for all time (or, so they think...) Various tribal wars are in progress in Africa and many are related to religious differences.

These wars are not purely religious wars per se. There are many side issues; but the main issues radiate from the central schism of opposing religious philosophies. On a bigger scale, for all Christian nations and all Muslim nations, it is an age-old continuance of the Crusades. But even this war is an offshoot of a much older grievance between Israelite (Jew) and Ishmaelite (Arab, Muslim). For the Israelite and the Ishmaelite have an age-old land dispute based on the actions of their forefather, Abraham, when he had a child by Hagar who was not his wife... (a religious and legal issue).

Every major group of nations can be labeled by common philosophical goals. Whether those goals be based on a supreme being (a god or gods) or whether it be based on an atheistic communal model, they are all preferred models of lifestyles and objectives within each group. When these philosophies conflict on non-resolvable issues, war is inevitable.

And when an entire planet of nation-state "alliances" is dedicated to tearing itself apart, how can the planet's peoples defend themselves against a common global threat? Can we afford to fight amongst ourselves while the climate changes become more an more life-threatening; or, while we are faced with an imminent economic collapse on a global basis; or, while the earthquakes and volcanoes destroy more land and commerce; or, while the Sun moves toward an increased energy output which could burn our crops to dust; or, while our polar ice caps are melting to sink our coastal areas and with them over 60% of human settlements; or, while an (as yet) undiscovered (but statistically expected) meteor of life-extinction proportion and speed hurls toward our world; or, while the threat of new, horrible diseases grows from the climatic and warfare stimuli; or, while the possibility of alien invasion of Earth grows with every new admission of guilt from persons formerly under strict codes of secrecy?... Well? Can we afford to be divided while any one of these threats could and may soon destroy civilization as we have come to know it?

Of course not... However, we will not become a race of earthmen until one of these threats knocks us to our knees... which is where we should have been all along in reverence to God: **Elohim**, They, The Oneness, who created all we perceive.

Beware of an impostor. For while we have squabbled between ourselves, a wolf in sheep's clothing has been tricking our troubled tribes to prepare us to submit to the authority of one who will appear to bring peace and, then, claim to be God and is not...

You CAN Be Delivered From the Deception...

see Pages 123-127...

THE 2010 FINAL ADDITION

It has been thirty-three years since I first wrote this book and thirty-two years the first printing. In this final edition, I left the first 200 pages intact with the exception of some minor editing changes and included the 1992 and 1998 additions to the book which so many readers requested in recent years but which are no longer in print. Hopefully, mankind's awareness is sufficiently elevated that what I have to reveal will not be so difficult to believe or understand as it would have been had I tried in the earlier editions of this book.

As I begin this final edition, winter also begins here in the Rockies. A chill is in the air and snow is falling outside. And, like some mythical hobbit, I am sitting by a fireplace where flames cast flickering flashes onto the walls and the scent of hot chocolate wafts around me. How strange and incongruous this all seems as I begin to write of the evil, frightening…. and deceptive things attending the end of this age. I know this is the last edition of my book as the events I have so long warned about are already starting to happen and soon Americans will no longer the enjoy freedom of speech granted by the first amendment to the U.S. Constitution.

-Stan Deyo, 10 February 2010

Beware the Messengers of Light from the Heavens

The entire planet is encompassed by a dark and chaotic atmosphere. Evil abounds like never before in human history. Nations are either already at war with each other or are planning to start wars with each other. Nuclear weapons are being prepared in the Middle East for events that will lead the world to Armageddon… as prophesied in the Bible and related extra-biblical documents.

Global food shortages are growing by the day. The unpredictable climate changes (for whatever reason) are playing havoc with the food crops, food delivery systems and human settlements. The world economy is about to collapse as the US dollar weakens. Fatal, viral diseases (some man-made) are appearing all over the world and no cures are known. The failing global economy is producing massive unemployment – hence, the loss of homes and other assets. Religious differences between Islam and the rest of the world's major religions are laying the foundation of widespread, horrific religious conflicts. Geological threats loom as huge earthquakes like the one that flattened Haiti and massive volcanic eruptions become more frequent. An increase in near-Earth-orbit meteorites and thousands of untracked meteors of serious sizes is adding to the darkness enveloping the minds of mankind. And last, but most importantly, is the threat of alien invasion which looms increasingly at a background level in daily news. Leaks from highly qualified witnesses to the alien presences are forcing the governments to grudgingly (or so it seems) reveal more and more data from their classified archives.

Until this edition I have primarily addressed how a group of 'enlightened' persons (the "Illuminati") were using secretly acquired and developed technologies to herd the people of Earth into a New World Order - one to be controlled by those 'enlightened persons.'

I explained how the post WWII illuminists planned to unite the people of Earth under a global order that would have but one religion, one law, one economy, one leader and, eventually, a predominantly single race of earthmen. I explained how those illuminists were planning to fool the peoples of Earth with a fake 'alien' landing and its implied threat to the world. I explained that many advanced technologies would be held back from the people so they would be properly impressed with the fake alien arrival and the attendant 'advanced' technologies. I even shared basic technological information which I, myself, had learned during my association with those illuminists and afterwards when I expanded my own knowledge from other people who were no longer working for the Illuminists.

However, I did NOT explain how their plans would be altered by a real 'alien' presence pretending to be 'god' – one of which the Bible clearly warns. The Bible says the 'alien' presence will be revealed here on Earth in the last days of this age and that they will be given absolute control over many people in several regions by the world dictator (commonly called the 'antichrist') at that time. Later in this 2010 addition, I will give you an 'unsealed' or decrypted translation of eight verses from the Book of Daniel before I explain the 'alien agenda' that Satan is already spinning to the unsuspecting.

The last words given to Daniel by our Messiah (God) tell us the book was encrypted as this verse clearly states: Daniel 12:9 And He said, "Go, Daniel! For these words [are] closed up and sealed until the end-time." However you can skip over the detailed proofs and still get the bottom line on what I have decoded from the Book of Daniel.

Furthermore, because of my own involvement many decades ago in the development of flying saucer technologies for the Illuminists, I knew that a **great global deception** was being planned using occult technologies. Their planned deception gave me an outline to apply to those parts of those sealed words when trying to re-translate, unseal or decrypt them. When several words could be used for a given group of consonants by applying different vowels, I chose the vowels which would produce words that fit the outline I obtained from the Illuminists. What I have discovered is mind-blowing!

In the book of the prophet Daniel are eight verses which, when translated and decrypted, tell us these things. I might add here that Y'shua Ben Elohim (Jesus), Himself, pointed us to the writings of Daniel as most important for these end times. Jesus accepted Daniel as a full-blown prophet even though Daniel was not accepted by the Jewish sages as a major prophet since they say God never talked to Daniel directly. However, it should be pointed out that students of the Revelation of John in the New Testament can verify the Son of God (Jesus – hence "God" as those who see the Messiah see God) did indeed speak to Daniel while standing on the waters of the great Tigris River as recorded in Daniel 10-12. The same being with eyes of fire wearing a golden waistband and having arms and feet the color of polished bronze is described in the Revelation of John as that described by Daniel.

Daniel's name is Chaldean Aramaic which is akin to ancient Hebrew. His name literally means 'God's Judgement' although most translators make it, 'God is my Judge.' This is an important issue as the Book of Daniel holds keys to decoding the book of the Revelation of Y'shua Ben Elohim (Jesus) to the prophet John many years later on the Island of Patmos. The literal translation of 'Daniel' implies his book is a writing about God's Judgement of all the global empires of the world, their kings and their people until the end of this age and the establishment of a benevolent kingdom by God that will never end. We are in the end-time as evidenced by the rebirth of Y'srael (Israel) as nation in May of 1948.

The following four verses are from an interpretation of a dream by Daniel. He made this interpretation of a dream experienced by his, King Nebuchadnezzar. In his dream the king saw a tall statue of a man. The head was made of fine gold, the breasts and arms of silver, the belly and thighs of bronze, its legs of iron, feet and toes were each made of different metals with some toes made only of clay. God showed Daniel the statue represented the world kingdoms in order from the Babylonian Empire to the final world empire just before the return of the Messiah to set up His everlasting Kingdom.

The last world empire was represented by the feet and toes while the penultimate world empire was represented by the eastern and western 'legs' of the Roman Empire. Interestingly, the legs were made of iron (a strong metal) which were followed by the feet made of iron and clay (a fragile material). This showed the last global empire would be an extension or a rebirth (not a replacement) of the former Roman Empire - one expanded into ten regions of mixed strengths.

History tells us the Roman Empire atrophied and left the vestiges of its former power and structure sequestered within the 'Roman Curia' (Roman Court) of the 'Roman Catholic Church.' It is no coincidence the Catholic Church is still divided into two branches or legs: The Eastern and Western Catholic Churches.

One last 'world empire' or 'new world order' is being formed before our eyes. And, not surprisingly, the blueprint for that world order was created by the Club of Rome (see the map on page 200 of this book). You will see in their proposed division of the world they made ten divisions (dioceses) just like the dream Daniel interpreted. My decryption of the eight selected verses from the Book of Daniel will tell even more about this last major world empire or the 'Revived Roman Empire:" In the following verses found in the King James Bible I inserted the actual Hebrew words next to them for later reference as I decode and retranslate the original text.

The ancient Hebrew and Chaldean Aramaic scrolls had no written vowels. Today, modern versions of the scrolls include the vowels as the little dots and lines underneath the Hebrew consonants. They were added many centuries later by scribes and translators based upon **what they thought the vowels should be**. For example, in written English if we were to remove the vowels and write just the consonants it could create confusion to translators. Take the word, 'LiNe' and remove the vowels. You are left with 'LN'. Then pretend you are a translator 2,600 years in the future trying to decide what word you meant in a sentence by substituting various vowels between the two consonants. That future translator would have to decide between the following words: LiNe, LoNe, LeNd, LiNk, LeNs, LeNt, LuNg, LiNt, LaNe, LuNe LaNd and LuNa. He would ultimately have to choose the word that seems to best fit in context with the rest of what you wrote. This is what I have done in the translation of the eight verses which follow.

The first four verses I selected and my decryption of their meanings follow; however, should you wish to skip my detailed analysis you can skip down to the paragraph labeled, "*First Four Of The Eight Verses Expanded.*"

Dan. 2:41 And whereas thou sawest the feet and toes, part of [the] **potters'** (דִּי־-פֶּחָר) clay (חֲסַף), and part of **iron** (בַּרְזֶל), the kingdom shall be divided; but there shall be in it of the strength of the **iron** (בַּרְזֶל), forasmuch as thou sawest the **iron** (בַּרְזֶל) **mixed** (מְעָרַב) with **miry clay** (בַּחֲסַף טִינָא).

Dan. 2:42 And as the toes of the feet were part of **iron** (בַּרְזֶל), and part of **clay** (חֲסַף), so the kingdom shall be partly strong, and partly broken.

Dan. 2:43 And whereas thou sawest **iron** (בַּרְזֶל) **mixed** (מְעָרַב) with **miry clay** (בַּחֲסַף טִינָא), **they shall mingle** (מִתְעָרְבִין לְהֱוֹן) **themselves** with the seed of **men**: (אֲנָשָׁא) but they shall not **cleave** (דָבְקִין) one to another, even as iron is **not mixed** (לָא מִחְעָרַב) with **the clay** (חַסְפָּא).

Dan. 2:44 And in the days of these kings shall the God of heaven set up a kingdom, which shall never be destroyed: and the kingdom shall not be left to other people, but it shall break in pieces and consume all these kingdoms, and it shall stand for ever.

Decrypting the preceding four verses with the knowledge mankind has gained about the 'alien' presence on Earth today begins to open the 'sealed' words of Daniel to us. Daniel lived in Chaldea which was an eastern district of ancient Babylon. His first language was Hebrew yet after many years in Babylon he mixed Arabic, Chaldean Aramaic and some Sumerian words and idioms in his Hebrew writings in the first two chapters of his book. So, bearing this in mind one can see how difficult it can be to 'unseal' words which have undergone so many translation along the way to us in this time.

One thing really leaps out of these verses. Daniel used four different words which were translated as the 'clay' in these verses. Why he did so presents a puzzle we need to solve; so I will expand the translations of *'potters' clay,' 'clay,' 'the clay'* and *'miry clay.'* Then I will expand the translations of: *'iron,' 'man,' 'they,' 'mixed,' 'mingle'* and *'cleave.'*

The words חֲסַף דִּי־פֶחָר (chasahph di peychar) are translated as *'potters' clay'* in **Daniel 2:41**. These three words come from Sumerian, Chaldean Aramaic and Hebrew respectively. The Sumerian meaning of פֶחָר (peychar) is **'fear or dread'** as opposed to *'potter'*. The Hebrew meaning of 'di' (דִּי) is **'of'**. While the Chaldean Aramaic word חֲסַף (chasahph) is translated as **'clay'**. When we put these original language meanings together it reads, **'clay of fear or dread'** which means the clay regions of the kingdom are not only pliable like potters' clay but are also filled with fear… of, presumably, the **iron** regions of the kingdom.

The Chaldean Aramaic word חֲסַף (chasahph) is translated as *'clay'* in **Daniel 2:41** and **Daniel 2:43**; however, the Chaldean Aramaic word חַסְפָּא (chahsfah) translated as *'the clay'* in **Daniel 2:43** presents a troubling enigma. Taken as a whole the Chaldean Aramaic meaning of חַסְפָּא (chahsfah) is **'shards of earthen ware.'** However when one looks at סְפָא (safah) the root verb for the word in Chaldean Aramaic we find it means **'to feed'**. I am reminded of a passage from the Book of Enoch [26] which describes the behavior of the offspring of the fallen angels who were cast to Earth before the Flood of Noah.

*Enoch 7:1-5 And all the others together with them took unto themselves wives, and each chose for himself one, and they began to go in unto them and to defile themselves with them, and they taught them charms and enchantments, and the cutting of roots, and made them acquainted with plants. And they became pregnant, and they bare **great giants**, whose height was three thousand ells: Who consumed all the acquisitions of men. And when men could no longer sustain them, the **giants turned against them and devoured mankind**.*

---

26   The book of Enoch was written in the 2nd century B.C., and was popular for about 500 years, with Jews and early Christians alike. It is one of 15 works of the Jewish apocrypha. The Jews rejected the book of Enoch when they made a canon of their own scriptures late in the second century A.D.

During the first three hundred years of Christianity, early church leaders made reference to it. The early second century "Epistle of Barnabus" makes much use of the Book of Enoch. Second and Third Century leaders, including Justin Martyr, Irenaeus, Origin and Clement of Alexandria all referenced the book. Tertullian (160-230 A.D.) even called the Book of Enoch "Holy Scripture". The Ethiopic Church also added the Book of Enoch to its canon. It was widely known and read the first three centuries after Christ. This and many other books became discredited after the Council of Laodicea. And being under ban of the authorities, afterwards it gradually passed out of circulation.

Later theologians apparently disliked it because of its content regarding the nature and actions of fallen angels. The Reformers, influenced by the Jewish canon of Old Testament, also considered it as non-canonical and thus it was removed from the Protestant Bible. Catholics apparently do consider the book of Enoch as canonical, as one of 12 of the 15 they accept.

Perhaps the expanded translation of this word חַסְפָּא (chahsfah) for **'the clay'** should reflect that the regions occupied by mankind will be not only like **'shards of earthen ware'** but also like **'fodder to those beings of the iron regions.'**

The Chaldean Aramaic words בַּחֲסַף טִינָא (bahchahsaph tinah) (in **Daniel 2:41** and **Daniel 2:43**) are traditionally translated as **'miry clay'** or **'ceramic clay'.** The Chaldean Aramaic word, טִינָא (tinah), is traditionally translated as **'thin, fragile clay as found in clay baskets or clay pottery.'** The literal word for word translation of these two words in their Hebrew form is, **'with clay of the clay'** which does not make sense. However, there was a region in western Babylon named "Ayn at-Tinah" where a certain type of clay was mined to make **Nabataean**[27], **fine, painted pottery** - which was widely known throughout the Middle East. This type of clay produced a very thin and, hence, delicate or fragile pottery. So, in Daniel's usage of the word he would have emphasized the fragility of the clay pottery by using this word טִינָא (tinah) as an adjective. Thus, this phrase should read, **'with the especially thin clay from Tinah.'**

The Hebrew word for **'iron'** is בַּרְזֶל (bahretzel) which was adopted from the Chaldean Aramaic words, פַּרְזֶל (pahretzel) and בַּרְזְלָא (pahretzelah). Since the Chaldean Aramaic root verb, for the Chaldean Aramaic forms means **'to separate'** or **'to scatter'** we can enhance the definition of **'iron'** to **'iron that scatters or separates with force.'**

The Hebrew word, אֱנָשׁ (ehnash) meant **'mortal'** in Hebrew and **'man'** or 'mankind' in Chaldean Aramaic; but the word was traditionally translated as, simply, **'mankind.'** My expanded translation of the word means the same but adds more depth to it. Combining the Hebrew, **'mortal,'** with the Chaldean Aramaic, **'man'** or 'mankind' emphasizes mankind is mortal while **'they'** are not. Thus, we read, **'mortal mankind.'**

The Hebrew word מְעָרַב (mahrahb) commonly translated as **'mixed'** in Dan. 2:41 and **Dan. 2:43** comes from the root verb עֲרַב (ahrab) which can also be translated as **'to intermix'** or **'to have intercourse with'** or **'to enter into a negotiation with.'** So, the enhancement to the translation of this word here might also be, **'to have intercourse with** or **to intermix.'**

---

27   The Clay Sources: Several provenience studies concluded that Nabataean painted fine wares were only made in the Petra area [3, 4, 5], where there is abundance of clay in the ash-Sharah mountains to the north and east of Petra.

The analyses of clay samples from the Petra region, done under CERAMED, concluded that the site of 'Ayn at-Tinah is the most probable source used by the Nabataean potters [6, see also H. Bearat and F. Alawneh, "Preliminary Examination of the Nabataean Painted Pottery from Petra, Jordan"; and T.S. Akasheh, B. Khrisat, M.N. Na'es, K. Amr and D. Ferro,"Archaeometric Investigation of Nabataean Painted Fine Ware"; both in this volume).

The 'Ayn at-Tinah clay was deposited in shallow waters during semi-arid geological periods (during the Cenomanian 98.9-93.5m yrs ago). The clay has a high expansion/shrinkage rate, and is extremely plastic thus enabling the manufacture of the characteristic "egg-shell" Nabataean wares under proper manufacturing procedures. The high iron contents of the clay result in red coloured pottery that is also typical of the Nabataean wares.

Extracted from: Recovery and Reproduction Technology of Nabataean Painted Fine Ware by Khairieh 'Amr (1), Talal Akasheh (2) and Maram Na'es (2)

(1) The Jordan Museum, P.O. Box 831225 Amman 11183, Jordan. Fax (+962) (6) 4629312. Tel (+962) (6) 4629317. e-mail: khairiehamr@cyberia.jo.

(2) Queen Rania Institute for Tourism and Heritage, The Hashemite University, POB 330127, Zarqa. Jordan. Fax (+962) (5) 3826613. Tel (+962) (5) 3826600. e-mail: tsakasheh@cyberia.jo; maram103@yahoo. com.

The Chaldean Aramaic words, מִתְעָרְבִין לֶהֱוֹן (mitarabin leheon) are commonly translated as *'they shall mingle themselves.'* Just the word, מִתְעָרְבִין (mitarbin), itself, is from the Arabic verb, חַצַּר (chaztar) meaning, '**to call together.**' However, the root verb for מִתְעָרְבִין (mitarbin) is the Arabic word עֲרָב (arahb), which has several other derivative meanings in Arabic. עֲרָב (Arehb) means '**strangers**' and עֲרָבָה (arabah) means '**the heavens.**' So, the enhanced translation of *'they shall mingle themselves with the seed of men,'* might well be, '**the immortal strangers from the heavens shall have intercourse with mortal mankind**' where '**intercourse with**' may mean '**dialogue with**' or '**sex with.**'

The Chaldean Aramaic word, דָּבְקִין (dahbakeen) is traditionally translated as *'cleave'* which is fairly descriptive of what Daniel meant albeit in King James English. It would read better as, '**stick together.**'

First Four Of The Eight Verses Expanded:

*Dan. 2:41 And whereas you saw the feet and toes, part of fearful, fragile **clay**, and part of clay-smashing **iron**, the kingdom shall be divided; but there shall be in it of the strength of the clay-smashing **iron**, forasmuch as you saw the clay-smashing **iron mixed** together by negotiation with the thin, fragile **clay of Tinah**.*

*Dan. 2:42 And as the toes of the feet were part of clay-smashing **iron**, and part of broken shards of **clay** vessels, so the kingdom shall be partly strong, and partly broken.*

*Dan. 2:43 And whereas you saw clay-smashing **iron mixed** together by negotiation with the thin, fragile **clay of Tinah**, the clay-smashing **iron regions of the immortal strangers (gods) from the heavens shall be part of the same world kingdom with the regions of the seed of mortal mankind**: but **these mortals and immortals** shall not **stick together**, even as clay-smashing iron is not mixed with the weak, fragile, fodder-like **clay of Tinah** which is something to trample underfoot or to scatter or shatter.*

Dan. 2:44 And in the days of these kings shall the God of heaven set up a kingdom, which shall never be destroyed: and the kingdom shall not be left to other people, but it shall break in pieces and consume all these kingdoms, and it shall stand for ever.

Daniel is warning of a coming world kingdom in which some of the ten regions are controlled by immortal beings from the heavens. The first two verses above describe the global 'kingdom' in more of a regional level. However, *Daniel 2:43* is written on a more detailed level whereby one considers the references to iron and clay regions to be the beings that inhabit those regions – not the regions themselves. In that respect the verse is saying that although the "***immortal strangers (gods) from the heavens***" try to interbreed with the "***seed of mortal mankind***" they will not succeed in making strong, cross-species beings and that those offspring will be fragile and weak.

Furthermore one can see that those regions will coexist in the global kingdom but will not '**hangout with**' or '**be close to**' the human regions of that global kingdom. Also, it is possible that the **iron regions** may have low regard for the **clay regions** as we look upon our livestock today.

This second four verses of the eight verses from the Book of Daniel give us further information on who forms and dictates the kingdom of those ten toes.

*Daniel 11:36 And the king shall do according to his will; and he shall exalt himself, and magnify himself **above every god** (עַל־כָּל־אֵל), and shall speak marvelous things against the God of gods, and shall prosper till the indignation be accomplished: for that which is decreed shall be done.*

*Daniel 11:37 Neither shall he regard the God of his fathers, nor **the desire of women** (חֶמְדָּה), nor regard any god: for he shall magnify himself above all.*

*Daniel 11:38 But the god of **forces** (מָעֻזִּים) in his estate (עַל־כַּנּוֹ) shall he honor; and a god whom his fathers knew not he shall honor with gold, and silver, and with precious stones, and pleasant things.*

*Daniel 11:39 Thus shall he do in **the most strong holds** (לְמִבְצְרֵי מָעֻזִּים) with an **alien** (נֵכָר) **god** (אֱלוֹהַּ), whom he shall acknowledge. He shall increase with glory; and he shall cause **them** to rule **over many** (בָּרַבִּים), and **shall divide** (יְחַלֵּק) the land (אֲדָמָה) for gain (בִּמְחִיר).*

In Daniel 11:36, the phrase '**above every god**' is עַל־כָּל־אֵל (ahl-kawl-el) and it is intriguing by its very structure. There are three groups of two consonants all ending in 'ל' ('L' in English). In the Hebrew alphabet, the letter 'ל' (lamed) is the tallest letter. The 'ל' is located in the middle of the alphabet and when coupled with letter 'כ' (kaf) the pair is neither too far to the left or to the right - representing a balance. The letter also represents the number '30' which, coincidentally, was the age of Jesus when He began His adult ministry.

The ancient wise men of Israel said the letter 'ל' represented 'wisdom' for it pointed the way upwards to God and the Truth. Furthermore, one can plainly see the letter represents the serpent (which was changed to a snake) which led Adam and Eve to rebel and taste of the tree of knowledge (and hopefully, wisdom). The use of the three 'ללל' lameds in this phrase may also point to the coming Antichrist or Satan because of the use of three snakes which, for some unknown reason, reminds me of '666' which is the last Antichrist's number.

The 'ל' is also the first consonant and second letter (right to left) in many Hebrew and Chaldean forms of the word for 'God' or 'god' such as אֱלוֹהַּ אֱלָה אֵל הִים and so forth. So when Daniel used the phrase, '**above every god**' (עַל־כָּל־אֵל) he was accenting the fact by using three 'ללל' (lameds). The three lameds can also represent the 'Trinity' or the three-part 'God of gods' whom the Hebrews and Christians call, "God". This is especially interesting because at the time Daniel wrote this the '**Trinity**' had not been completed. Most Hebrews knew of basically two parts to God: God the Father (ha Elohim) and the Holy Spirit (Ruach ha Kodesh). The Messiah (Mashiach) had not yet been born into a mankind and His godly nature was still a mystery to most. Daniel was blessed by God in that he was allowed to see and converse with the third member of the 'Trinity' - 'The Messiah' - in His pre-incarnate form as He stood on the water of the Tigris in front of Daniel.

As I said before this is one of the reasons that Daniel was not considered a major prophet by the sages of Israel. They classified the major prophets as those whom God had spoken to directly. They did not recognize the being who spoke with Daniel in **Daniel 10-12** as God or the Son of God. That 'Being' was clearly described as 'Jesus The Messiah' and 'The Son of God' in the Christian book, *"The Revelation of John"* which would explain why only a few Jewish sages would have been able to recognize that GOD did indeed speak directly to Daniel.

I have not yet decoded any other hidden meanings in this phrase; but I do find it poetic in nature due to the rhyming of the last letters. When I analyzed this phrase numerically I found three two-letter groupings representing the numbers '70,30'...'20,30'...'1,30' or summed by twos '100'...'50'...'31' or all totaled '181.'

In **Daniel 11:37** the meaning of the phrase חֶמְדַּת נָשִׁים (teh-mah-daht) *'the desire of women'* is a little less obvious. Daniel 11:37 connects three grammatical objects of equal relevance. The first is 'God of his fathers' and the last is 'any god'. So, what would *'the desire of women'* have to do with gods? In Daniel's time the women of Israel all hoped their son might be the promised Messiah of Israel – yet they did not know He would be half human and half God (Son of The God of gods). Daniel met Him at the Tigris River and was shown His Godly status; so, the proper translation of *'the desire of women'* according to Daniel should be **'The Promised Messiah (Son of God) of Israel'.**

It is interesting to note that most ancient cultures determined an entity to be a 'god' if they exhibited such advanced technology as to be defined only as 'magic' or if the entity was immortal. Sometimes 'gods' were invented to explain the powerful working of weather and geological processes. In essence, gods were thought to all-powerful, omniscient and immortal. However, in the Jewish, Christian and Islamic monotheistic faiths, they gave reverence to the Chief of the gods – not the 'gods' He created. In contrast, the polytheistic Greek and Roman religions worshipped not only the Chief of the 'gods' but the lower 'gods' as well.

In **Daniel 11:38**, the word *'forces'* מָעֻזִּים (mah-u-zeem) and can be translated, **'fierce forces.'** And, in the same verse, the phrase *'in his estate'* עַל־כַּנּוֹ (ahl-kah-no) can be translated as, **'in his protected place.'**

The phrase *'most strong holds'* in Daniel 11:39 is לְמִבְצְרֵי מָעֻזִּים (lah-mi-bah-ahtz-ray mah-u-zeem) can also be translated as **'inaccessible fortresses.'** The *'alien god'* whom he (the king of Earth or the Antichrist of the last days) honors will have technology that will allow them to pass through Earth, water and space with equal ease; which is reminiscent of how the 'grey aliens' pass through solid walls in many UFO contactee or abduction cases. This might mean fortresses underwater, buried underground or off the Earth. However, wherever they are they will be out of reach for mankind at the current technological stage of development.

The phrase *'with an alien god'* in **Daniel 11:39** is עִם־אֱלוֹהַ נֵכָר (eem-ehloah-naykar) really holds has a surprising meaning when I apply my translating technique. The word עִם normally translated as *'with'* comes from the Hebrew verb עָמַם meaning, **'to hide or conceal'.** The word אֱלוֹהַ still means, *'god.'* However, the word נֵכָר (naykar) from the verb נָכַר (neekar) can mean **'to seem alien'** or **'to feign oneself a alien.'** So I decrypt the phrase, *'with an alien god,'* to ,**'with a hidden, immortal being pretending to be a god.'**

The next phrase to upgrade is *'over many'* (בְּרַבִּים). When I divided this phrase into its components I discovered that רַבִּים is also the plural form of רַב which means **'defender'** So, with the בְּ meaning *'over',* the revised phrase becomes, **'over the defenders.'**

The last clause of Daniel 11:39 is traditionally translated as *'and shall divide* (יְחַלֵּק) *the land* (אֲדָמָה) *for gain* (בִּמְחִיר).' However, Daniel uses הָא־ אֶרֶץ (וְאֶתָא) - a completely different word for *'land'* in several other verses of his book. The Hebrew word אֲדָמָה (ah-dahm-ah) which Daniel put in this verse means something completely different; and in light of my expanded translations of the preceding verses it makes more sense. The Hebrew word אֲדָמָה is derived from the verb אָדַם (ah-dahm) which means **'men'** and with the ה at the end it means **'feminine or weak men'** which can also be stated as **'weak mankind or humans'**. The word יְחַלֵּק (ya-chah-leyq) does mean **'shall divide'** but the expanded meaning of it is **'shall divide by lot.'** The words, for gain (בִּמְחִיר) can also be translated as **'in the future.'**

Last Four Of The Eight Verses Expanded:

Daniel 11:36 And the king shall do according to his will; and he shall exalt himself, and magnify himself above every god, and shall speak marvelous things against the God of gods, and shall prosper till the indignation be accomplished: for that which is decreed shall be done.

Daniel 11:37 Neither shall he regard the God of his fathers, nor the **Son of the 'God of gods'**, *nor regard any god: for he shall magnify himself above all.*

Daniel 11:38 But the god of **fierce forces**[28] **in his protected place** *shall he honor; and a god whom his fathers knew not shall he honor with gold, and silver, and with precious stones, and pleasant things.*

Daniel 11:39 Thus shall he do in inaccessible fortresses with **a hidden, immortal being pretending to be a god,** *whom he shall acknowledge. He shall increase with glory; and he shall make them to rule* **over the defenders;** *and* **he shall separate the weak humans by lot in the future.**

So, in summary, the eight verses I have more correctly translated from the Book of Daniel tell us about the short-term future ahead of us. It will be one where mortal man is deceived into accepting a tyrannical world empire designed by truly evil, immortal beings posing as 'gods' and staying hidden from public access.

If you have persevered through my laborious decrypting efforts and you are a fan of *Stargate* then you might agree with me on two points. First, it is an interesting coincidence that the decrypting expert on ancient written languages in the *Stargate* series is named, *"Daniel"* and, second, that it can take many days or longer to render a proper translation of ancient writings in a dead language.

In the third section of this book I mentioned (almost in passing) that Biblical prophecy tells us, "messengers of light" will arrive on Earth at the end of this age as part of a **Huge Deception.** The Bible tells us those messengers (angels of light) and their leader will be cast down to Earth in the last days of this age. Furthermore, it tells us to beware their coming as they will not have our best interests in mind. They will pose as aliens or "beings from off our world." They will claim to be our creators and our elder brothers. They will claim they have come back to Earth in peace to help us solve our many global problems. They will offer to help mankind establish a planet of peace where we can all live together in harmony.

However, as far-fetched as it may sound, the Bible tells us even though those beings were alive when mankind was created, they are NOT our creators. They, themselves, were created long before by the same Creator (The Elohim) who made

---

[28]   It should be noted here that I found another VERY interesting tidbit in Daniel 11:31. Although it does not relate directly to the "Alien Deception" it is of major importance to students of end-time prophecy. Daniel 11:31 states, "And arms shall stand on his part, and they shall pollute the sanctuary of strength, and shall take away the daily sacrifice, and they shall place the abomination that maketh desolate." Most scholars say this "abomination of desolation" is some sort of a profaning object placed in the 3rd Temple at Jerusalem. It is thought it would be some sort of pig or other unclean thing. HOWEVER, I believe it to be an atomic bomb due to the exact translation of the original verse. The original Hebrew is הַשִּׁקּוּץ מְשֹׁמֵם (ha-she-kootz me-show-mahm) which should be translated as 'the contamination which lays waste'. To me this really says a nuclear device of some sort which is put in Jerusalem as a device of 'fierce forces' to force agreement with the antichrist's will. Failure to do so will result in the nuclear annihilation of Jerusalem and all the holy sites of the world's major religions.

mankind and all the other sentient beings in the heavens. Furthermore at some point in those distant ages, one third of those sentient beings rebelled against the The Elohim; and their leader was Lucifer (the light bearer) as I discussed earlier in this book.

The Torah tells us Lucifer or the "dragon" (a flying reptile with scales) was punished for leading Adam and Eve to rebel against The Elohim in the Garden of Aden ('Eden' in English). According to the modern translations that "dragon" was changed into a serpent (a scaled reptile that could only fly in our atmosphere). One could say the Aden history uses a metaphor when discussing Lucifer's fall from grace. He was originally able to travel between the heavens as a powerful and beautiful creature but after he caused Adam and Eve to cut themselves off from The Elohim, he was limited to Earth. Lucifer was thereafter referred to as, "Satan" (challenger or adversary) in the ancient writings because he challenged the authority of The Elohim.

Satan may seem stupid to those who have not studied these things; however, he is anything but that. He was the wisest and most radiantly beautiful being The Elohim created in the beginning of this saga. In fact he was like a secretary of state serving The Elohim.

Thousands of years have passed since the time in the Garden of Aden. Mankind has been fruitful and filled the Earth as The Elohim instructed in the Genesis account. The time has finally come for Satan and his adherents to make one last attempt to convince as many human souls as possible to accept Satan as their 'god.' They hope to win this trial by doing so; however, they do so using deception and lies.

As I said, Satan was given a brilliant mind which means he was and still is way ahead of our most brilliant minds here on Earth. This means he would have read and understood in depth the Bible, The Torah, The Qu'ran and every other 'holy' book on the planet through the ages. **Importantly**, he would also know what events the Christians, Jews, Muslims, Buddhists, Hindus and all other religions on Earth are expecting at the end of this age – especially the return of our Creator.

Therefore, it is important we realize that Satan knows what we are all expecting to witness. He knows a significant portion of mankind could recognize him for what he really is if he were to appear on the scene without the appearance of an obviously evil race of aliens before himself. He wants to come as a savior to mankind and 'mother Earth' in these threatening times. He knows the wise among mankind will be expecting him to arrive posing as the promised Messiah or "Light" of the world before the real Messiah comes back to Earth to establish His Kingdom of true peace and harmony. They will know to warn others about his ruse and will do so until they are removed from the Earth by the True Messiah. If he did not counterfeit the second coming Satan would not be able to fool the entire planet.

He would have but one gambit (chess move) left to deceive mankind. He would have to create a fake alien arrival of one or more types of beings to threaten mankind; and then he would have to arrive with his followers to 'save' mankind from the evil aliens who had been afflicting them. So clever will his deception be that if the faithful followers of Y'shua Ben Elohim were left on earth at that time they, too, would be fooled. If God's people were not warned in advance it would be hard to argue after the fact that yet a third alien arrival was yet to come and that they would really be the good guys.

The TV series, "V", is a great example of the coming deception. In the TV series, aliens looking like humans arrive proclaiming peace and safety. Underneath their fake human exteriors they are upright lizards - complete with scaled, reptilian skin. In the series a portion of these alien lizards take sides with mankind to overthrow their own kind who are evil and deceitful. Mankind is grateful and the series ends with mankind happily going into the future with the 'good serpents'... What people don't see in the TV series is the last book, "The Second Generation," written by Kenneth Johnson - the author of the series. In that book he introduces a third alien race who arrive some twenty-three years later to help mankind overthrow the original aliens who had virtually hoodwinked and enslaved humanity in that short time. However, the parallels to what the Bible warns us about end here. The True Messiah and His people will not be some third alien race from "out there." They will be The Son of our Creator and His People from a concentric (some call it a parallel) Universe.

Evidence will be given to all of us who serve the True Messiah and who seek to avoid the great deception that is approaching its climax now. Stand alert; stand holy; and as The Bible teaches, "Be gentle as doves yet clever as serpents." All is not and will not be as it seems in the media...

The Schauberger Legacy

In "Appendix 6" of "The Cosmic Conspiracy" I explained about lowering the pressure in front of an object while increasing the pressure behind it to produce forward motion in a given fluid or gas. This type of 'recycled entrainment' propulsion can be applied in many ways and in many types of fluids or fluid-like media. In "Appendix 6" I only addressed a theoretical craft utilizing such a method with electrodynamic propulsion. However, after I wrote the book I began to experiment with less complex forms of recycled entrainment propulsion using props, MHD (magneto-hydro-dynamic) fields, high voltage fields and turbines.

Along the way I read about the work of an Austrian research scientist named Viktor Schauberger (1885-1958) whose research and development of novel methods of propulsion in fluids led him on a very dangerous path of discovery through two world wars and the demands of Adolf Hitler - whom he managed to resist and live.

He came from a long line of foresters stretching back about 400 years. So, it is no surprise that in his early life he served as the head warden of the forest and hunting territories in Upper Austria. From what he observed as a warden in the wild he was able to invent a number of practical flotation devices for logging as well as a new type of turbine. He published a paper for Wasserwirtschaft (the Austrian Journal of Hydrology) on his observations of the effects of temperature on the movement and oxygenation of water. He then delved into using capillary action to produce electricity directly from falling water. When he was about 50 years old he began to investigate using implosion technology on water to produce power and propulsion in a device he named, "the repulsine".

However, the portion of his research that I found most fascinating was his observations of trout under various conditions

Brown Trout

of the streams in which they lived. Schauberger studied the trout's ability to jump up high waterfalls with little apparent effort. Within this phenomenon he saw evidence for his theory that the trout exploited some hitherto unknown source of energy within the water.

As Schauberger wrote, *"It was spawning time one early spring moonlight night I was sitting beside a waterfall waiting to catch a dangerous fish poacher. What then occurred took place so quickly that I was hardly able to comprehend. In the moonlight falling directly onto the crystal clear water, every movement of the fish, garnered in large numbers, could be observed. Suddenly the trout dispersed, due to the appearance of a particularly large fish which swam up from below to confront the waterfall. It seemed as if it wished to disturb the other trout and danced in great twisting movements in the undulating water, as it swam quickly to and fro. Then, as suddenly, the large trout disappeared in the jet of the waterfall which glistened like falling metal. I saw it fleetingly under a conically-shaped stream of water, dancing in a wild spinning movement the reason for which was at first not clear to me. It then came out of this spinning movement and floated motionlessly upwards. On reaching the lower curve of the waterfall, it tumbled over and with a strong push reached behind the upper curve of the waterfall. There, in the fast- flowing water, with a vigorous tail movement it disappeared. Deep in thought I filled my pipe, and as I wended my way homewards, smoked it to the end. I often subsequently saw the same sequence of play of a trout jumping a high waterfall. After decades of similar observations, like rows of pearls on a chain, I should be able to come to some conclusion. But no scientist has been able to explain this phenomenon to me."*

After due consideration of Schauberger's observation and his complaint, I believe I have the correct scientific assessment of what he observed. The fish was observed to be in rapid motion *"to and fro"* under the waterfall before it was observed inside the conical field of water inside the waterfall. It was positioning itself to generate the vortex of water around itself for the ascent to the top of the waterfall.

At first, the fish was described as *"dancing in a wild spinning movement."* Inside that cone of water. That *"conically-shaped stream of water"* around the fish was visible to Schauberger because the water inside that cone or vortex was spinning fast enough to create an inertial wall between itself and surrounding falling water. The difference in density of the two produced a reflective medium which he could see in the moonlight. The fish created that cone or vortex of water by swimming back and forth in a slightly elliptical path while he arched his tail and back to cause the water around him to start to re-circulate in a tornado-like shape as shown in this illustration:

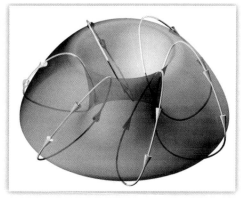

In a glass jar of water in which the water has been stirred in a circular motion the centrifugal force of the spin toward the wall of the glass forces water to rise up on the outside of the jar while reducing the height of the water toward the center of the jar.

In most chemical laboratories one will find a magnetic stirring device which can spin a plastic-coated spin magnet inside jar of water without having to physically stir the water with a spoon or similar instrument. If one had access to such a device another stirring test like the following could be performed.

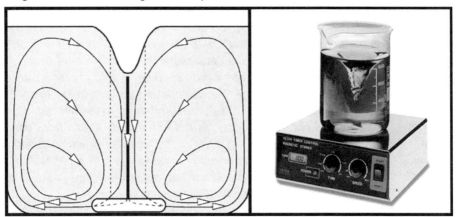

Fill a small jar to the brim with water, drop a stirring magnet into the water, seal the jar tightly and then turn on the magnetic stirring device. Since the stirring magnet would be inducing spin into the bottom of the sealed jar of water no 'tornado' of air could be pulled down the center of the water. The centrifugal force would create convection of water around and above the spinning magnet.[29] A doughnut-shaped field of water would form around and above the magnet (as the following illustration shows). Of course if the device were left to run for a few minutes it would eventually cause the water near the top of the jar to spin as well.

Such a closed-system of spinning water is much akin to the special case of the salmon or European Brown Trout (see fish illustration) which Schauberger observed rising up the waterfall inside a cone of water. In the laboratory, there were two dynamic forces acting on the water and one static 'force' formed by the wall of the jar which was also acting on the water to deflect it upwards. One of those two forces was the spin force from the spinning magnet and the other was gravity.

In Schauberger's waterfall observation, the same two forces were acting on the water around the fish: the falling water in the waterfall which was accelerated by gravity and the spinning forces were created by the fish. The momentum of all that

_____
29   Vortex flow generated by a magnetic stirrer
    Gábor Halász, Balázs Gyüre, and Imre M. Jánosia_von Kármán Laboratory for Environmental Flows, Eötvös University, Pázmány P. s. 1/A, H-1117 Budapest, Hungary
    K. Gábor Szabób_ Department of Fluid Mechanics, Budapest University of Technology and Economics, H-1111, Bertalan L. u. 4-6, Budapest, Hungary
    Tamás Tél von Kármán Laboratory for Environmental Flows, Eötvös University, Pázmány P. s. 1/A, H-1117 Budapest, Hungary
    Received 6 March 2007; accepted 14 July 2007. We investigate the flow generated by a magnetic stirrer in cylindrical containers by optical observations, particle image velocimetry measurements, and particle and dye tracking methods. The tangential flow is that of an ideal vortex outside a core, but inside downwelling occurs with a strong jet in the very middle. In the core region dye patterns remain visible over minutes indicating inefficient mixing in this region. The results of quantitative measurements can be described by simple relations that depend on the stirring bar's rotation frequency. The tangential flow is similar to that of large atmospheric vortices such as dust devils and tornadoes. © 2007 American Association of Physics Teachers. DOI: 10.1119/1.2772287   http://karman3.elte.hu/janosi/pdf_pub/AJP07vortex.pdf

water in the waterfall formed a dynamic, elastic, inertial wall to reflect the water inside the spinning vortex made by the fish upwards and then back down again to form a spinning conical flow around the fish. The fish contorted its body as shown in the images on the next page.

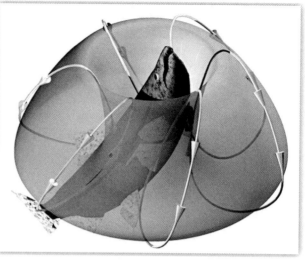

At the same time the fish angled all its fins, (tail, paired pectoral fins, dorsal, paired ventral, anal and adipose fins) and opened its gills to deflect water around and along its body in such a fashion as to stir the water around itself against that 'wall' of the waterfall into a vortex. This made the fish appear to be dancing inside the conical-shaped stream around it. As the fish deflected the energy of the falling water into the conical field of spinning water, air bubbles in the turbulent waterfall were trapped in the space between the waterfall and the spinning vortex made by the fish. The moonlight was thus reflected off the air bubbles to allow Schauberger to see 'the conical-shaped stream of water around the fish as it prepared to be drawn up the waterfall.

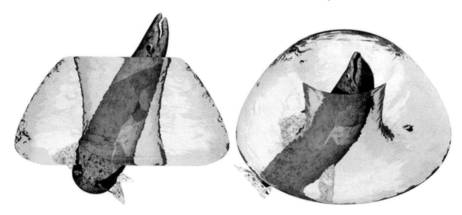

As soon as enough momentum had been redirected into the vortex surrounding the fish, the air bubbles dispersed and the fish straightened its body in the center of the spinning 'conical doughnut-shaped' capsule of water around it and allowed itself to be drawn upwards against the falling water.

In effect, the fish was 'tacking into the wind' (or in this case into the falling water stream) using the energy of the falling stream to draw itself up the waterfall. Now that is nothing short of impressive! The physics principles of this process are waiting to be applied to boats, submarines, aircraft, land vehicles and a host of other devices.

Fish are known to use vortices[30] in the shape of a torus to propel themselves through the water with greater ease than just stroking the water. "By essentially hitching a ride and letting these vortices propel them along, the scientists say, fish can swim against a current with considerably less exertion than is required in calmer settings." [31]

Who Will Announce Return Of The Fake, Alien 'Gods'?

It would be remiss of me to close this final edition of my book without giving some clues as to how the 'alien deception' might be thrust upon the unsuspecting people of Earth. It is no secret that the Catholic Church is holding meetings and aiming policy

---

30   AQUATIC PROPULSION IN FISHES BY VORTEX RING PRODUCTION. http://darwin.bio.uci.edu/~edrucker/home/text from MS word/asz'98_abstract.htm
    E.G. Drucker* and G.V. Lauder. Brown University, Providence, RI and University of California, Irvine. Society of Integrative and Comparative Biology, 1998 Annual Meeting.
    The difficulty in directly measuring properties of unsteady fluid flow surrounding freely-swimming fishes has led to qualitative analyses of hydrodynamic structures in the wake and steady-state models (e.g. "lift-based" and "drag-based") for estimating swimming thrust. To gain new insights into mechanisms of un-steady aquatic force production, we used Digital Particle Image Velocimetry to visualize quantitatively the structure of the wake behind the pectoral fins of slowly swimming bluegill sunfish (Lepomis macrochirus, 20 cm total body length, TL). Fish swam in a flow tank at 0.5, 1.0, and 1.5 TL/s against a current seeded with 12 μm silver-coated glass beads. Particle motion was illuminated by a 3 W argon laser focused into a 1-2 mm-thick light sheet in three separate perpendicular planes (horizontal, parasagittal, frontal) and imaged using high-speed video (250 fields/s). Use of two video cameras simultaneously permitted a three-dimensional reconstruction of the flow. Cross-correlation analysis of pairs of consecutive video images allowed calculation of planar flow velocity fields and changes in fluid angular velocity (vorticity). Abduction of the pectoral fin resulted in fluid flow with opposite-sign vorticity around the dorsal and ventral fin edges. This downstroke produced a vortex ring (in the shape of a torus) 2-5 cm in diameter that traveled laterally, posteriorly, and ventrally. Each pectoral fin generated one vortex ring per complete downstroke/upstroke cycle. A momentum jet directed through the center of each ring is the source of both "lift-based" and "drag-based" thrust forces for swimming. The deep body of bluegill separates left- and right-side vortices and, by limiting hydrodynamic interactions between the two fins, is expected to enhance maneuverability.

31   Source: Massachusetts Institute Of Technology http://www.sciencedaily.com/releas-es/2004/01/040105072113.htm **Scientists Pinpoint How Fish Save Energy By Swimming In Schools**
    "Using a tank designed to mimic a turbulent waterway, scientists have found that fish use a unique and clever swimming motion to harness the energy of eddies in flowing water. By essentially hitching a ride and letting these vortices propel them along, the scientists say, fish can swim against a current with considerably less exertion than is required in calmer settings. The study provides the first direct view of the technique used by fish to exploit the energy of eddies generated by the swimming movements of other members of a school..."
    "This study reveals mechanisms through which the energy contained in the eddies of turbulence can be extracted, with great reduction in the required swimming effort, said Michael S. Triantafyllou, a professor in MITís Department of Ocean Engineering. Since turbulence is present everywhere in moving water or air -- in the ocean, the lakes, and streams, as well as the atmosphere -- many opportunities arise for engineer-ing applications of these new principles of energy extraction.
    First author James C. Liao, a graduate student in Harvard's Department of Organismic and Evolutionary Biology, likens the technique fish use to swim against a turbulent current to tacking, the back-and-forth mo-tion that allows a sailboat to move forward passively, if somewhat indirectly, against a breeze..."
    "They discovered a previously unknown swimming motion, dubbed the Karman gait, that exploits the energy of eddies to minimize fish muscle activity. Viewed from below, the bodies of fish swimming in turbu-lent waters flutter gently, like a flag flapping in the breeze, a motion that entails far less muscle activity than swimming through still waters.
    This new link between the use of eddies and decreased muscle activity supports a hydrodynamic explana-tion for fish distributions in schools and in current-swept habitats, Liao said. Our work is the first to establish a causal mechanism to explain the hydrodynamic benefits of fish schooling, and resolves the paradox of how wild fish can move upstream against currents faster than they have been reported to swim in the laboratory. Additional authors of the paper are George V. Lauder, a professor of biology at Harvard, and MIT alumnus David N. Beal (S.M. 1997, Ph.D. 2003). The work was supported by grants from Sigma Xi, the American Museum of Natural History, the Robert A. Chapman Memorial Scholarship at Harvard, the National Sci-ence Foundation, and MIT Sea Grant."

toward forming 'a common dialogue' between the religions of the world in an effort to bring world peace. It has been doing so openly for over thirty years and is intensifying its efforts every year. The Catholic Church is falling victim to a movement to unite the religions of the world under one encompassing doctrine.[32]

On 13 November, 2009 the Australian ABC news released a story entitled, *"Alien Life Is Possible: Vatican."* According to the article, the Pope's chief astronomer, Jesuit priest Father Jose Gabriel Funes and also director of the Vatican Observatory, conceded other intelligent beings could exist in outer space. This conclusion was drawn by the astronomer and other scientific experts called in by the Vatican to study the possibility of extraterrestrial life and its implications for the Church.

According to many mainstream news stories I found on the internet in early 2010, several leading Catholic priests believe aliens are visiting Earth. Some suggested the Fatima visions were created by - a 'flying saucer' or 'UFO.' Although these priests do not speak for the Vatican, they do represent a ground swell in the Catholic community.

When the 'alien' presence is released to the world (as I discussed earlier) it will shake the very foundations of the religious faiths of people all over the world. To maintain control of the various religions the leaders of these religions will have to explain the alien existence in terms of their respective faiths. The secular population of the world which account for about 16% of the people will not have a problem accepting the presence of aliens - for the most part. Christianity represents about 33% of the global population while Islam accounts for another 21% and Hinduism for 14%. By aiming carefully crafted propaganda at these three faiths and the secular population, 84% of the world could be convinced that the initial 'alien' presence is a good thing. So, I believe when the alien presence is announced officially, the major religions will rapidly support their presence. They will do so to maintain order within their ranks and out of fear the power the 'aliens' might wield on those who do not accept them and their plan for organizing Earth into a global government.

The man who makes the official announcement of the 'alien' presence will have the support of the 'aliens.' He will use technology given to him by the 'aliens' to bring order to the planet (albeit an unstable order). He will have the support of the world religion. Presently several candidates for the job have been suggested by various Biblical scholars. These include: Barack Obama, Javier Solana, Prince Charles of Wales, Prince Hassan of Jordan, "the Maitreya" and another called, "the 12th Imam Mahdi."

Look for this mystery man by seeing who arises from the political arena of the planet. He will be a deceiver and a pathological liar. He will have the support of the major religious leaders of the world. He will also ratify an agreement or a treaty between all the nations of the world and the 'aliens'. BUT, most importantly.....

The True Messiah can be discerned using the following verses (if nothing else):

Matt. 24:23-27   *Then if any man shall say unto you, Lo, here is Christ, or there; believe it not. For there shall arise false Christs, and false prophets, and shall show great signs and wonders; insomuch that, if it were possible, they shall deceive the very elect Behold, I have told you before. Wherefore if they shall say unto you, Behold, he is in the desert; go not forth: behold, he is in the secret chambers; believe it not. **For as the lightning comes out of the east, and shines even unto the west; so shall also the coming of the Son of man be.***

---

32   http://www.cuttingedge.org/NEWS/n1094.cfm
     http://www.united-religions.org/
     http://www.parliamentofreligions.org/index.cfm

The following concerns the artistic coding I used in the cover painting. It was originally on the inside of both the front and back covers; but put them here in this Millennium Edition.

The Artistic Code Behind My Cover Painting

Due to numerous inquiries we have received regarding the cryptic symbolisms on my front cover. I have included the following seven items for all to witness...
Signed: Stan Deyo, on the 4th of the 7th, 1998.

ITEM 1: The title, *"The Cosmic Conspiracy,"* was constructed using the code on page 78 to give the following cryptic messages. Firstly, the pattern of the number of letters in the title is a message. That pattern is '3-6-10' which transliterates to 'Godhead-man of sin-ordinal perfection.' By philosophically combining the three parts of that message a sentence forms, 'Sinful man stands between GOD and the harmony of true unity.' If one sums the three numbers '3, 6 and 10', one obtains the sum of '19' whose two digits sum to '10' which self-sums to '1' or 'unity'.

In the title, three words were used as in many other places throughout the book to represent the Triune Godhead of the Judeo-Christian ethic. The word 'THE' was given superior position because its number represents that Triune Godhead. The word 'COSMIC' was given a lower position because its number represents that which GOD created. The word 'CONSPIRACY' was given the lowest position even though its number represented 'ordinal perfection'. This was to show the Satanic conspiracy fits into the perfection of GOD's will even though the conspiracy, itself, is a violation of the Law. Its position is. However, the lowest in the mind of GOD.

The title words are arranged in a triangle whose apparent apex represents repentant mankind looking upward for that final union between GOD and those of His created beings who have accepted His Law of their own free will. Though the apex of the title pattern is not visible, its presence is discerned by the enlightened eye.

ITEM 2: The Great Pyramid without a capstone represents the Great Pyramid at Gizeh and that which it, itself, represents. The missing capstone symbolizes that the collective body of Christian believers has not yet united in the visible form with its Chief... their Messiah... the Christ. Ancient Masonic legend says. "He who seeks the Chief Cornerstone rejected by the builders dwells in the Lodge of Wisdom illumined by Ner Tamid" (the Eternal Flame). An informal transliteration of this phrase gives 'He who seeks the Christ who was once rejected by Israel is a wise and enlightened being.'

The crumbling, alabaster, covering stones reveal the final stages of the long and arduous process of exposing the Conspiracy. The Roman style of lettering was used to denote how similar the ancient Roman Empire and the coming New World Order will be. The partially revealed Roman numerals 'DCCLX' indicate there is yet more revelation to follow.

ITEM 3: The pyramidal crystal containing a rose is an impostor apex. The building material of the main pyramid and of the impostor cornerstone are different. The true Son of Man (the Christ) would be depicted by non-transparent material which would have light emanating from it; also, it would be a 'double cubic stone' to lock into place.

The impostor (the antichrist), however, is represented by this floating crystal which is quite transparent to those with discernment. As the impostor tries to 'steal the throne' the real Messiah (see item 7) will return to prevent it.

Although the rose may suggest beauty and sweet smells. remember these things conceal thorns for the undiscerning mind. The sinister eye in the rose represents the eye of Satan (the Phoenix) watching from deep within his disguise to the unsuspecting.

ITEM 4: The three dark flying saucers represent the imminent arrival of the unholy 'trinity' of Satan's grand deception as they are 'cast down to Earth'. The faint white glow of ionized air over each craft represents the superficial "light" of these dark messengers. The craft's glowing red underside represents the fires of Hell from whence came these dark beings.

The red and green channels of light underneath the craft represent the two extremes of thought in human affairs generated by the machinations of these dark beings (the dark Illuminati). These two extremes are loosely known as 'the left' (pro-socialist) and 'the right' (pro-fascist). In reality, as in this picture the two are part of the same dictatorial source which churns the water. The water represents mankind in chaos (see item 6).

The spiral vortex shows that one should not take something unless someone else gives it first. In essence, if 'the left' gain then 'the right' lose. The process should. in reality. be reversed so that the initial act is a giving. If such were the case in my drawing then the spiral vortex would be inverted as a "giving first" sign.

ITEM 5: The star-studded sky serves as a reminder to the great and catastrophic changes which prophecy predicts will soon strike Earth. Stars appear to move from their places as our lens-like atmosphere is parted like a scroll and splashed about from the effects of a colliding comet and the impact of a large, burning meteoric 'mountain'.

The semi-darkness represents not only the pre-dawn state of human consciousness - but also the semi-darkness which will strike Earth when our Sun temporarily reduces most of its light output (in our direction). The lack of vegetation amongst the rocks below. however, reminds us that the Sun will also appear temporarily seven times brighter than its present value (in our direction) causing wide-spread drought.

ITEM 6: The troubled sea represents mankind in its current, chaotic state. It is being churned by the 'dark Illuminati'. Notice that the pyramid is an enduring edifice which sits above the sea unmoved by the machinations of the dark ones. Someday soon through the mystical action of the return of the true Messiah, these waters will calm and become the "Great, Westward-Flowing River of Life".

ITEM 7: The Cross of Light (The Great Light) approaches... His life-giving Light radiating in all directions-without end. His coming: a ray of hope for the repentant and a two-edged sword of judgment to the unrepentant. Look up! Witness the signs! He Descends to Earth and soon!:

The SON of The Most High CREATOR...

LORD of Hosts...
The True MESSIAH...
KING of Kings...
YESHUA HA MASHIACH

Amen

Index

catastrophic changes 277
Cathie, Captain Bruce 93
catholic 232, 235
Catholic 259
Catholic Church 275
cavities 246, 247
cavity resonance 42
Cayce, Edgar 47, 231
celestial body 144
Celestial Heaven 127
celestial observations 177
centaur 110
Central Intelligence Agency
    (CIA) 8, 22, 30, 49,
    235, 238, 239
centrifugal force 271
centrifugal turbine fan 188
centuries 62, 119, 144, 258
century 32, 49, 51, 130,
    143, 144, 227
Century Association 94
ceramic 242
Ceylon 199
chained in the bottom-
    less pit 137
Chaldea 263
Chaldean 261
Chaldean Aramaic 263
challenger 111, 134
channeled 232, 242
channeler 232
channelers 231
channeling 231
channels 277
chaos 73, 74, 78, 79, 83,
    101, 103, 112, 130,
    136, 233, 277
chaotic 78, 277
Chapel Hill, N.C. 143
Charbury 64
charged clouds 41, 42
charged particle 32,
    48, 53, 168
charged particles 32, 48, 53
chariot 257
charismatic 231
Charles Thomson 209, 257
Charles University in
    Prague 143
Charlottesville 257
chemical fuels 143
Cherenkov radiation 176
Cheyenne Mountain 42
Chief Superintendent of the
    Aeronautical Research
    Lab 28, 170, 171
child of GOD 127
children of Israel 116,
    128, 257

Chile 198
China 50, 67, 133,
    194, 229, 234
Chinese 50, 67, 69, 91
Chinese Cultural
    Reformations 69
chlorine ions 167
Christ 69, 73, 86, 112, 116,
    117, 118, 119, 120,
    121, 123, 124, 125,
    126, 127, 128, 135,
    136, 138, 229, 276
Christian 63, 79, 101,
    106, 129, 139, 227,
    229, 232, 259, 276
Christian conversions 63
Christianity 65
Christians 70, 122, 124,
    129, 138, 139, 140,
    229, 231, 259, 269
Church 123, 124, 126,
    128, 136, 137, 138
church age 132
Church on Earth 126
CIA 235, 238, 239
circular rainbows 108
citizen of the new world 131
city of Babel 115
civilian 8, 30, 82, 234
civilian UFO research 8, 82
civilisation 130, 139
civilization 17, 86, 92,
    247, 248, 259
clairvoyant 232
clandestine 240
Clarke, Dudley 146
Clarke Electronics
    Laboratories 17
Clark, Evans 101
classified 8, 11, 18, 30, 38,
    42, 125, 239, 241, 246
clay 262
clear number disk 107
cleave 263
Cleveland, Ohio 248
climate 49, 230, 259
climatic 259
cloning on Earth 110
close of the great
    tribulation 136
cloud 8, 46, 132, 229, 257
clouding of the Sun 132
clouds 235
clouds in heaven 135
Club of Rome 85, 86, 87,
    88, 89, 90, 99, 100, 130,
    193, 194, 197, 211, 262
coal 49, 90
coastal areas 259

coastal flooding 47
Cohen, Benjamin V. 101
coils 44, 145, 188, 190, 242
colliding 277
colliding beams of
    deuterons 37
collision 233, 234
collision course with
    the Earth 132
Colombia 198
Colorado 41, 42, 53
Colorado test 42
colour of dead blood 134
Columbia 64
Columbia University 64
combatants 103,
    120, 135, 235
comet 277
Comforter 126
coming deception 269
coming Deliverer 112
commercial air trans-
    portation 22
committee 16, 100, 146, 257
commodities 90, 130, 228
commodity 130, 230
common language 115
communal 139, 228, 259
communism 102
communist 73, 89,
    99, 101, 235
communist doctrines 73
communistic revolutions 101
complex wave form 108
computer 25, 28, 55, 57,
    85, 87, 88, 89, 99,
    131, 195, 196, 230
computers 25, 86
concentrated energy 176
concentric 270
concentric layers 19
concrete 17, 227
condensers 155, 159
conditioning 8, 37, 82,
    83, 101, 243
conflict 53, 87, 91, 101,
    103, 110, 112, 113,
    118, 124, 194
conflict resolution 87,
    113, 194
Confronting the Future 88
congregations 232
Congress 69, 70, 90, 257
Congressional 257
conically-shaped stream
    of water 271
Connolly, Sir Willis 100
conquest of space 145
conscious 14, 29, 110, 227

implants 230
impostor 259, 276
imprisonment in their
    abyss 137
Impulsive Plasma Cur-
    rent Sheets 38
inaccessible fortresses 267
incarnated into a hu-
    man body 118
incompressible fluid
    space 176
India 49, 199, 259
Indian 24, 42, 47, 49,
    53, 91, 227
Indian deaths by famine 49
Indian floods 53
Indian Ocean 91
Indonesia 199
industrialists 241
indwelt by the spirit
    of Satan 133
inertial 19, 172, 176, 177,
    178, 179, 184, 186
inertial curl 176
inertial wall 271
infiltrators 63
infinitely short travel time 22
initiate witches 74
inner shell 36
Institute for Advanced Study
    at Princeton 143
Institute of Advanced
    Math 144
Institute of Pure Physics 143
Institute of the Aeronauti-
    cal Sciences 21
Institutional 194
instructor pilot 7, 8, 10
insulating plate 161
insulation 18, 188
insulator 157, 158,
    159, 190, 242
INTEL 147
Intelligence 12, 29, 49, 241
intelligence personnel 22
intelligence war 9
intelligent beings 275
intelligent life 248
intense heat 12
Interavia article 18
intercessor 121
interests of national
    security 238
Interlinear Greek
    Translation 123
interlocked conspiracies 64
intermarriage 235
International Club of
    Rome 100

international disarmament 47
international economic
    law 89
international order 88
Internet 228, 237
Invaders 243
invading 235
invasion 55, 83, 130,
    227, 235, 239, 240
invasion of Earth 83
invasion of Israel 130
invasions 228, 231
inventor 16, 17, 143
inventors 40, 81
invincible 247
invisibility 192
invisible 13, 25, 52, 86, 94,
    133, 147, 234, 240
invisible college 86, 240
ion accelerators 188
ion electron beam
    reactors 145
ionic 162, 163
ionization potentials 186
ionized 14, 168, 242, 277
ionosphere 41, 44
ionospheric 'tornadoes' 53
ions 55, 161, 167,
    168, 188, 190
Iran 89, 91, 198, 230, 259
Iraq 91, 198, 230
Ireland 197, 230
Irian Jaya 50
iron 262
Iron Curtain 14
irreversible 233
Isaiah, the Book of 122, 136
Ishmaelite 259
Islam 260
Islamic 259, 267
island 10, 42, 132
Israel 63, 89, 91, 116, 118,
    124, 125, 126, 128, 130,
    132, 133, 134, 135, 230,
    234, 257, 259, 266, 276
Israel in the Old Testa-
    ment 125
Israelite 259
Italy 197

J

Jabal Tuwayq 243
Jacobins 63, 103
Jah 62
James Lovell 209, 257
Japan 85, 133, 194, 197, 230
Japanese 91
Jastrow 204
Jefferson 69, 70, 257

Jeffersonian Era 80
Jefferson, Thomas
    69, 70, 257
Jerimiah, the Book of 122
Jerusalem 117, 128, 137
Jesuits 65, 66
Jesuits, the Society
    of Jesus 65
Jesus 65, 69, 112, 117, 118,
    120, 121, 126, 127, 128,
    135, 138, 227, 228, 229,
    231, 232, 248, 261
Jesus Christ 69, 120,
    127, 128, 135, 138
Jesus is Lord 121
Jesus the Christ 112,
    118, 120, 121, 126
jet engines 145
Jew 64, 66, 106, 130,
    132, 133, 232, 259
Jewish 64, 66, 106, 130, 232
Jewish blood 130
Jewish Elders 66
Jewish sages 261, 266
Jews 66, 68, 73, 125,
    231, 238, 269
John Morin Scott 209, 257
John Rutledge 209, 257
John Schwarz 205
Jones, Jim 231
Jones, Raymond F. 17
Jordan 91, 145, 198
Joseph 99, 102, 117
Joshua the Book of 61, 122
joules 212
JPL 249
Judean province 116
judge 68, 231, 258
judged 120, 227
Jupiter 58, 248, 249

K

Kabbalistic Brotherhood 62
Kabbalistic gematria 9, 73
Kabbalistic mystery
    school 61
Kabbalistic society
    of Israel 63
Kabbalistic teachings 61
Kamchatka 50, 233
Kant 102
Kaplan, Martin 142
Kennedy 120, 238
Kenneth Johnson 270
Keyhoe, Donald 21
KGB 22
Khan, Ghengis 234
kill 112, 116, 117,
    118, 229, 238

The Cosmic Conspiracy Final Edition 2010 © Stan Deyo 2010

military 11, 12, 15, 17, 20, 24, 45, 73, 91, 99, 142, 144, 155, 239, 258
military significance 155
Milky Way Galaxy 48
Millenium 135
millenium ages 136
millennial Kingdom 136
millennial nations 137
Millennium 227, 276
Minerval 67
mingle 263
miniature 'flying saucer' 145
Minihan, Captain Kenneth A. 12
miracles 134, 231
miry clay 263
misery 86, 233
missile 239
missiles 12, 142, 143, 145, 233
missing capstone 74, 276
mists 114, 229
M.I.T. 85, 86, 144
model Universe 111, 114
modern mystery code 78
modern mysticists 60, 62
molecular lattice 34
molten 250
Monaco 197
monetary 78, 83, 131, 233
monetary numbering system 78
monetary system 233
money 20, 131, 155, 230, 239, 240
Mongol 234
monotheistic faiths 267
Monticello 69, 257
Montserrat 233
moon 249, 250
Moon 53, 114, 132, 134, 135, 231
moonlight 271
Moon will not give its light 135
Moore, William 240
moral decay 79
Moray 203
Morgan, J.P. 64
Morgan, William 68
Morin Scott, John 257
Morley 100, 172, 174, 177
Morley, Edward W. 172
Morningsky, Robert 231
Morocco 198
morphology 106, 107
mortal 264
mortal beings 111

mortal mankind 265
mortal men 136
Moses 61, 63, 257
Moses dividing the Red Sea 209
Mosshiahni 227, 228, 229
most strong holds 267
mother ship 246
mothership 247
motherships 246
motive power 163
motor 37, 91, 161, 162, 163
motto 257
mountains will be levelled 135
Mount Ararat 115
Mount Ararat in Turkey 115
Mount St Helens 205
movie 243
movies 82, 83, 243
Mozer, Professor F. 20
Mozer's approach 155
Mozer's quantum mechanical approach 154
multinational corporations 10
Munich 67
Muslim 229, 259
Muslims 231, 259, 269
mutated 230
mutual annihilation 33
mutual enhancement 33
mysterious 28, 51, 60, 123, 145, 146, 230, 237, 239
mysterious floating metal ball 145
mystery 26, 37, 61, 62, 64, 66, 73, 78, 79, 85, 94, 109, 126, 140, 142, 144, 146
Mystery Babylon 59
mystery schools 61, 64, 66, 85
mystical 69, 70, 73, 88, 94, 103, 116, 118, 121, 127, 133, 232, 277
mystical action 121, 127, 277
mystically 258
mysticism 60, 73, 88, 92, 257
mysticists 60, 62, 65, 73, 88
Mystic Order 68
myth 233

N

Nabataean 264
naivet√© 237
Napoleonic order 211
NASA 92, 239, 240, 248, 249
National Advisory Committee for Aeronautics 146

National Oceanic and Atmospheric Administration 53
national security affairs 99, 238, 239
national survival 146
National Worker's Party 102
nation of Israel 128
nations 103, 116, 133, 135, 137, 139, 229, 230, 234, 235, 236, 241, 259
nations of Earth 133, 236
nation-state 259
NATO 211
natural disaster 88, 129
natural gas 212
Nature 160
Naval 241
navigation 26, 242
Nazareth 228
NAZI 102
near-Earth-orbit meteorites 260
necromancy 126, 232
negative mass 20, 154, 155, 156, 164, 166
negative thermions 162
negative weight 18
Nelson, John H. 52
NEPA findings 157
Nepal 199
Nephilim 106, 110, 111, 113
Neptune 248
Ner Tamid 276
neutralized bodies 154, 164, 165, 166
neutrino 156
New Age 233
New American Standard Bible 140
New Ark 127, 139
new 'believers' 134
new birth 121
New Boston, N.H. 143, 144
new covenant 115
new Earth 137
New Energy 81
New England 144
new Heaven 137
New Horizons in Propulsion 38
New Jersey 38
new Nazism 88
new order of the ages 258
new source of energy 248
Newsweek 92
new technology 16, 81, 242
New Testament 110, 125, 140, 261

tidal waves 46
Tigris 111
Tigris River 261
time is reference 175
time lag 172
Timeless Consciousness 109
times of crisis 131
Tinah 264
tissue damage 43
titanate 242
tithes 232
Titus, the Book of 121
Today's 'ark' 126
Togo 199
to have intercourse with 264
to intermix 264
Torah 122, 269
tornado-like shape 271
toroidal 242
toroidal entrainment 203
torpedo propulsion 203
torus 274
to save Israel 135
total investment 195
totalitarian 73, 101
tower, gigantic 115
Townsend Brown's saucers 20, 158
traces 12, 239
tracking 233
translational force 144
translation of the church 124, 140
transliterates 276
transliteration 276
transmission of power 19
transmitting electricity 40
transparent 67, 276
travel into outer space 145
treachery 114
Treasurer of Loans 257
tremendous 13, 42, 124, 142, 230, 231
triangle 70, 73, 74, 79, 107, 257, 276
triangular 74, 79, 246, 249
tribulation 123, 124, 126, 128, 129, 130, 131, 132, 133, 134, 135, 136, 139, 227, 229
Tribulation 227, 229
tribulations 231
Trimble jr, George S. 142, 144
trinity 73, 78, 277
Trinity 266
triune Elohim 121
Triune Godhead 276

Trojan 247
Trojan Horse 247
Tropic of Cancer 42, 47
Tropic of Capricorn 42
trout 271
Trucial Oman 198
trumpet 126, 135
Trump, Professor John G. 18
Truth 26, 42, 119, 120, 121, 129, 135, 140, 232, 241, 250
tsunamis 46
tuberculosis 230
Tunisia 198
turbine engine 158
turbine fan 188
turbines 37, 179, 270
turbulence 51, 181, 184, 188
Turkey 99, 115, 197
TV 228, 243
TV series, "V" 269
twelve hundred consular staff 25
twelve signs of the zodiac 119
twelve tribes of Israel 132
Twentieth Century Fund, Inc. 100
twenty 7, 10, 16, 18, 20, 21, 44, 65, 74, 82, 102, 132, 146, 227
twisting movements 271
Two 'ancient witnesses' 63
two metallic plates 162
tyrants 257

U

UFO 6, 7, 8, 10, 11, 12, 13, 21, 26, 30, 31, 82, 83, 229, 231, 240, 241, 242, 243, 275
UFO/alien 229
UFO films 30
UFO phenomena 13
UFO records 26
UFOs 8, 9, 10, 13, 14, 21, 23, 31, 188, 240
UFO sightings 8, 10, 13, 82
UFO situation 7, 30
UFO story 7
Uganda 199
ultraviolet 46, 230
ultraviolet radiation to bombard 46
Ulyanov, Vladimir Illich 102
underground 24, 31, 57, 67, 242, 243
underground antenna 24
underground cities 57

underneath 25, 42, 230, 246, 247, 277
undersea 24, 247
Underwood, Professor E.J. 100
underworld of the dead 118
undesirables of the new age 129
undulating water 271
unearthed 239
unemployment 260
U.N. General Assembly 101
unholy 277
unidirectional forces exerted 162
unification 235
unified 127, 155, 156, 157, 161, 178, 186, 235, 248
Unified Field Law 178
unified field links 157
unified field problem 248
Unified Field Theory 178
uniform 26, 33, 46, 166, 186, 242
uniform acceleration 242
United Nations 13, 45, 80
United States 20, 21, 24, 42, 45, 46, 49, 68, 69, 70, 73, 78, 90, 130, 142, 143, 144, 145, 146, 147, 155, 197, 229, 230, 241, 246, 257, 258
United States Air Force 146
United States Geological Survey 90
United States of America 70, 197, 257, 258
unity 276
Unity 78, 89
Universal Energy Systems, Inc. 38
universal gravitation 15, 142, 143, 144, 145, 146
University of California 49, 143
University of Indiana 143, 144
University of Ingolstadt 65
University of North Carolina 37, 143, 239
unrepentant 133, 135, 227, 277
unseal 263
unsealed 261
upper ionosphere 44
Upper Volta 199
Urantia 83, 231
URL 228
Uruguay 198

William Barton 257
William Churchill
    Houston 209, 257
Williams, Sir John 171
Willis, Eric H. 23
wind, great rushing 137
winged eagle 88
Winston Salem, N.C. 143
Winterhaven, Project 155
wireless broadcast of
    electricity 192
wireless electrical
    broadcasting 40
wisdom 26, 60, 62, 63, 70,
    74, 107, 229, 233, 266
wise men of the East 116
witchcraft 231
witness 24, 113, 122,
    138, 276, 277
Witness 277
witnessed 13, 16, 132, 241
witnessing 235
Witty, David B. 21
wizard 17, 146
woes 91, 230
Wold 160
wolf in sheep's clothing 259
women of Israel 267
Word of GOD 123, 126,
    139, 140, 231
Words 62
work of GOD 125
world affairs 93
world dictator 261
world dictatorship 86,
    131, 211
world economic progress 89
world economies 130
world economy 139
world federal state 86
world Freemasonry 69
world government 101
world kingdoms 261
world news 138
world peace 60, 80, 89, 275
World population model 195
world's central banks 90
world system 87, 124,
    136, 194, 195
World System 194
World War II 83
worldwide state of
    emergency 130
worst plagues 133
wrath 132
Wright Air Development
    Centre 37
Wright Brothers 145, 146
Wright Field 146

WWI 246
WWII 22, 44, 73, 236,
    241, 242, 246
WWII Nazi bird 73

X

XVI 121

Y

Yahweh 62, 106, 107, 111,
    112, 113, 114, 115,
    116, 117, 118, 119,
    120, 121, 122, 139
Yates, William B. 144
Year of Beginning 258
Yemen 198
YESHUA 277
Yiddish 238
Y'shua Ben Elohim
    (Jesus) 261
Y'srael (Israel) 261
Yugoslavia 89, 197, 241
Yugoslavian 40

Z

Zaire 199
Zambia 199
Zatzkis 160
Zealand 30, 47, 50,
    93, 171, 197
Zionists 101
zodiac, twelve signs
    of the 119